Biocatalysis and Pharmaceuticals

Biocatalysis and Pharmaceuticals

A Smart Tool for Sustainable Development

Special Issue Editor

Andrés R. Alcántara

MDPI • Basel • Beijing • Wuhan • Barcelona • Belgrade

Special Issue Editor
Andrés R. Alcántara
Complutense University
Spain

Editorial Office
MDPI
St. Alban-Anlage 66
4052 Basel, Switzerland

This is a reprint of articles from the Special Issue published online in the open access journal *Catalysts* (ISSN 2073-4344) from 2018 to 2019 (available at: https://www.mdpi.com/journal/catalysts/special_issues/biocatalysis_pharmaceuticals).

For citation purposes, cite each article independently as indicated on the article page online and as indicated below:

LastName, A.A.; LastName, B.B.; LastName, C.C. Article Title. *Journal Name* **Year**, *Article Number*, Page Range.

ISBN 978-3-03921-708-3 (Pbk)
ISBN 978-3-03921-709-0 (PDF)

Contents

About the Special Issue Editor

Andrés R. Alcántara (Professor). During his Ph.D. thesis at the University of Córdoba (Spain), thanks to an FPI Scholarship and all through his postdoctoral stay (working contract as Visiting Research Fellow) at the University of Kent at Canterbury, U.K., (three periods between 1989 and 1991, under the supervision of the late Professor A. Williams), his research line has been focused on Biotransformations and Applied Biocatalysis in Organic Chemistry, specifically, in the preparation of enantiomerically pure compounds as chemical drug precursors, using mainly proteases, lipases and alcohol dehydrogenases, either native or immobilized. On the other hand, his work fits perfectly within Sustainable Chemistry, since he has applied these biocatalysts mainly in sustainable bio-solvents, which generates a very beneficial synergy from the point of view of the sustainability of the processes. This background has reported him more than 100 scientific papers, reviews, monographs and book chapters and more than 100 communications in national and international congresses. He has also actively participated in more than 30 funded projects, contracts and research activities by different organizations, being the Principal Investigator in 6 of them. Currently, he is the Director of TransBioMat, a new research group also at UCM, still in the process of evaluation, together with researchers from the Institute of Biofunctional Studies. In this Complutense University, where he has developed his research work (briefly reviewed so far) and also his teaching work from 1989 to the present, he is Full Professor of Organic Chemistry since March 2018. Finally, as one of the first partners of SEBiot (Spanish Society of Biotechnology), he was, consecutively, Coordinator of the Section of Applied Biocatalysis (2004–2007), Member of the Board of Directors (September 2004–September 2006), Secretary Elect (September 2006–September 2008) and, from 2008 to 2012, Secretary General. Last December 2018 he was elected member of the Scientific Board of the Section on Applied Biocatalysis (ESAB) of the European Federation of Biotechnology (EFB).

Editorial

Biocatalysis and Pharmaceuticals: A Smart Tool for Sustainable Development

Andrés R. Alcántara

Department of Chemistry in Pharmaceutical Sciences, Section of Organic and Pharmaceutical Chemistry, Faculty of Pharmacy, Complutense University of Madrid, Plaza de Ramón y Cajal, s/n, E-28040 Madrid, Spain; andalcan@ucm.es; Tel./Fax: +34-91-394-1820

Received: 16 September 2019; Accepted: 19 September 2019; Published: 23 September 2019

1. Background

Biocatalysis is the term used to describe the application of any type of biocatalyst (enzymes, as isolated preparations of wild-type or genetically modified variants, or whole cells, either as native cells or as recombinant expressed proteins inside host cells) in a given synthetic schedule [1]. One type of applied biocatalysis, also called a biotransformation [2], takes advantage of the excellent enzymatic precision inherent to its use, in terms of chemoselectivity, regioselectivity, or stereoselectivity. The use of biotransformations has increased considerably in recent decades, complementing classical chemical synthesis in multiple industries, mainly for the preparation of pharmaceuticals [1,3–18], fine chemicals [19–21] or food products [22–24]. Additionally, and based on the principles and metrics of green chemistry [25–29] and sustainable chemistry [30–37], biocatalysis fits perfectly into this framework; in fact, biocatalyzed procedures are highly efficient, economical, and generate less waste than conventional organic syntheses [38–46]. As such, the interest in the application of biocatalysis within the pharma industry is not surprising, as this industry is by far the biggest waste producer [43,46–50]. Furthermore, as biotransformations are generally conducted under approximately the same temperature and pressure conditions, the possibility of carrying out coupled cascade processes is enabled, providing additional economic and environmental advantages [51–58]. Finally, most biotransformations can be easily developed in standard multipurpose batch reactors without requiring costly specific devices, such as high-pressure equipment [59,60], allowing the costless implementation of continuous processes [21,44,61–63].

In recent decades, the increase of the impact of biocatalysis within the pharma industry has not been linear. In fact, it was relatively minor until the last two decades, when a clear increase (the third wave of biocatalysis [64]) was caused by the popularization of the genetic manipulation of biocatalysts by directed evolution [15,65–75], a method involving fast generation of enzyme mutants using new molecular biology techniques combined with selective pressure via screening conditions. This technique, recognized by the Nobel Prize in Chemistry being awarded to pioneer Frances Arnold in 2018 [76,77], has allowed the redesign of enzymes to fulfill industrial requirements in terms of specificity, activity, and robustness, while keeping or even increasing its outstanding precision. Thus, with this potent tool in our hands, we are facing what has been called the 4th wave of biocatalysis [1,78,79], which will be fully implemented within the pharma industry when the speed of the overall process needed to create an improved biocatalyst (rational directed evolution in a design–make–test cycle, combining multiple disciplines in a continuous industrialized workflow) is improved by at least 200%–1000% [1,18,80,81].

2. The Present Issue

In this Special Issue, in which I have been honored to act as Guest Editor, different articles have been published covering very diverse areas of this fascinating discipline—the implementation of biocatalytic tools focused on pharmaceuticals. It comprises three reviews and eight research articles.

In the first review by Bastida and coworkers [82], the application of glycosaminoglycan (GAG)-degrading enzymes, specifically chondroitin sulfate (CS) lyases (CSases), is presented as an effective tool for the preparation of bioactive molecules possessing diverse therapeutic applications. Thus, initially the types, structures, and mechanisms of different CSases are described, and afterward, the application of these enzymes for the synthesis of low molecular weight chondroitin sulphate (LMWCS; used for osteoarthritis treatment, cardiocytoprotection, anticoagulant and antithrombotic activities, etc.) is reviewed, clearly illustrating the potential of these enzymes for the sustainable development of CS-based pharmaceutical products.

In the second review, presented by Alcántara and coworkers [83], different chemoenzymatic methods (using nitrilases, ketoreductases, and aldolases) for the synthesis of the lateral chain of statins—possessing two stereogenic centers, the absolute configuration of which is vital for the therapeutic activity of these drugs—are presented and compared. Statins—inhibitors of 3-hydroxy-3-methylglutaryl coenzyme A (HMG-CoA) reductase—are the largest selling class of drugs prescribed for the pharmacological treatment of hypercholesterolemia and dyslipidemia. In fact, the statin market involves a huge amount of money, which could potentially increase due to the recently described statins' pleiotropic effects (beneficial effects for cardiovascular health, regulation of the immune system, anti-inflammatory and immunosuppressive properties, prevention and treatment of sepsis, treatment of autoimmune diseases, osteoporosis, kidney and neurological disorders, or even in cancer therapy), which are also noted.

In a similar field, Rhimi and coworkers present in this Special Issue a review of the microbial bioreduction of cholesterol to coprostanol [84], a metabolite poorly absorbed by the human intestine, allowing it to have an impact on cholesterol metabolism and modulation of serum cholesterol levels. This biotransformation is still poorly understood, as few studies are available examining cholesterol-metabolizing bacteria and their associated genes. Thus, by understanding the molecular characterization of these bacterial pathways (currently, three different pathways are proposed), as presented in this review, it could be possible to design new hypocholesterolemic strategies that could be complementary to the previously mentioned prescription of statins.

As mentioned before, eight research articles are included in this Special Issue. In the article presented by Plou and coworkers, a controlled enzymatic hydrolysis of chitosan or chitin with different enzymes was employed to obtain three types of chitooligosaccharides (COS), with molecular weights ranging 0.2–1.2 kDa—namely, fully deacetylated (fdCOS), partially acetylated (paCOS), and fully acetylated (faCOS) chitooligosaccharides [85]. Subsequently, the chemical composition of these biopolymers was established, as well as the anti-inflammatory activity of the three COS mixtures. In this case, this was done by measuring their ability to reduce the level of tumor necrosis factor (TNF) in murine macrophages (RAW 264.7) after stimulation with a mixture of lipopolysaccharides (LPS). Results showed that fdCOS and faCOS chitooligosaccharides effectively displayed anti-inflammatory activity, therefore proving their potentiality.

Sulfuretin is a naturally occurring aurone displaying a remarkable spectrum of biological activities (including against acquired lymphedema, anti-Parkinson's disease activity, antioxidant action, therapeutic benefits in bone disease and regeneration, as well as neuroprotective effects). Although sulfuretin glucosides are important sources of innovative drugs, few glucosides of sulfuretin have been observed in nature, meaning the preparation of sulfuretin glycosides is an attractive research field. In this sense, in the article presented by Kong and coworkers [86], the glycosyltransferase (GT)-catalyzed glycodiversification of sulfuretin is described, specifically using a flavonoid GT (named OcUGT1) for the glucosylation of sulfuretin with UDP-Glc. In this article, ten glycosylated products (three monoglucosides, five diglucosides, and two triglucosides) are characterized; the three monoglucosides were identified as sulfuretin 3′-, 4′-, and 6-glucoside, while the major diglucoside was assigned as sulfuretin 4′,6-diglucoside. The exact structures of the other four diglucosides (traces) were not well characterized, but they were inferred to be sulfuretin 3′,6-diglucoside, sulfuretin 3′,4′-diglucoside, and two disaccharide glucosides. Finally, the structural identification of the remaining two triglucosides was not performed because of their small amount, although one of them was deduced to be sulfuretin 3′,4′,6-triglucoside, according to the previously reported catalytic behavior of OcUGT1. Remarkably, at least six of the ten sulfuretin glucosides are described for the first time, making this a pioneering article in describing the simultaneous production of monoglucosides, diglucosides, and triglucosides of sulfuretin from a single glycosyltransferase.

The use of laccases for catalyzing polymerization of bioactive phenolic compounds has become very attractive because of the enhanced physicochemical and biological properties of the obtained products, which are generally used as nutraceuticals. In this context, the article presented by Eibes and coworkers [87] describes the influence of enzyme activity on rutin (also named rutoside, quercetin-3-*O*-rutinoside, or sophorin) oligomerization using low (1000 U/L) and high (10,000 U/L) initial activities of laccase from *Trametes versicolor* in a food-compatible reaction medium; rutin oligomers with the best characteristics were obtained in the reaction with the lowest laccase activity, significantly improving the apparent aqueous solubility and xanthine oxidase inhibitory activity compared to its control reaction, without compromising the antioxidant activity. The thermal stability of rutin oligomers was negatively affected by increasing the enzyme concentration, and by comparing the antioxidant capacity of similar mean molecular mass oligomers produced with different laccase activities, these authors concluded that higher enzyme dosages promoted the formation of multiple intermolecular bonds between rutin units, which negatively affected their antioxidant activity. Remarkably, this is the first study focusing on the effect of laccase activity upon the products obtained in enzymatic oligomerization of the rutin flavonoid as a key parameter to enhance and tailor their physicochemical and biological properties.

The article presented by Otero and coworkers [88] describes different extraction methods of high-value hydrophilic spirulina biocomponents (peptides of therapeutic interest), using four selective enzymatic degradations of spirulina biomass, catalyzed by two proteases and endo- and exoglucanases, illustrating the usefulness of biocatalysts in this applied field. The four enzyme-assisted extraction processes were optimized, determining best experimental conditions (pH, temperature, enzymatic loading or duration of enzymatic pre-treatment), and scaled up, showing a superior behavior compared to those extractions not employing enzymes. The best results for hydrophilic extraction were obtained using Alcalase® (a serine endo-peptidase from *Bacillus licheniformis*, mainly subtilisin A), because of its effective degradation of membrane proteins, lipoproteins, and peptidoglucan under very mild conditions.

Sterically demanding 2,2-diaryl-2-hydroxy carboxylic acids are valuable chiral building blocks for the synthesis of antimuscarinic agents. When facing a hypothetical enzymatic kinetic resolution of these bulky substrates, there are two main drawbacks: First of all, esters having α-quaternary or α- tertiary centers display high steric hindrance, hampering their approach to the enzymatic active site, (only pig liver esterase (PLE) has been proven active for this purpose). Additionally, esters of 2-hydroxy-2-(3-hydroxyphenyl)-2-phenylacetate (the key precursors for preparation of antimuscarinic agents) display poor stereo discrimination because of the two aromatic groups directly bound to the stereocenter, which differ only by the presence of a meta substituent on one of the two aromatic rings. For this difficult task, Pinto, Carzaniga, and coworkers [89] have described a very smart approach, using a double enzymatic hydrolysis: In fact, because in a first PLE-catalyzed hydrolysis of the quaternary α-hydroxyester, only moderate enantioselectivity (80% ee) was obtained, these authors chemically re-esterified the enantiomerically enriched α-hydroxyacid and carried out a second enzymatic hydrolysis. With this methodology (optimized by choosing suited co-solvents (DMSO) and additives (β-cyclodextrins)), it was feasible to prepare the desired optically pure α,α-diaryl-α-hydroxyacid on a multi-milligram scale.

Speranza, Ubiali, and coworkers present in this Special Issue a paper describing the chemoenzymatic synthesis of ribavirin, tecadenoson, and cladribine (nucleoside analogues, well-established drugs in clinical practice, mainly used as anticancer and antiviral agents) via "one-pot, one-enzyme" transglycosylation; that is, the transfer of the carbohydrate moiety from a nucleoside donor to a heterocyclic base [90]. For this purpose, purine nucleoside phosphorylase from *Aeromonas hydrophila* (AhPNP) was the biocatalyst used, using 7-methylguanosine iodide and its 2′-deoxy counterpart as sugar donors. Good conversions (49%–67%) were achieved in all cases under screening conditions. Similarly, 7-methylguanine arabinoside iodide was prepared for the purpose of synthesizing the antiviral vidarabine via a novel approach, although neither the phosphorolysis of the sugar donor nor the transglycosylation reaction were observed. Finally, this strategy was used to prepare two other ribonucleosides structurally related to ribavirin and tecadenoson, namely, acadesine (5-aminoimidazole-4-carboxamide-1-β-*D*-ribofuranoside, also named AICA-riboside or AICAR) and 2-chloro-N^6-cyclopentyladenosine (CCPA), leading to a moderate yield (52%) for the latter compound only. This study clearly paves the way for the development of a new synthesis of the target Active Pharmaceutical Ingredientes APIs at a preparative scale, and contributes to the understanding of the specific substrate requirements of AhPNP.

The stereoselective synthesis of enantiopure amines is an interesting task because of their role as intermediates in pharmaceutical synthesis, as well as for a variety of other chemical products. There are several biocatalyzed approaches leading to these optically pure amines using different enzymes (amine dehydrogenases, imine reductases, reductive aminases, or amine transaminases). Two contributions in this Special Issue focus on this area. Campos, Gotor-Fernández, and coworkers apply a broad panel of commercially available amine transaminases (ATAs) for the biotransanimation of 3,4-dihydro-2*H*-1,5-benzoxathiepin-3-one (the chemical synthesis of which is reported via some previously undescribed intermediates) to furnish the correspondent enantiopure amines [91]. The optimization of the reaction conditions (enzyme loading, temperature, and reaction times) using ATA03 from *Neosartorya fischeri* and ATA07 from *Mycobacterium vanbaalenii* (leading to the (*S*)-amine), as well as TA-P1-G05 for the (*R*)-counterpart, allowed a milligram-scale synthesis of the pure amines, which are useful as building blocks for the preparation of antiproliferative drugs.

In the second article producing enantiopure amines, Hollmann and coworkers report a methodology based on the aerobic photooxidation (mediated by water-soluble sodium anthraquinone-2-sulfonate (SAS) and heterogeneous graphitic carbon nitride (g-C_3N_4)) of primary and racemic secondary alcohols to the corresponding aldehydes and prochiral ketones, which are subsequently transformed into the (chiral) amines with commercial aminotransaminases (one (*R*) and four (*S*)-ATAs), using isopropylamine as the sacrificial amine donor [92]. The system worked in a one-pot, one-step fashion; as time-consuming intermediate isolation and purification steps were omitted, the required amount of organic solvents was minimized, with a concomitant reduction of waste. The productivity was significantly improved by switching to a "one-pot, two-step" procedure. A wide range of aliphatic and aromatic compounds was transformed into the enantiomerically pure corresponding amines via the photo-enzymatic cascade, providing good yields and excellent ee values.

In conclusion, these eleven papers clearly illustrate the versatility of biocatalysis in the preparation of bioactive compounds. Once again, let me reiterate what an honor it has been for me to act as Guest Editor of this Special Issue. I would like to thank all of the authors participating in it, as well as all of the reviewers for providing me with their valuable comments. Finally, I cannot refrain from thanking all of the staff of the *Catalysts* Editorial Office, especially Caroline Zhan, Associated Editor, whose efforts made this Special Issue possible.

Funding: This research received no external funding.

Conflicts of Interest: The authors declare no conflict of interest.

References

1. Truppo, M.D. Biocatalysis in the Pharmaceutical Industry: The Need for Speed. *ACS Med. Chem. Lett.* **2017**, *8*, 476–480. [CrossRef] [PubMed]
2. Faber, K. *Biotransformations in Organic Chemistry: A Textbook*, 7th ed.; Springer International Publishing AG: Cham, Switzerland, 2018. [CrossRef]
3. Bezborodov, A.M.; Zagustina, N.A. Enzymatic Biocatalysis in Chemical Synthesis of Pharmaceuticals (Review). *Appl. Biochem. Microbiol.* **2016**, *52*, 237–249. [CrossRef]
4. Hoyos, P.; Pace, V.; Hernáiz, M.J.; Alcántara, A.R. Biocatalysis in the Pharmaceutical Industry. A greener future. *Curr. Green Chem.* **2014**, *1*, 155–181. [CrossRef]
5. Lopez-Iglesias, M.; Mendez-Sanchez, D.; Gotor-Fernandez, V. Native Proteins in Organic Chemistry. Recent Achievements in the Use of Non Hydrolytic Enzymes for the Synthesis of Pharmaceuticals. *Curr. Org. Chem.* **2016**, *20*, 1204–1221. [CrossRef]
6. Patel, R.N. Applications of Biocatalysis for Pharmaceuticals and Chemicals. In *Organic Synthesis Using Biocatalysis*; Stewart, J.D., Ed.; Elsevier: Amsterdam, The Netherlands, 2016; pp. 339–411. [CrossRef]
7. Patel, R.N. Pharmaceutical Intermediates by Biocatalysis: From Fundamental Science to Industrial Applications. In *Applied Biocatalysis: From Fundamental Science to Industrial Applications*; Wiley-VCH GmbH & Co. KGaA: Weinheim, Germany, 2016; pp. 367–403. [CrossRef]
8. Patel, R.N. Green Processes for the Synthesis of Chiral Intermediates for the Development of Drugs. In *Green Biocatalysis*; John Wiley & Sons, Inc.: Hoboken, NJ, USA, 2016; pp. 71–114. [CrossRef]
9. Hoyos, P.; Pace, V.; Alcántara, A.R. Chiral Building Blocks for Drugs Synthesis via Biotransformations. In *Asymmetric Synthesis of Drugs and Natural Products*; Nag, A., Ed.; CRC Press: Boca Raton, FL, USA, 2018; pp. 346–448.
10. Alcántara, A.R. Biotransformations in Drug Synthesis: A Green and Powerful Tool for Medicinal Chemistry. *J. Med. Chem. Drug Des.* **2018**, *1*, 1–7.
11. Sun, H.; Zhang, H.; Ang, E.L.; Zhao, H. Biocatalysis for the synthesis of pharmaceuticals and pharmaceutical intermediates. *Bioorg. Med. Chem.* **2018**, *26*, 1275–1284. [CrossRef] [PubMed]
12. Rosenthal, K.; Lutz, S. Recent developments and challenges of biocatalytic processes in the pharmaceutical industry. *Curr. Opin. Green Sustain. Chem.* **2018**, *11*, 58–64. [CrossRef]
13. Patel, R.N. Biocatalysis for synthesis of pharmaceuticals. *Bioorg. Med. Chem.* **2018**, *26*, 1252–1274. [CrossRef]
14. Groger, H. Enzyme catalysis in the synthesis of pharmaceuticals. *Bioorg. Med. Chem.* **2018**, *26*, 1239–1240. [CrossRef]
15. Li, G.Y.; Wang, J.B.; Reetz, M.T. Biocatalysts for the pharmaceutical industry created by structure-guided directed evolution of stereoselective enzymes. *Bioorg. Med. Chem.* **2018**, *26*, 1241–1251. [CrossRef] [PubMed]
16. Adams, J.P.; Brown, M.J.B.; Diaz-Rodriguez, A.; Lloyd, R.C.; Roiban, G.D. Biocatalysis: A Pharma Perspective. *Adv. Synth. Catal.* **2019**, *361*, 2421–2432. [CrossRef]
17. Hughes, D.L. Biocatalysis in drug development-highlights of the recent patent literature. *Org. Process Res. Dev.* **2018**, *22*, 1063–1080. [CrossRef]
18. Devine, P.N.; Howard, R.M.; Kumar, R.; Thompson, M.P.; Truppo, M.D.; Turner, N.J. Extending the application of biocatalysis to meet the challenges of drug development. *Nat. Rev. Chem.* **2018**, *2*, 409–421. [CrossRef]
19. Green, A.P.; Turner, N.J. Biocatalytic retrosynthesis: Redesigning synthetic routes to high-value chemicals. *Perspect. Sci.* **2016**, *9*, 42–48. [CrossRef]
20. Hoyos, P.; Hernáiz, M.J.; Alcántara, A.R. Biocatalyzed Production of Fine Chemicals. In *Reference Module in Life Sciences*; Moo-Young, M., Ed.; Pergamon: Oxford, UK, 2017. [CrossRef]
21. Thompson, M.P.; Penafiel, I.; Cosgrove, S.C.; Turner, N.J. Biocatalysis using immobilized enzymes in continuous flow for the synthesis of fine chemicals. *Org. Process Res. Dev.* **2019**, *23*, 9–18. [CrossRef]
22. Bilal, M.; Iqbal, H.M.N. State-of-the-art strategies and applied perspectives of enzyme biocatalysis in food sector—Current status and future trends. *Crit. Rev. Food Sci. Nutr.* **2019**, 1–15. [CrossRef] [PubMed]
23. Bilal, M.; Iqbal, H.M.N. Sustainable bioconversion of food waste into high-value products by immobilized enzymes to meet bio-economy challenges and opportunities—A review. *Food Res. Int.* **2019**, *123*, 226–240. [CrossRef] [PubMed]

24. Raveendran, S.; Parameswaran, B.; Ummalyma, S.B.; Abraham, A.; Mathew, A.K.; Madhavan, A.; Rebello, S.; Pandey, A. Applications of microbial enzymes in food industry. *Food Technol. Biotechnol.* **2018**, *56*, 16–30. [CrossRef]
25. Li, C.-J.; Anastas, P.T. Green Chemistry: Present and future. *Chem. Soc. Rev.* **2012**, *41*, 1413–1414. [CrossRef]
26. Anastas, P.; Eghbali, N. Green Chemistry: Principles and Practice. *Chem. Soc. Rev.* **2010**, *39*, 301–312. [CrossRef]
27. Sheldon, R.A. Fundamentals of green chemistry: Efficiency in reaction design. *Chem. Soc. Rev.* **2012**, *41*, 1437–1451. [CrossRef] [PubMed]
28. Dunn, P.J. The importance of Green Chemistry in Process Research and Development. *Chem. Soc. Rev.* **2012**, *41*, 1452–1461. [CrossRef] [PubMed]
29. Erythropel, H.C.; Zimmerman, J.B.; de Winter, T.M.; Petitjean, L.; Melnikov, F.; Lam, C.H.; Lounsbury, A.W.; Mellor, K.E.; Jankovic, N.Z.; Tu, Q.S.; et al. The Green ChemisTREE: 20 years after taking root with the 12 principles. *Green Chem.* **2018**, *20*, 1929–1961. [CrossRef]
30. Marion, P.; Bernela, B.; Piccirilli, A.; Estrine, B.; Patouillard, N.; Guilbot, J.; Jerome, F. Sustainable chemistry: How to produce better and more from less? *Green Chem.* **2017**, *19*, 4973–4989. [CrossRef]
31. Smith, D.J. The Past, Present, and Future of Sustainable Chemistry. *ChemSusChem* **2018**, *11*, 5–10. [CrossRef] [PubMed]
32. Kummerer, K. Sustainable Chemistry: A Future Guiding Principle. *Angew. Chem. Int. Ed.* **2017**, *56*, 16420–16421. [CrossRef] [PubMed]
33. Anastas, P.T.; Zimmerman, J.B. The United Nations sustainability goals: How can sustainable chemistry contribute? *Curr. Opin. Green Sustain. Chem.* **2018**, *13*, 150–153. [CrossRef]
34. Horváth, I.T. Introduction: Sustainable Chemistry. *Chem. Rev.* **2018**, *118*, 369–371. [CrossRef]
35. Halpaap, A.; Dittkrist, J. Sustainable chemistry in the global chemicals and waste management agenda. *Curr. Opin. Green Sustain. Chem.* **2018**, *9*, 25–29. [CrossRef]
36. Hogue, C. Differentiating green chemistry from sustainable chemistry. *Chem. Eng. News* **2019**, *97*, 19.
37. Noce, A.M. Green chemistry and the grand challenges of sustainability. In *Green Chemistry Education: Recent Developments*; Benvenuto, M.A., Kolopajlo, L., Eds.; Walter De Gruyter Gmbh: Berlin, Germany, 2019; pp. 1–11. [CrossRef]
38. Patel, R.N. *Green Biocatalysis*; John Wiley & Sons, Inc.: Hoboken, NJ, USA, 2016. [CrossRef]
39. Sheldon, R.A. Biocatalysis and Green Chemistry. In *Green Biocatalysis*; John Wiley & Sons, Inc.: Hoiboken, NJ, USA, 2016; pp. 1–15.
40. Sheldon, R.A.; Woodley, J.M. Role of Biocatalysis in Sustainable Chemistry. *Chem. Rev.* **2017**, *118*, 801–838. [CrossRef] [PubMed]
41. Sethi, M.K.; Chakraborty, P.; Shukla, R. Biocatalysis—A Greener Alternative in Synthetic Chemistry. In *Biocatalysis: An Industrial Perspective*; The Royal Society of Chemistry: Cambrige, UK, 2018; pp. 44–76. [CrossRef]
42. Shoda, S.; Uyama, H.; Kadokawa, J.; Kimura, S.; Kobayashi, S. Enzymes as Green Catalysts for Precision Macromolecular Synthesis. *Chem. Rev.* **2016**, *116*, 2307–2413. [CrossRef] [PubMed]
43. Sheldon, R.A.; Brady, D. Broadening the Scope of Biocatalysis in Sustainable Organic Synthesis. *ChemSusChem* **2019**, *12*, 2859–2881. [CrossRef] [PubMed]
44. Guajardo, N.; Domínguez de María, P. Continuous biocatalysis in environmentally-friendly media: A triple synergy for future sustainable processes. *ChemCatChem* **2019**, *11*, 3128–3137. [CrossRef]
45. Foley, A.M.; Maguire, A.R. The Impact of Recent Developments in Technologies which Enable the Increased Use of Biocatalysts. *Eur. J. Org. Chem.* **2019**, *2019*, 3713–3734. [CrossRef]
46. Faber, K.; Fessner, W.D.; Turner, N.J. Biocatalysis: Ready to Master Increasing Complexity. *Adv. Synth. Catal.* **2019**, *361*, 2373–2376. [CrossRef]
47. Roschangar, F.; Colberg, J.; Dunn, P.J.; Gallou, F.; Hayler, J.D.; Koenig, S.G.; Kopach, M.E.; Leahy, D.K.; Mergelsberg, I.; Tucker, J.L.; et al. A deeper shade of green: Inspiring sustainable drug manufacturing. *Green Chem.* **2017**, *19*, 281–285. [CrossRef]
48. Sheldon, R.A. The E factor 25 years on: The rise of green chemistry and sustainability. *Green Chem.* **2017**, *19*, 18–43. [CrossRef]
49. McElroy, C.R.; Constantinou, A.; Jones, L.C.; Summerton, L.; Clark, J.H. Towards a holistic approach to metrics for the 21st century pharmaceutical industry. *Green Chem.* **2015**, *17*, 3111–3121. [CrossRef]

50. Roschangar, F.; Sheldon, R.A.; Senanayake, C.H. Overcoming barriers to green chemistry in the pharmaceutical industry—The Green Aspiration Level [trade mark sign] concept. *Green Chem.* **2015**, *17*, 752–768. [CrossRef]
51. Lin, J.-L.; Palomec, L.; Wheeldon, I. Design and Analysis of Enhanced Catalysis in Scaffolded Multienzyme Cascade Reactions. *ACS Catal.* **2014**, *4*, 505–511. [CrossRef]
52. Sigrist, R.; da Costa, B.Z.; Marsaioli, A.J.; de Oliveira, L.G. Nature-inspired enzymatic cascades to build valuable compounds. *Biotechnol. Adv.* **2015**, *33*, 394–411. [CrossRef] [PubMed]
53. O'Reilly, E.; Turner, N.J. Enzymatic cascades for the regio- and stereoselective synthesis of chiral amines. *Perspect. Sci.* **2015**, *4*, 55–61. [CrossRef]
54. Schrittwieser, J.H.; Velikogne, S.; Hall, M.; Kroutil, W. Artificial Biocatalytic Linear Cascades for Preparation of Organic Molecules. *Chem. Rev.* **2017**, *118*, 270–348. [CrossRef] [PubMed]
55. Bornscheuer, U.T. (Chemo-) enzymatic cascade reactions. *Z. Naturforsch. C* **2019**, *74*, 61–62. [CrossRef] [PubMed]
56. Ruales-Salcedo, A.V.; Higuita, J.C.; Fontalvo, J.; Woodley, J.M. Design of enzymatic cascade processes for the production of low-priced chemicals. *Z. Naturforsch. C* **2019**, *74*, 77–84. [CrossRef] [PubMed]
57. Groger, H.; Schallmey, A.; Schwab, H.; Kourist, R. Multi-enzyme cascades as synthetic tool for biocatalysis. *J. Biotechnol.* **2019**, *294*, 88. [CrossRef] [PubMed]
58. Rudroff, F. Whole-cell based synthetic enzyme cascades-light and shadow of a promising technology. *Curr. Opin. Chem. Biol.* **2019**, *49*, 84–90. [CrossRef]
59. Woodley, J.M. Scale-Up and Development of Enzyme-Based Processes for Large-Scale Synthesis Applications. In *Biocatalysis in Organic Synthesis*; Faber, K., Fessner, W.D., Turner, N.J., Eds.; Georg Thieme: Stuttgart, Germany, 2015; pp. 515–546. [CrossRef]
60. Ramesh, H.; Nordblad, M.; Whittall, J.; Woodley, J.M. Considerations for the Application of Process Technologies in Laboratory- and Pilot-Scale Biocatalysis for Chemical Synthesis. In *Practical Methods for Biocatalysis and Biotransformations 3*; John Wiley & Sons, Ltd.: Hoboken, NJ, USA, 2016; pp. 1–30. [CrossRef]
61. Rogers, L.; Jensen, K.F. Continuous manufacturing—The Green Chemistry promise? *Green Chem.* **2019**, *21*, 3481–3498. [CrossRef]
62. Lindeque, R.M.; Woodley, J.M. Reactor Selection for Effective Continuous Biocatalytic Production of Pharmaceuticals. *Catalysts* **2019**, *9*, 262. [CrossRef]
63. Britton, J.; Majumdar, S.; Weiss, G.A. Continuous flow biocatalysis. *Chem. Soc. Rev.* **2018**, *47*, 5891–5918. [CrossRef] [PubMed]
64. Bornscheuer, U.T.; Huisman, G.W.; Kazlauskas, R.J.; Lutz, S.; Moore, J.C.; Robins, K. Engineering the third wave of biocatalysis. *Nature* **2012**, *485*, 185–194. [CrossRef]
65. Wang, X.Y.; Wang, B.Y.; Wang, Z.W.; Chen, T.; Zhao, X.M. The Research Progress of Protein Directed Evolution. *Prog. Biochem. Biophys.* **2015**, *42*, 123–131.
66. Porter, J.L.; Boon, P.L.; Murray, T.P.; Huber, T.; Collyer, C.A.; Ollis, D.L. Directed evolution of new and improved enzyme functions using an evolutionary intermediate and multidirectional search. *ACS Chem. Biol.* **2015**, *10*, 611–621. [CrossRef]
67. Currin, A.; Swainston, N.; Day, P.J.; Kell, D.B. Synthetic biology for the directed evolution of protein biocatalysts: Navigating sequence space intelligently. *Chem. Soc. Rev.* **2015**, *44*, 1172–1239. [CrossRef] [PubMed]
68. Renata, H.; Wang, Z.J.; Arnold, F.H. Expanding the Enzyme Universe: Accessing Non-Natural Reactions by Mechanism-Guided Directed Evolution. *Angew. Chem. Int. Ed.* **2015**, *54*, 3351–3367. [CrossRef]
69. Denard, C.A.; Ren, H.Q.; Zhao, H.M. Improving and repurposing biocatalysts via directed evolution. *Curr. Opin. Chem. Biol.* **2015**, *25*, 55–64. [CrossRef] [PubMed]
70. Porter, J.L.; Rusli, R.A.; Ollis, D.L. Directed Evolution of Enzymes for Industrial Biocatalysis. *ChemBiochem* **2016**, *17*, 197–203. [CrossRef]
71. Davis, A.M.; Plowright, A.T.; Valeur, E. Directing evolution: The next revolution in drug discovery? *Nat. Rev. Drug Discov.* **2017**, *16*, 681–698. [CrossRef]
72. Bornscheuer, U.T.; Hauer, B.; Jaeger, K.E.; Schwaneberg, U. Directed evolution empowered redesign of natural proteins for the sustainable production of chemicals and pharmaceuticals. *Angew. Chem. Int. Ed.* **2019**, *58*, 36–40. [CrossRef]
73. Zeymer, C.; Hilvert, D. Directed evolution of protein catalysts. In *Annual Review of Biochemistry*; Kornberg, R.D., Ed.; Annual Reviews: Palo Alto, CA, USA, 2018; Volume 87, pp. 131–157.

74. Arnold, F.H. Directed Evolution: Bringing New Chemistry to Life. *Angew. Chem. Int. Ed.* **2018**, *57*, 4143–4148. [CrossRef]
75. Reetz, M.T. *Directed Evolution of Selective Enzymes: Catalysts for Organic Chemistry and Biotechnology*; Wiley-VCH: Weinheim, Germany, 2017.
76. Illanes, A. Dr. Frances Arnold is awarded with the Nobel Prize in Chemistry 2018: Good news for biocatalysis. *Electron. J. Biotechnol.* **2018**, *36*, A1. [CrossRef]
77. Jones, C.W. Another Nobel Prize for Catalysis: Frances Arnold in 2018. *ACS Catal.* **2018**, *8*, 10913. [CrossRef]
78. Poppe, L.; Vertessy, B.G. The Fourth Wave of Biocatalysis Emerges The 13th International Symposium on Biocatalysis and Biotransformations. *ChemBiochem* **2018**, *19*, 284–287. [CrossRef]
79. Bornscheuer, U.T. The fourth wave of biocatalysis is approaching. *Philos. Trans. R. Soc. A Math. Phys. Eng. Sci.* **2018**, *376*, 7. [CrossRef]
80. Sheldon, R.A.; Brady, D. The limits to biocatalysis: Pushing the envelope. *Chem. Commun.* **2018**, *54*, 6088–6104. [CrossRef]
81. Woodley, J.M. Accelerating the implementation of biocatalysis in industry. *Appl. Microbiol. Biotechnol.* **2019**, *103*, 4733–4739. [CrossRef]
82. Benito-Arenas, R.; Zarate, S.G.; Revuelta, J.; Bastida, A. Chondroitin sulfate-degrading enzymes as tools for the development of new pharmaceuticals. *Catalysts* **2019**, *9*, 322. [CrossRef]
83. Hoyos, P.; Pace, V.; Alcántara, A.R. Biocatalyzed synthesis of statins: A sustainable strategy for the preparation of valuable drugs. *Catalysts* **2019**, *9*, 260. [CrossRef]
84. Kriaa, A.; Bourgin, M.; Mkaouar, H.; Jablaoui, A.; Akermi, N.; Soussou, S.; Maguin, E.; Rhimi, M. Microbial reduction of cholesterol to coprostanol: An old concept and new insights. *Catalysts* **2019**, *92*, 167. [CrossRef]
85. Santos-Moriano, P.; Kidibule, P.; Miguez, N.; Fernandez-Arrojo, L.; Ballesteros, A.O.; Fernandez-Lobato, M.; Plou, F.J. Tailored enzymatic synthesis of chitooligosaccharides with different deacetylation degrees and their anti-inflammatory activity. *Catalysts* **2019**, *9*, 405. [CrossRef]
86. Yuan, S.; Xu, Y.L.; Yang, Y.; Kong, J.Q. OcUGT1-catalyzed glucosylation of sulfuretin yields ten glucosides. *Catalysts* **2018**, *8*, 416. [CrossRef]
87. Muniz-Mouro, A.; Gullon, B.; Lu-Chau, T.A.; Moreira, M.T.; Lema, J.M.; Eibes, G. Laccase activity as an essential factor in the oligomerization of rutin. *Catalysts* **2018**, *8*, 321. [CrossRef]
88. Verdasco-Martin, C.M.; Echevarrieta, L.; Otero, C. Advantageous preparation of digested proteic extracts from *Spirulina platensis* biomass. *Catalysts* **2019**, *9*, 145. [CrossRef]
89. Pinto, A.; Serra, I.; Romano, D.; Contente, M.L.; Molinari, F.; Rancati, F.; Mazzucato, R.; Carzaniga, L. Preparation of sterically demanding 2,2-disubstituted-2-hydroxy acids by enzymatic hydrolysis. *Catalysts* **2019**, *9*, 113. [CrossRef]
90. Rabuffetti, M.; Bavaro, T.; Semproli, R.; Cattaneo, G.; Massone, M.; Morelli, C.F.; Speranza, G.; Ubiali, D. Synthesis of ribavirin, tecadenoson, and cladribine by enzymatic transglycosylation. *Catalysts* **2019**, *9*, 355. [CrossRef]
91. Gonzalez-Martinez, D.; Fernandez-Saez, N.; Cativiela, C.; Campos, J.M.; Gotor-Fernandez, V. Development of biotransamination reactions towards the 3,4-dihydro-2H-1,5-benzoxathiepin-3-amine enantiomers. *Catalysts* **2018**, *8*, 470. [CrossRef]
92. Gacs, J.; Zhang, W.Y.; Knaus, T.; Mutti, F.G.; Arends, I.; Hollmann, F. A photo-enzymatic cascade to transform racemic alcohols into enantiomerically pure amines. *Catalysts* **2019**, *9*, 305. [CrossRef]

Review

Chondroitin Sulfate-Degrading Enzymes as Tools for the Development of New Pharmaceuticals

Raúl Benito-Arenas [1], Sandra G. Zárate [2], Julia Revuelta [1,*] and Agatha Bastida [1,*]

1 Departamento de Química Bio-orgánica, IQOG-CSIC. c/Juan de la Cierva 3, E-28006 Madrid, Spain; rbenito@iqog.csic.es

2 Facultad de Tecnología, Carrera de Ingeniería Química, Universidad Mayor Real y Pontífice de San Francisco Xavier de Chuquisaca, Regimiento Campos 180, Casilla 60-B, Sucre, Bolivia; zarate.sandra@usfx.bo

* Correspondence: julia.revuelta@iqog.csic.es (J.R.); agatha.bastida@csic.es (A.B.); Tel.: +34-9-1562-2900 (J.R. & A.B.)

Received: 27 February 2019; Accepted: 19 March 2019; Published: 1 April 2019

Abstract: Chondroitin sulfates are linear anionic sulfated polysaccharides found in biological tissues, mainly within the extracellular matrix, which are degraded and altered by specific lyases depending on specific time points. These polysaccharides have recently acquired relevance in the pharmaceutical industry due to their interesting therapeutic applications. As a consequence, chondroitin sulfate (CS) lyases have been widely investigated as tools for the development of new pharmaceuticals based on these polysaccharides. This review focuses on the major breakthrough represented by chondroitin sulfate-degrading enzymes and their structures and mechanisms of function in addition to their major applications.

Keywords: lyases; chondroitin sulfates; glycosaminoglycan; polysaccharides

1. Introduction

Proteoglycans consist of a central protein core with *O*-linked glycosaminoglycan (GAG) side-chains. They can be categorized into four main groups based on differences between the repeating disaccharide units comprising GAGs: heparan sulfate (HS), chondroitin sulfate (CS), dermatan sulfate (DS) and hyaluronic acid (HA) (Figure 1A).

CSs are a family of highly sulfated polysaccharides that have recently acquired relevance in the pharmaceutical industry due to their interesting therapeutic applications [1]. Structurally, these are linear polysaccharides and their basic unit is a disaccharide composed of a D-glucopyranosyluronic acid (◊) or L-idopyranosiduronic acid (◊) glycosidically linked (β (1→3)) to an *N*-acetyl-D-galactosamine residue (□) in which the hydroxyl groups undergo sulfation at one or more positions. The disaccharide subunits found in natural CSs are shown in Figure 1B.

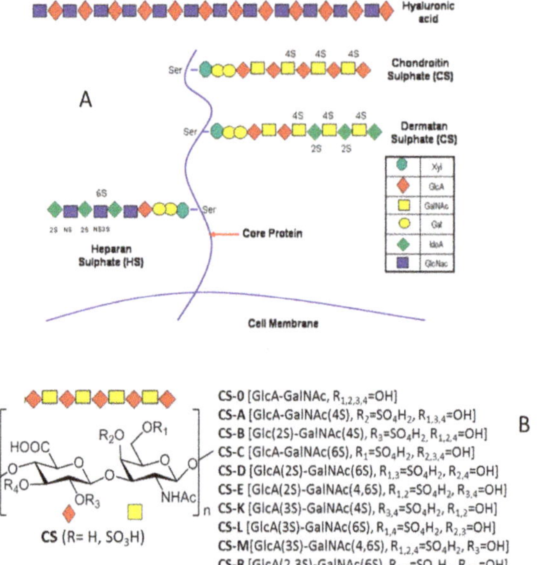

HOOC, R_2O, OR_1, R_4O, OR_3, NHAc, n

CS (R= H, SO_3H)

CS-0 [GlcA-GalNAc, $R_{1,2,3,4}$=OH]
CS-A [GlcA-GalNAc(4S), R_2=SO_4H_2, $R_{1,3,4}$=OH]
CS-B [Glc(2S)-GalNAc(4S), R_3=SO_4H_2, $R_{1,2,4}$=OH]
CS-C [GlcA-GalNAc(6S), R_1=SO_4H_2, $R_{2,3,4}$=OH]
CS-D [GlcA(2S)-GalNAc(6S), $R_{1,3}$=SO_4H_2, $R_{2,4}$=OH]
CS-E [GlcA(2S)-GalNAc(4,6S), $R_{1,2}$=SO_4H_2, $R_{3,4}$=OH]
CS-K [GlcA(3S)-GalNAc(4S), $R_{3,4}$=SO_4H_2, $R_{1,2}$=OH]
CS-L [GlcA(3S)-GalNAc(6S), $R_{1,4}$=SO_4H_2, $R_{2,3}$=OH]
CS-M[GlcA(3S)-GalNAc(4,6S), $R_{1,2,4}$=SO_4H_2, R_3=OH]
CS-R [GlcA(2,3S)-GalNAc(6S), $R_{3,4}$=SO_4H_2, $R_{1,2}$=OH]

B

Figure 1. (**A**) The general structure of proteoglycans. The chemical structures of glycosaminoglycans (GAGs) are shown. (**B**) The natural sulfation patterns of the chondroitin sulfates (CSs).

Their ubiquity in the human body and their essential functions for life have aroused great interest in their medical applications [2]. In addition to their well-known applications in the treatment of osteoarthritis [3] and thrombosis [4], other potential pharmaceutical applications have been proposed [5]. Table 1 summarizes the physiological functions of CSs in animal cells and tissues and their potential medical/pharmacological applications.

Table 1. Physiological functions in animal cells and tissues and potential medical/pharmacological applications of CSs.

Physiological Function	Reference
Cell–cell/cell–matrix interactions	[6]
Immune modulation	[7,8]
Host–pathogen interactions	[9,10]
Anticoagulant activities	[11,12]
Potential Therapeutic Application	
Anti-inflammatory	[13,14]
Antiviral	[15,16]
Antimalarial vaccine	[17,18]
Anticancer	[19,20]
Antiparasitic	[21]
Biomarker	[22]
Liver regeneration	[23]
Repair of the central nervous system	[24,25]
Neuroprotective	[26]
Wound healing	[27]

Additionally, CSs are widely used in other pharmacological applications such as coating materials for implants, hydrogels in controlled release applications, components of 3D-constructs such as tissue engineering scaffolds and even as biosensors in diagnostic devices [8,28]. Table 2 summarizes these applications of CS.

Table 2. Applications of CS in 2D and 3D systems.

Glycosaminoglycan Conjugate	Type of Application	Target Tissue/Application Method	Reference
Biotinylated Hyaluronic Acid(HA)CS	Material coating	In vitro biosensor	[29]
Collagen/CS	Implant coating	Osseointegration	[30,31]
Gelatin methaclylate/CS methacrylate	Hydrogel	Cartilage regeneration	[32]
Cross-linked thiolated HA, CS, Heparine, gelatin	Biodegradable hydrogel	Drug/Growth factor (GF) delivery, tissue regeneration	[33–35]
Cross-linked CS–tyramine	Hydrogel	Drug delivery	[36,37]
Chemically cross-linked HA/CS	Hydrogel matrix/particles	Wound dressing, skin regeneration	[38–41]
Photochemically cross-linked HA/CS	Hydrogel	Cell encapsulation, cartilage repair	[42]
Collagen-CS	Porous scaffold	Neovascularization, tissue (bone) regeneration	[43,44]
Genipin cross-linked HA, CS	Porous scaffold	Cartilage regeneration	[45]
Carbodiimide-cross-linked HA, CS, Dermatan sulfate (DS), Chitosan gelatin	Porous scaffold	Cartilage regeneration	[46,47]
Electrospun collagen–CS	Porous mesh	Artificial extracellular matrix(ECM), cartilage regeneration	[48–50]
Chitosan–CS	Nanoparticles	GF delivery, bone regeneration	[51]

However, CSs present not only positive outcomes with regard to their pharmaceutical applications—numerous downsides have also been pointed out. On the one hand, CSs extracted from natural sources have a high structural diversity in terms of their molecular weight (M_w) and degree of sulfation. On the other hand, the preservation of the functionality of CSs while maintaining their biocompatibility is a challenging task and must be addressed depending on the specific application.

To provide solutions to these drawbacks, new approximations have been described during the last decade, with the enzymatic modification of CSs being one of the most widely employed [52,53]. In this context, glycosaminoglycan lyases (GAGLs) can be combined with separation methods for the preparation of CS oligosaccharides for biological evaluations as well as for disaccharide analysis and polysaccharide sequencing [54].

These enzymes have important therapeutic value for the treatment of diseases related to GAGs [55]. Hence, chondroitinase ABC, for example, is being tested in clinical trials for the treatment of spinal cord injury [56]. The same CSase inhibits melanoma invasion, proliferation and angiogenesis [57], and has also been applied as a subretinal injection [58] and for the treatment of intervertebral disc protrusion [59]. On the other hand, HAase has been successfully used as an adjuvant for infiltration anesthesia due to the increased membrane permeability induced by the hydrolysis of HA [60]. Finally, some GAG-degrading enzymes are used as an adjuvant therapy in cancer, in which their administration to reduce the progression of metastatic breast cancer is well tolerated without adverse events [61,62].

Despite the great interest in these direct medical applications, in this review we focused only on the GAG-degrading enzymes, specifically CSases as biocatalytic tools for the development of new pharmaceuticals based on GAGs.

2. Types, Mechanism and Structure of CS Lyases

GAG-degrading enzymes are widely distributed in nature and are structurally diverse depending on whether they are produced by eukaryotic or prokaryotic organisms. These enzymes catalyze the depolymerization of GAGs and are classified according to their enzymatic mechanism in two categories;

hydrolases and lyases. They act with an extremely high degree of stereospecificity. Additionally, a classification based on amino acid sequence similarities has been proposed (http:www.cazy.org).

Mammalian enzymes are hydrolases and their mechanism is the same as glucosidases in which the glycosyl–oxygen (C1–*O*) bond is hydrolyzed by the addition of a water molecule [63], affording saturated oligosaccharide products. In contrast, bacterial enzymes degrade GAGs either through hydrolysis or by a β-elimination reaction (lyases). The latter degrades GAGs by cleaving the oxygen–aglycone (*O*–C4) linkages on the non-reducing side of uronic acids yielding unsaturated C4–C5 products [64–66]. The mechanism of action of these enzymes is shown in Figure 2. In the first step, the negative charge on the C5 carboxylate group is neutralized presumably by interaction with a positively charged arginine or calcium ion, thereby reducing the acidity of the C5 proton. Next, the proton at C5 of the GlcA is abstracted by a His residue, leading to the elimination of the 4-*O*–glycosidic bond and the formation of a double bond between C4–C5 of the uronic acid. Finally, an acidic residue of the protein (Tyr) donates a proton to the *O*-leaving group of the glucosamine, reconstituting the hydroxyl functional group at the reducing end of the cleaved bond and releasing the products.

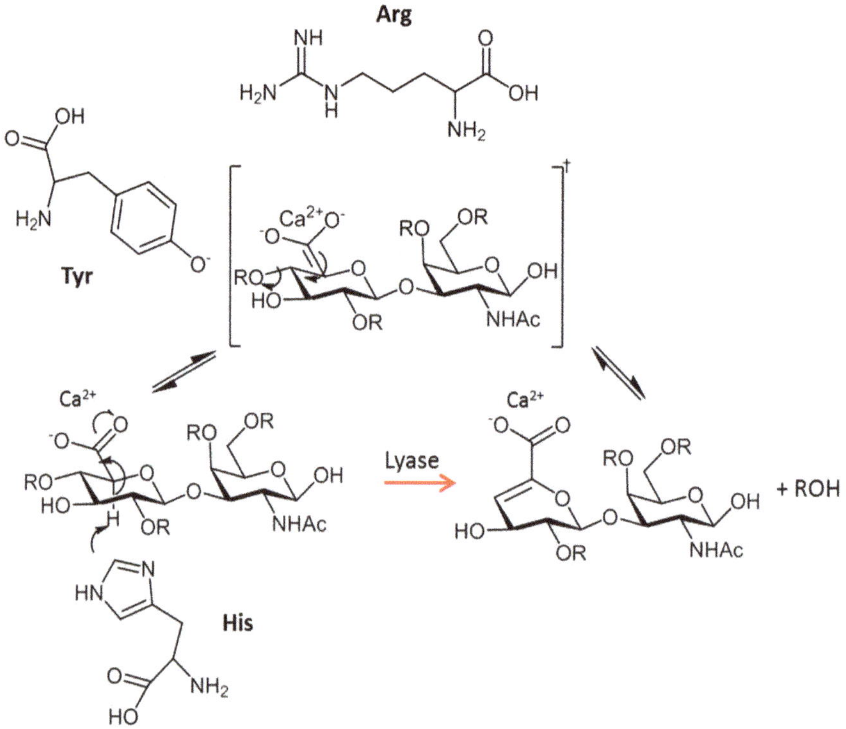

Figure 2. The catalytic mechanism of the GAG-degrading enzymes by elimination cleavage (lyases).

These enzymes catalyze reactions with an extremely high degree of stereospecificity, with extensive variation in specificity among lyases for different GAG classes [54]. Accordingly, these enzymes can be divided into three groups depending on the composition of the repeating disaccharide unit: GlcA-β(1-4)-GlcNAc for heparinases (Hsases) and hyaluronidases (HAases), and GlcA-β(1-3)-GlcNAc for chondroitinases (CSases) [67]. In this review we focused on those that degrade polysaccharides in which the uronic acids are β(1→3) linked to N-acetyl-D-galactosamine (CS/DS) or to N-acetyl-D-glucosamine (HA) (Figure 3). Some exceptions have been found to this stereospecificity. Hence, an HSase isolated from *Erpobdellidae* (*Nephelopsis obscura* and *Erpobdella* punctata), for example, degrades HA by hydrolysis [68,69] while HAase lyase from *Streptococcus pneumoniae* degrades HA/CS by β-elimination [66].

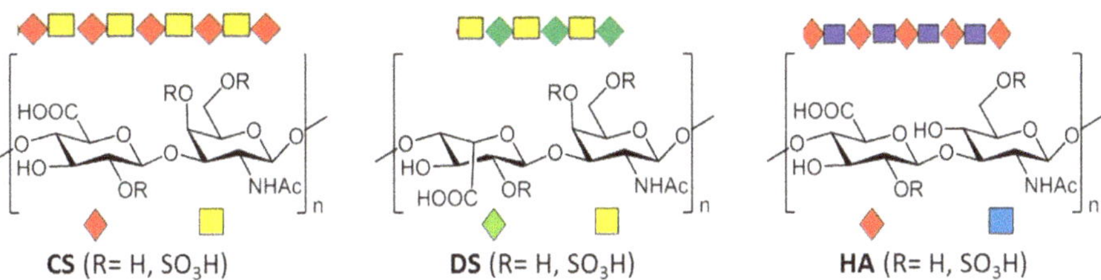

Figure 3. The chemical structure of GAG disaccharide-repeat building blocks of CS, dermatan sulfate (DS) and hyaluronic acid (HA).

In addition to the GAG-class specificity, the majority of these enzymes degrade glycosidic bonds with absolute uronic acid epimer specificity towards either GlcA or IdoA, as well as being dependent on their sulfation pattern. Thus, CSase AC specifically degrades the glucoronic acid-containing glycosidic bonds present in CS-A (chondroitin 4-sulfate) and CS-C(chondroitin 6-sulfate). In fact, the nomenclature for the CSases has been established based on this sulfation pattern specificity. Hence, CSase AC, for instance, cleaves CS-A and CS-C but not DS (chondroitin 2,4-Disulfate, CS-B).

Finally, the GAG-degrading enzymes can present endolytic or exolytic modes of action. In the former, the cleavage occurs in the middle of the GAG chain, yielding a mixture of disaccharides, tetrasaccharides and longer oligosaccharides. In the latter, the enzyme degrades the chain from the end, releasing only disaccharide products.

Table 3 and the following text summarize the findings regarding the classes, mechanisms and structures of the CS-specific lyases.

Table 3. CS-degrading enzymes (lyases).

Lyase	Substrate	PDB	Species	Catalytic tetrad	Ref.
CSase ABC II (exo) CSase ABC II (exo) CSase ABC I (endo)	CS, DS, HA Tetra-CS, Tetra-DS CS, DS, HA	2Q1F 1HNO	*Bacteroides thetaiotaomicron* *Proteus vulgaris* *Proteus vulgaris*	Glu^{628}–Tyr^{461}–His^{454}– Arg^{514} His^{453}–Tyr^{460} His^{501}–Tyr^{508}–Arg^{560}– Glu^{653}	[70–74]
CSase AC II (Exo) CSase AC I (Endo) CSase AC^{Y253A}	CS, HA CS, HA	1RWF;1RWG;1RWH;1RW9;1RWA;1RWC 2WA; 2XO3 1CB8 1HMW;1HM2;1HM3;1HMU	*Arthrobacter aurescens* *Streptomyces coelicor A3* *Flavobacterium heparinum* *Pedobacter heparinus*	His^{225}–Tyr^{234}–Arg^{288}–Glu^{371}	[75–78]
CSase B (Endo)	DS	1DBO;1DBG;1OFL;1OFM	*Flavobacterium heparinum*	Lys^{250}–Arg^{271}–His^{272}	[79,80]
HAase	HA, CS	1OJO;1OJM; 1OJN;1OJP	*Streptococcus pneumoniae*		[81]

2.1. Chondriotinases ABC

Chondriotinases ABC (EC 4.2.2.4) with endo activity (CSase ABC I) [73] or exo activity (CSase ABC II D) [70] are catalysts with a high degree of stereospecificity. CSases ABC can degrade chondroitin, CS (A and C), DS and HA independent of their sulfation pattern by β-elimination, producing unsaturated disaccharides and tetrasaccharides. They are not active against keratan sulfate, HS and heparin. In rare cases, a single enzyme is able to degrade both uronic acid isomers (IdoA and GlcA) efficiently.

2.1.1. CSase ABC I Endolyase

CSase ABC I endolyase from *Proteus vulgaris* (EC:4.4.4.20) presents three domains (ID:1HNO); an N-terminal domain with a fold similar to that of carbohydrate-binding domains, a middle domain with an $(\alpha/\alpha)_5$ fold (typical for CSase AC) and a C-terminal domain with β-sheet folding (typical for CSase B) (Figure 4). The substrate-binding site is in the middle domain, a wide-open cleft with two structural folds that evolved to perform these reactions in an epimer-specific fashion [70–73] (Figure 4). The putative catalytic residues of CSase ABC I from *P. vulgaris* are His^{501}, Tyr^{508}, Arg^{560} and Glu^{653}, which were identified by site-directed mutagenesis [74]. The His^{501} residue plays the critical role of the proton abstraction of C5 from the IdoA/GlcA moiety during catalysis, the Glu^{653} residue is involved in a hydrogen bonding network in the active site, the Tyr^{508} residue is essential in the protonation of the leaving group in the GAG and the Arg^{560} residue near the IdoA/GlcA is able to stabilize the carbanion intermediate formed during catalysis, making the C5 proton more labile (Figure 4). This enzyme presents a shared identity of over 90% in the following species: *Shigella* sp. FC1655, *Proteus mirabilis*, *Klebsiella pneumoniae*, *Proteus hauseri* and *Proteus penneri*. This enzyme is able to promote functional recovery in the injured central nervous system via its role in the disruption of the normal organization of the extracellular matrix.

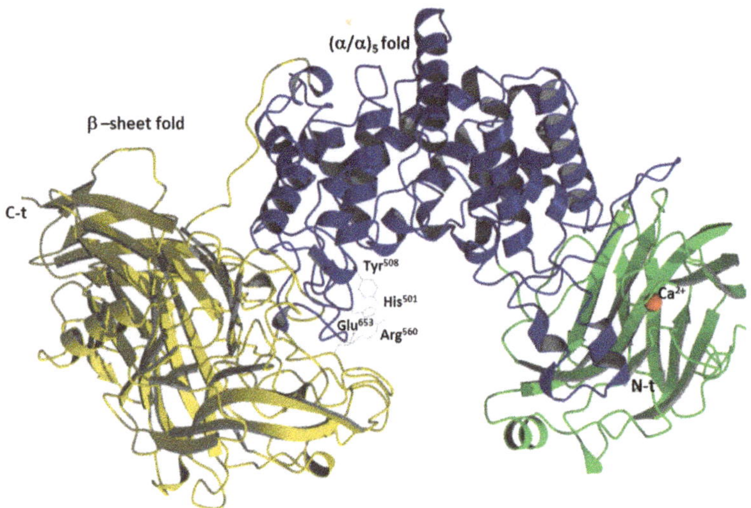

Figure 4. CSase ABC I from *P. vulgaris*. The N-terminal domain is colored green, the middle (catalytic) domain is in blue and the C-terminal domain is in yellow. The Ca^{2+} ion is shown as a red sphere. The catalytic tetrad is shown by the stick (His^{501}, Tyr^{508} and Arg^{560}).

2.1.2. CSase ABC II Exolyase from *Bacteroides thetaiotaomicron*

CSase ABC II exolyase from *Bacteroides thetaiotaomicron* (gene *BT_3324*) is a broad-specificity lyase which degrades CSs and DS to yield only disaccharide products. This enzyme has a preference for CS-A over CS-C and exhibits low activity against HA. The enzyme presents three structural domains; an N-t domain which adopts a β-jellyroll fold (carbohydrate binding), a central domain which adopts an $(\alpha/\alpha)_5$ incomplete toroid and a C-t domain which contains four antiparallel β-sheets. Glu^{628}, Tyr^{461}, His^{454} and Arg^{514} contribute to the catalytic tetrad and one structural Ca^{2+} ion is located in the N-t domain [70]. Tyr^{461} in a deprotonated state acts as the catalytic base abstracting the C5-bound proton from glucuronic acid, His^{454} serves as a catalytic base and Glu^{628} plays a part in positioning both His^{454} and Arg^{514} in the precise orientation necessary for effective CS/DS degradation, but is not directly involved in catalysis. Similar proteins with a shared identity of 90% are found in other species, including: *Bacteroides faecis CAG:32, Bacteroides thetaiotaomicron CAG:40, Bacteriodes sp. AR20* and *Klebsiella oxytoca*.

2.1.3. CSase ABC II Exolyase from *Proteus vulgaris*

CSase ABC II exolyase from *Proteus vulgaris* has a broad specificity of GAG activity which preferentially degrades the tetra- and hexasaccharide derivatives of CS and DS produced by the CS ABC endolyase, to yield the respective disaccharides. This enzyme is inhibited by Ni^{2+}. The catalytic residues are His^{453} (proton acceptor) and Tyr ^{460} (proton donor).

2.1.4. Other CSases

Recently, CSase ABC from *Acinetobacter sp. C26* has been described, with the finding that its activity increases in the presence of a number of different ions (Na^+, K^+, Mn^{2+}) and is strongly inhibited by other kinds of ions (Cu^{2+}, Hg^{2+}, Al^{3+}) [82].

CSase ABC displays a more open cleft in the central domain (substrate binding), which is different from the other lyases.

2.2. CSases AC

CSases AC with endo activity (from *Flavobacterium heparinum*) [83] or exo activity (from *Arthrobacter aurescens*) [83] degrade CS (CS-A or CS-C) and HA, generating unsaturated disaccharides. These enzymes show a low level of homology with several hyaluronate lyases, although they share its fold. CSase AC from *Flavobacterium heparinum* is composed of two domains; an α-helical domain (N-t) within enzymatic activity residues and a β-sheet domain (C-t). These enzymes are sensitive to the 5-epimerization of the GlcA moiety, so they can only degrade CSs containing a domain with an $(\alpha/\alpha)_5$ toroid fold. The crystal structures of the lyase–GAG complex showed that His^{225} is the candidate for the catalysis and the Tyr–His–Glu–Arg residues are present in the catalytic center. Four residues—His^{225}, Tyr^{234}, Arg^{288} and Glu^{371}—are near the catalytic site of chondroitin AC lyase from *Flavobacterium heparinium* (Figure 5) [78]. The His^{225} residue is a candidate for the general base and the removes the proton attached to C5 of the glucuronic acid, the Tyr^{234} residue is able to protonate the leaving group and the Arg^{288} residue contributes to charge neutralization and stabilization of the enolate anion intermediate. These enzymes could have endolytic or exolytic activity depending on the microorganism (Table 3), while the activity itself is independent of metal ions [78]. The structural alignment of CSases AC and CSase ABC shows that the Tyr–His–Arg–Glu catalytic tetrads of CSases AC have counterparts in CSase ABC [78].

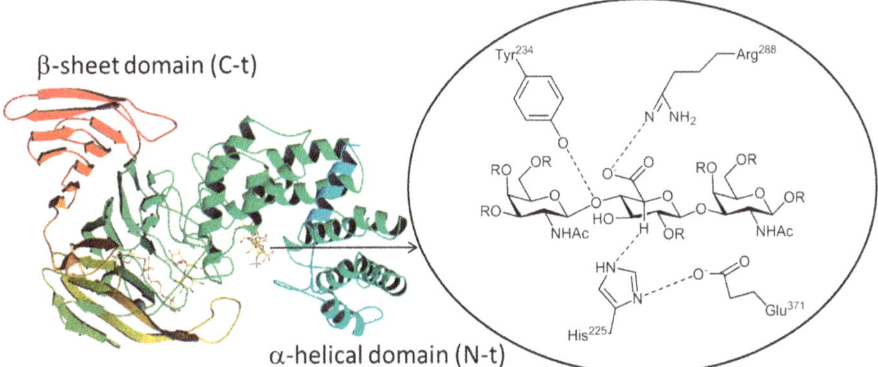

Figure 5. Ribbon stereo drawing of the chondroitinase AC I from *Flavobacterium heparinum* showing the N-t domain ($(\alpha/\alpha)_5$ (green), β-sheet sandwich C-t domain (orange)) and active site.

2.3. CSase B

CSase B from *Flavobacterium heparinum* [83] cleaves endolytically on the GAG DS, generating oligosaccharides, tetrasaccharides and unsaturated 4-sulfate-disaccharides. This enzyme is often used in studies to assess the structural characterization and antithrombin activity of DS by chromatographic techniques. The first crystallographic study of this enzyme with DS facilitated the identification of the subsites in the active site [79,80]. Later, establishing the structure of the lyase–CS complex provided a more complete picture of the active site of the enzyme including the identification of the catalytic residues Lys^{250}, Arg^{271} and His^{272} [79,84]. Mutation of the amino acid Lys resulted in the inactivation of the enzyme, which is attributed to the role of the residue in stabilizing the carbanion of C5 formed during catalysis. The 3D X-ray structure revealed the presence of a divalent ion coordinated by conserved acidic residues (one asparagine and two glutamates) and utilized for charge neutralization of the acidic group of iduronic acid [84] (Figure 6). The activity of this enzyme is inhibited by Co^{2+}, Fe^{2+} and Ba^{2+} ions. CSase B adopts a β-helical fold, typical of several polysaccharide lyases and hydrolases (Figure 6).

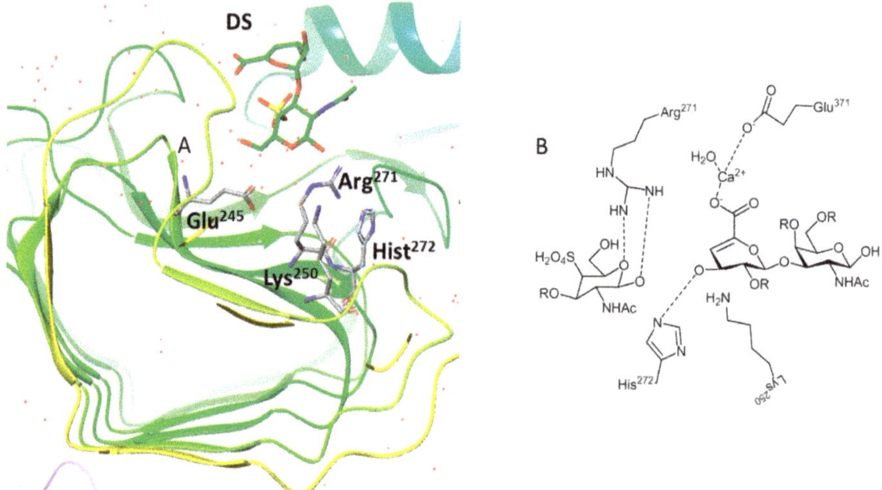

Figure 6. (**A**) Ribbon stereo drawing of Chondroitinase B from *Flavobacterium heparinum* (pdb ID:1egu); (**B**) the active site of the enzyme CSase B. Figure generated by Maestro and ChemDraw.

2.4. HAases

HAases from *Streptococcus pneumoniae* degrade hyaluronan and chondroitin/CSs only in four positions of the sulfate group based on the β-elimination mechanism [85]. These lyases show low levels of homology with chondroitinases of type AC. The enzymatic activity residues are within the N-terminal domain. The enzyme molecule is composed of two domains, the catalytic domain having a $(\alpha/\alpha)_5$ barrel fold (N-t) and the C-t domain comprising an antiparallel β-sandwich (Figure 7). The function of the C-t domain is the modulation of the oligosaccharide substrate access to the catalytic cleft present in the N-t domain. Any changes in the binding mode of the protein were detected when used as disaccharides of CS-O, CS-A and CS-C.

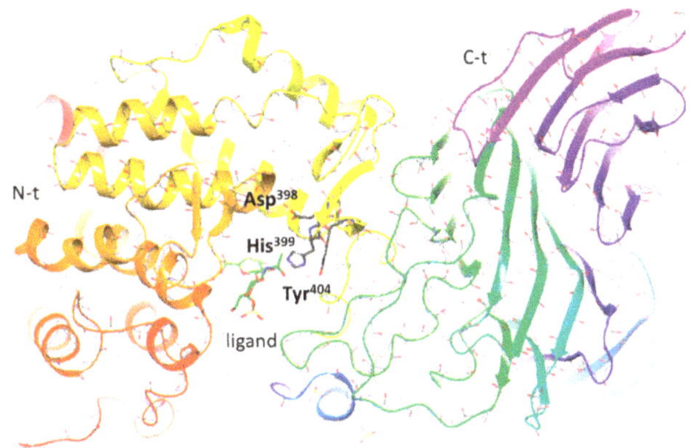

Figure 7. Ribbon stereo drawing of the hyaluronidase from *Streptococcus pneumoniae* (pdb ID:1ojo). Figure generated by Maestro.

The HAase from *Vibrio sp. FC509* degrades hyaluronan and CS variants except CS-E (GlcA-GalNAc(4,6S)), while the desulfation of the GalNAc unit abrogates its activity on the β(1-4) linkage between the disaccharide units. This protein uses a general acid–base catalysis mechanism as with the other lyases [62].

3. Applications of CSases

3.1. Synthetic Applications: Preparation of Low Molecular Weight Chondroitin Sulphate (LMWCS) as Therapeutic Agents

The preparation of low molecular weight chondroitin sulphate (LMWCS) has mainly been accomplished through acidic, basic or oxidative treatment or by enzymatic depolymerization, allowing oligosaccharides with different molecular weights and degrees of polydispersity to be obtained depending on the employed degradation conditions [86] (Table 4). In the case of the chemical procedures, fairly drastic conditions can cause undesirable reactions and partial or total desulfation of the obtained oligosaccharides, modifying the biological properties of the resulting oligosaccharides [87].

Table 4. The M_w of natural and depolymerized CS.

SAMPLE [a]	MOLECULAR WEIGHT (Da)				
	Natural	Depolymerized CS/Low Molecular Weight of CS			
		HCl/H_20/60 °C	NaOH/MAOS [b]/60 °C	Enzymatic	H_2O_2(30%)
CS-B	18.150	1.826	8.085	2.583	1.561
CS-S	31.300	1.217	4.269	1.994	1.511
CS-P	10.070	2.690	3.663	2.920	2.191

[a] CS-B, from bovine cartilage; CS-S, from shark cartilage; CS-P, from porcine cartilage. [b] Microwave-assisted organic synthesis.

On the contrary, enzymatic depolymerizations are more specific and allow better control of the processes, as well as being environmentally friendly. Additionally, the polysaccharide substrates and enzymes are relatively inexpensive, meaning the oligosaccharides can be prepared in large quantities at a low cost. For this reason, several specific (CSase AC I, CSase AC II and CSase ABC) and non-specific enzymes (HAases) have been employed in the preparation of LMWCS.

These oligosaccharides have therapeutic applications, the depolymerization of natural CSs being a strategy that can be used to remove their main limitations in many medical applications. It is well understood that CSs of natural origin have high polydispersity, varying significantly in chain length even when isolated from a single source [88,89]. Furthermore, their high molecular weights preclude their use in many medical uses, impacting not only their biological activities but also their equally important pharmacological properties [90]. For these reasons, their degradation products (LMWCS) have been found to be much more useful than native CSs.

Several studies have demonstrated that changes in the M_W cause different immune responses and that the use of long-chain CSs can even result in the cancellation of the anti-inflammatory activity [91,92]. In a similar way, LMWCS has demonstrated a superior effect on collagen-induced arthritis as compared to that of intact CSs [93]. To this must be added that even though the M_W is similar, the biological effect is further augmented in the case of CSs with a narrow range of molecular weights, i.e., polysaccharides with low polydispersity [94]. An explanation for these results is that only LMWC derivatives reach the bloodstream, as absorption through the gastrointestinal tract is a M_W-dependent process [95,96]. In fact, it has been reported that LMWCS administered orally for osteoarthritis treatment is more readily absorbed and hence arrives at the joint and is distributed into the cartilage more effectively than native CSs [97]. Finally, the M_W also has an important effect on the elimination of exogenously administered CSs. The use of LMWCS prevents its hepatic accumulation and improves its renal filtration [98,99].

Accordingly, the preparation of LMWCS for osteoarthritis treatment has attracted much attention in recent years, especially with the cloning, expression and characterization of new GAG lyases such as chondroitin lyase AC II from *Arthrobacter aurescens* [100] and the chondroitinase ABC I from *Proteus vulgaris* with a maltose-binding protein [101] or with glyceraldehyde-3-phosphate dehydrogenase [102] as fusion proteins.

LMWCS 4-sulfated polysaccharides have been prepared by the degradation of CS-A from bovine trachea tissue using a bovine testicular hyaluronidase [103]. These 4-sulfated polysaccharides (CS-A) are known for their antioxidant activity, a capacity that is closely related to the treatment of diseases such as cancer, cardiovascular and cerebrovascular diseases and ischemia, as well as aging processes [104]. In a similar way, the digestion of CS-A from cow cartilage with an extracellular chondroitinase ABC produced by *Sphingomonas paucimobilis* afforded CS-A oligosaccharides. These promote in vitro cardiocytoprotection, decreasing the well-known damage induced by isoproterenol and accelerating the recovery of myocardial cells [105].

Recently, the preparation of low molecular weight fucosylated chondroitin sulfates (LMWfCSs) has attracted much attention [106] (Figure 8). Fucosylated CSs (fCSs) are CSs with fucose branches invariably extending from the 3-*O*-position of the GlcA. These have demonstrated anticoagulant and antithrombotic activities as substitutes for heparin [107] and have attracted considerable attention as potential antitumor drugs [108,109] and for their application as a treatment of hyperlipidemia [110,111]. Unfortunately, native fCSs can also cause side effects such as the activation of FXII, platelet aggregation [112], hypertension and spontaneous bleeding in humans [87], limiting their therapeutic applications. However, LMWfCSs have demonstrated that in addition to the retention or even enhancement of biological activities in comparison with native fCS, depolymerized polysaccharides exhibit negligible adverse effects [113].

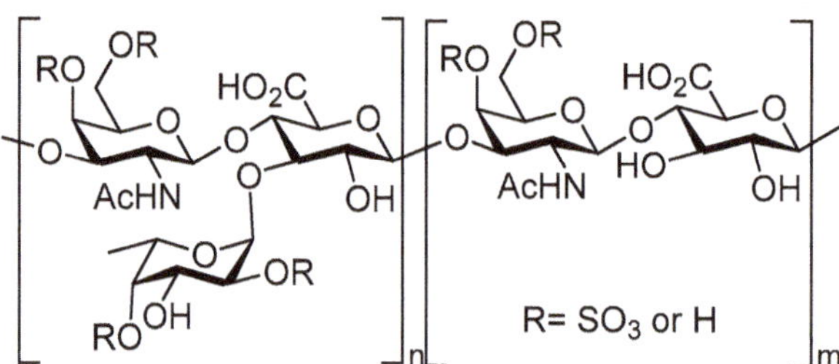

Figure 8. Chemical structure of fucosylated CS (fCS).

3.2. Analytical Applications

3.2.1. Oligosaccharide Mapping

Knowledge of the oligosaccharide sequences that contain GAGs is an important prerequisite for a better understanding of their biological roles and the development of pharmaceuticals based on them. Oligosaccharide mapping is an approach comparable to the peptide mapping of proteins that has been applied to GAGs in order to provide information on which oligosaccharide sequences are the bioactive domains.

The mapping technique involves specific enzymatic scission of polysaccharide chains, followed by high-resolution separation of the degradation products by chromatographic methods and the final analysis of the obtained oligosaccharides [114]. In this context, the use of enzymes that can specifically isolate certain domains through the selective digestion of other domains has been reported. Hence for example, a lyase from the marine bacterium *Vibrio sp. FC509* has been used to isolate several CS-E oligosaccharides and their interaction with herpes simplex virus receptors has been analyzed [115]. As such, the bioactive domain in the binding to the virus has been established [115], opening the door to the development of new antiviral drugs (Figure 9).

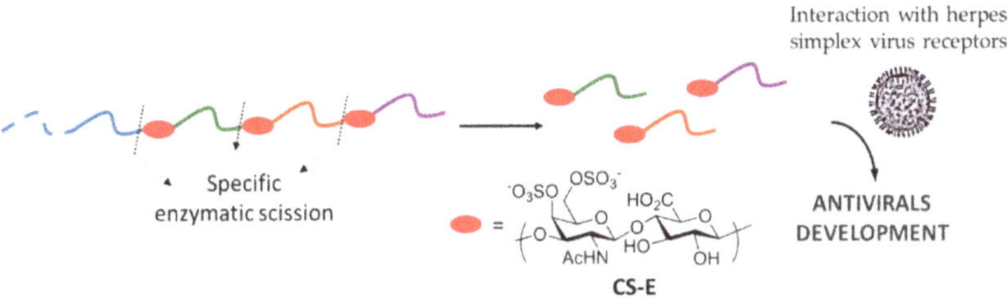

Figure 9. Oligosaccharide mapping for the discovery of new antivirals.

In a similar way, CS from shark fin cartilage has been digested with chondroitinase AC-I from *Flavobacterium heparinum*, an enzyme which cannot act on the galactosaminidic linkages bound to GlcA(2-*O*-sulfate)-GalNAc(6-*O*-sulfate) disaccharides (CS-D units) (Figure 10A) [116]. This digestion afforded five novel hexasaccharide sequences in addition to three previously reported sequences containing D-D-tetrasaccharide motifs (Figure 10B), which might be useful for the establishment of useful structure–function relationships in neuroglycobiological fields with regard to novel biomarkers.

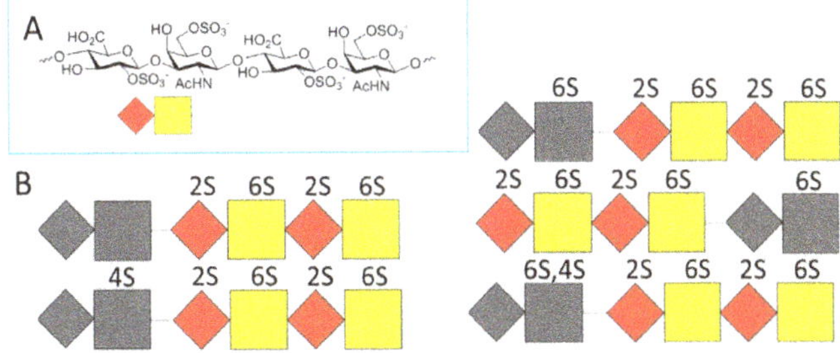

Figure 10. (**A**) The chemical structure of D-D-tetrasaccharide motifs. (**B**) The general structure of novel hexasaccharide sequences.

3.2.2. Compositional Analysis of GAGs

Another important application of GAG lyases is the compositional analysis of polysaccharides, both from natural sources and semi-synthetic products. Normally, a small quantity of the polysaccharide is exhaustively depolymerized using the proper lyase; then, the CS disaccharides obtained from this process are analyzed by High Performance Liquid Chromatograpgy (HPLC) and quantified by calculating the total peak areas of the disaccharides derived from a CS calibration curve.

As an example of the analysis of CS from natural sources, very recently a comprehensive disaccharide analysis of polysaccharides from different shellfish was performed to better understand the GAG structures in marine organisms [117]. According to the obtained results, the degree of sulfation in CSs depends on the species and, surprisingly, ΔDi-diSE is present in most shellfish (Figure 11).

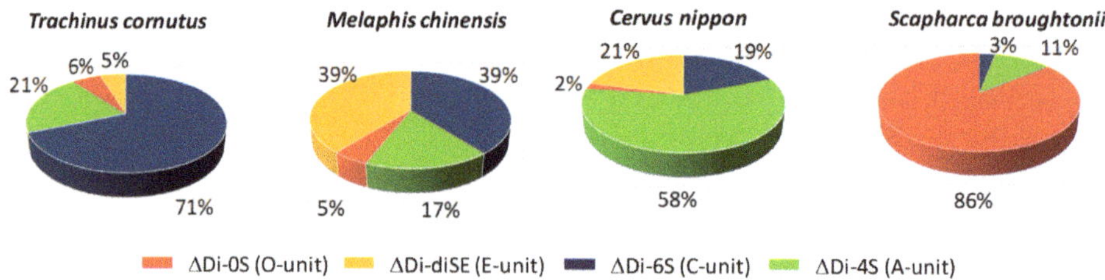

Figure 11. Unsaturated disaccharide composition of CSs from different types of shellfish.

Finally, enzymatic digestion followed by disaccharide analysis using HPLC is a very useful procedure for evaluation of the composition of semi-synthetic CSs. For example, we employed this method to determine the composition of a library of polysaccharides recently prepared in order to establish novel structure–function relationships [118] (Figure 12). Concretely, we showed that the particular sulfate distribution within the disaccharide repeating-unit plays a key role in the binding of growth factors, modulating the surface charge of the helical structure that, interestingly, has a significant influence on the binding capacity of CSs with several Grow Factors [119]. These findings provide additional strategies in the development of new CSs as growth factor binders for a broad range of therapeutically relevant applications [120].

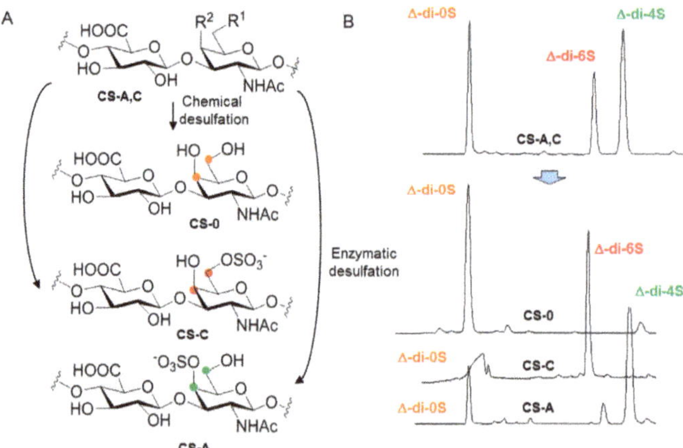

Figure 12. (**A**) The semi-synthesis of CS-0, CS-A and CS-C. (**B**) The analysis of disaccharides by HPLC after CSase ABC digestions.

4. Conclusions and Future Perspectives

The types, structure and activity of CS-degrading enzymes were the focus of this review. These lyases can be employed as tools for the development of new pharmaceuticals based on CS structures. During the last decades, these structures have been demonstrated to be involved not only in structural functions, but also in modulating numerous biological processes, such as angiogenesis, cell differentiation, growth and migration. These findings provide a promising avenue in the development of new pharmaceuticals based on CSs for a broad range of therapeutically relevant applications, including new drugs, drug carriers and medical implants.

However, it should be note that their development still faces challenges, which includes the discovery of more efficient and universal methods to synthesize LMWCSs with precisely controllable structures. Moreover, their interactions with growth factors and cells as well as their biological processes still need to be studied further. Thus, among other things, deep insight into CSs sequences as well as their relation to biological functions is required.

In our opinion, chondroitin sulfate-degrading enzymes will gradually become more present in the pharmaceutical industry in the next years, not only as excellent drug candidates [55–62] but also as important tools for the sustainable development of CS-based pharmaceutical products.

Author Contributions: Conceptualization, A.B. and J.R.; resources, S.Z. and R.B.-A; writing—review and editing, J.R. and A.B.

Funding: The authors thank the Ministerio de Economía y Competitividad (grants MAT2015- 65184-C2-2-R and CTQ2016-79255-P) for their support.

Conflicts of Interest: The authors declare no conflict of interest.

References

1. Volpi, N. Therapeutic applications of glycosaminoglycans. *Curr. Med. Chem.* **2006**, *13*, 1799–1810. [CrossRef] [PubMed]
2. Stick, R.V.; Williams, S. Glycoproteins and proteoglycans. In *Carbohydrates: The Essential Molecules of Life*; Stick, R.V., Williams, S., Eds.; Elsevier Science: Amsterdam, The Netherlands, 2008; pp. 369–412.
3. Hermann, W.; Lambova, S.; Muller-Ladner, U. Current treatment options for osteoarthritis. *Curr. Rheumatol. Rev.* **2018**, *14*, 108–116. [CrossRef] [PubMed]
4. Mourão, A.P. Perspective on the use of sulfated polysaccharides from marine organisms as a source of new antithrombotic drugs. *Mar. Drugs* **2015**, *13*, 2770–2784. [CrossRef] [PubMed]
5. Yamada, S.; Sugahara, K. Potential therapeutic application of chondroitin sulfate/dermatan sulfate. *Curr. Drug Discov. Tech.* **2008**, *5*, 289–301. [CrossRef]
6. Milstone, L.M.; Houghmonroe, L.; Kugelman, L.C.; Bender, J.R.; Haggerty, J.G. Epican, a heparan/chondroitin sulfate proteoglycan form of CD44, mediates cell-cell adhesion. *J. Cell Sci.* **1994**, *107*, 3183–3190. [PubMed]
7. Stabler, T.V.; Huang, Z.; Montell, E.; Verges, J.; Kraus, V.B. Chondroitin sulphate inhibits NF-κB activity induced by interaction of pathogenic and damage associated molecules. *Osteoarthritis Cartilage* **2017**, *25*, 166–174. [CrossRef]
8. Wu, F.F.; Zhou, C.H.; Zhou, D.D.; Ou, S.Y.; Liu, Z.J.; Huang, H.H. Immune-enhancing activities of chondroitin sulfate in murine macrophage RAW 264.7 cells. *Carbohydr. Polym.* **2018**, *198*, 611–619. [CrossRef] [PubMed]
9. Rogerson, S.J.; Chaiyaroj, S.C.; Ng, K.; Reeder, J.C.; Brown, G.V. Chondroitin sulfate A is a cell-surface receptor for Plamodium-falciparum infected erythrocytes. *J. Exp. Med.* **1995**, *182*, 15–20. [CrossRef] [PubMed]
10. Pereira, M.A.; Clausen, T.M.; Pehrson, C.; Mao, Y.; Resende, M.; Daugaard, M.; Kristensen, A.R.; Spliid, C.; Mathiesen, L.; Knudsen, L.E.; et al. Placental Sequestration of *Plasmodium falciparum* malaria parasites is mediated by the interaction between VAR2CSA and Chondroitin sulfate A on Syndecan-1. *PloS Pathog.* **2016**, *12*, e1005831. [CrossRef]
11. Mourao, P.A.S.; Pereira, M.S.; Pavao, M.S.G.; Mulloy, B.; Tollefsen, D.M.; Mowinckel, M.C.; Abildgaard, U. Structure and anticoagulant activity of a fucosylated chondroitin sulfate from echinoderm.Sulfated fucose branches on the polysaccharide account for its high anticoagulant action. *J. Biol. Chem.* **1996**, *271*, 23973–23984. [CrossRef]
12. Glauser, B.F.; Pereira, M.S.; Monteiro, R.Q.; Mourao, P.A.S. Serpin-independent anticoagulant activity of a fucosylated chondroitin sulfate. *Thromb. Haemos.* **2008**, *100*, 420–428. [CrossRef]

13. Ustyuzhanina, N.E.; Bilan, M.I.; Panina, E.G.; Sanamyan, N.P.; Dmitrenok, A.S.; Tsvetkova, E.A.; Ushakova, N.A.; Shashkov, A.S.; Nifantiev, N.E.; Usov, A.I. Structure and anti-inflammatory activity of a new unusual fucosylated chondroitin sulfate from *Cucumaria djakonovi*. *Mar. Drugs* **2018**, *16*, 389. [CrossRef] [PubMed]
14. Lovu, M.; Dumais, G.; du Souich, P. Anti-inflammatory activity of chondroitin sulfate. *Osteoarthritis Cartilage* **2008**, *16*, S14–S18.
15. Bergefall, K.; Trybala, E.; Johansson, M.; Uyama, T.; Naito, S.; Yamada, S.; Kitagawa, H.; Sugahara, K.; Bergström, T. Chondroitin sulfate characterized by the E-disaccharide unit is a potent inhibitor of herpes simplex virus infectivity and provides the virus binding sites on gro2C cells. *J. Biol. Chem.* **2005**, *280*, 32193–32199. [CrossRef] [PubMed]
16. Mycroft-West, C.J.; Yates, E.A.; Skidmore, M.A. Marine glycosaminoglycan-like carbohydrates as potential drug candidates for infectious disease. *Biochem. Soc. Trans.* **2018**, *46*, 919–929. [CrossRef] [PubMed]
17. Goel, S.; Muthusamy, A.; Miao, J.; Cui, L.W.; Salanti, A.; Winzeler, E.A.; Gowda, D.C. Targeted disruption of a ring-infected erythrocyte surface antigen (RESA)-like export protein gene in Plasmodium falciparum confers stable chondroitin 4-sulfate cytoadherence Capacity. *J. Biol.l Chem.* **2014**, *289*, 34408–34421. [CrossRef] [PubMed]
18. Alkhalil, A.; Achur, R.N.; Valiyaveettil, M.; Ockenhouse, C.F.; Gowda, D.C. Structural requirements for the adherence of Plasmodium falciparum-infected erythrocytes to chondroitin sulfate proteoglycans of human placenta. *J. Biol. Chem.* **2000**, *275*, 40357–40364. [CrossRef] [PubMed]
19. Asimakopoulou, A.P.; Theocharis, A.D.; Tzanakakis, G.N.; Karamanos, N.K. The biological role of chondroitin sulfate in cancer and chondroitin-based anticancer agents. *In Vivo* **2008**, *22*, 385–389. [PubMed]
20. Borsig, L.; Wang, L.; Cavalcante, M.C.M.; Cardilo-Reis, L.; Ferreira, P.L.; Mourão, P.A.S.; Esko, J.D.; Pavão, M.S.G. Selectin blocking activity of a fucosylated chondroitin sulfate glycosaminoglycan from sea cucumber: Effect on tumor metastasis and neutrophil recruitment. *J. Biol. Chem.* **2007**, *282*, 14984–14991. [CrossRef]
21. Merida-de-Barros, D.A.; Chaves, S.P.; Belmiro, C.L.R.; Wanderley, J.L.M. Leishmaniasis and glycosaminoglycans: a future therapeutic strategy? *Parasit. Vectors* **2018**, *11*, 536. [CrossRef] [PubMed]
22. Vallen, M.J.E.; van Tilborg, A.A.G.; Tesselaar, M.H.; ten Dam, G.B.; Bulten, J.; van Kuppevelt, T.H.; Massuger, L. Novel single-chain antibody GD3A10 defines a chondroitin sulfate biomarker for ovarian cancer. *Biomark. Med.* **2014**, *8*, 699–711. [CrossRef] [PubMed]
23. Yamaguchi, K.; Tamaki, H.; Fukui, S. Detection of oligosaccharide ligands for Hepatocyte growth factor/Scatter factor (HGF/SF), Keratinocyte growth factor (KGF/FGF-7), RANTES and Heparin cofactor II by neoglycolipid microarrays of glycosaminoglycan-derived oligosaccharide fragments. *Glycoconjugate J.* **2006**, *23*, 513–523. [CrossRef] [PubMed]
24. Djerbal, L.; Lortat-Jacob, H.; Kwok, J. Chondroitin sulfates and their binding molecules in the central nervous system. *Glycoconjugate J.* **2017**, *34*, 363–376. [CrossRef] [PubMed]
25. Miller, G.M.; Hsieh-Wilson, L.C. Sugar-dependent modulation of neuronal development, regeneration, and plasticity by chondroitin sulfate proteoglycans. *Exp. Neurol.* **2015**, *274*, 115–125. [CrossRef] [PubMed]
26. Ju, C.; Gao, J.; Hou, L.; Wang, L.; Zhang, F.; Sun, F.; Zhang, T.; Xu, P.; Shi, Z.; Hu, F.; et al. Neuroprotective effect of chondroitin sulfate on SH-SY5Y cells overexpressing wild-type or A53T mutant α-synuclein. *Mol. Med. Rep.* **2017**, *16*, 8721–8728. [CrossRef] [PubMed]
27. Zou, X.H.; Jiang, Y.Z.; Zhang, G.R.; Jin, H.M.; Hieu, N.T.M.; Ouyang, H.W. Specific interactions between human fibroblasts and particular chondroitin sulfate molecules for wound healing. *Acta Biomater.* **2009**, *5*, 1588–1595. [CrossRef]
28. Kowitsch, A.; Zhou, G.Y.; Groth, T. Medical application of glycosaminoglycans: A review. *J. Tissue Eng. Regen. Med.* **2018**, *12*, E23–E41. [CrossRef] [PubMed]
29. Altgarde, N.; Nileback, E.; de Battice, L.; Pashkuleva, I.; Reis, R.L.; Becher, J.; Moller, S.; Schnabelrauch, M.; Svedhem, S. Probing the biofunctionality of biotinylated hyaluronan and chondroitin sulfate by hyaluronidase degradation and aggrecan interaction. *Acta Biomater* **2013**, *9*, 8158–8166. [CrossRef] [PubMed]
30. Kliemt, S.; Lange, C.; Otto, W.; Hintze, V.; Möller, S.; von Bergen, M.; Hempel, U.; Kalkhof, S. Sulfated Hyaluronan Containing Collagen Matrices Enhance Cell-Matrix-Interaction, Endocytosis, and Osteogenic Differentiation of Human Mesenchymal Stromal Cells. *J. Proteome Res.* **2013**, *12*, 378–389. [CrossRef]

31. Korn, P.; Schulz, M.C.; Hintze, V.; Range, U.; Mai, R.; Eckelt, U.; Schnabelrauch, M.; Moller, S.; Becher, J.; Scharnweber, D.; Stadlinger, B. Chondroitin sulfate and sulfated hyaluronan-containing collagen coatings of titanium implants influence peri-implant bone formation in a minipig model. *J. Biomed. Mater. Res. Part A* **2014**, *102*, 2334–2344. [CrossRef] [PubMed]
32. Levett, P.A.; Melchels, F.P.W.; Schrobback, K.; Hutmacher, D.W.; Malda, J.; Klein, T.J. A biomimetic extracellular matrix for cartilage tissue engineering centered on photocurable gelatin, hyaluronic acid and chondroitin sulfate. *Acta Biomater.* **2014**, *10*, 214–223. [CrossRef] [PubMed]
33. Cai, S.S.; Liu, Y.C.; Shu, X.Z.; Prestwich, G.D. Injectable glycosaminoglycan hydrogels for controlled release of human basic fibroblast growth factor. *Biomaterials* **2005**, *26*, 6054–6067. [CrossRef] [PubMed]
34. Elia, R.; Newhide, D.R.; Pedevillano, P.D.; Reiss, G.R.; Firpo, M.A.; Hsu, E.W.; Kaplan, D.L.; Prestwich, G.D.; Peattie, R.A. Silk-hyaluronan-based composite hydrogels: A novel, securable vehicle for drug delivery. *J. Biomater. Appl.* **2013**, *27*, 749–762. [CrossRef]
35. Shu, X.Z.; Ahmad, S.; Liu, Y.C.; Prestwich, G.D. Synthesis and evaluation of injectable, in situ crosslinkable synthetic extracellular matrices for tissue engineering. *J. Biomed. Mater. Res. Part A* **2006**, *79A*, 902–912. [CrossRef]
36. Jin, R.; Lou, B.; Lin, C. Tyrosinase-mediated in situ forming hydrogels from biodegradable chondroitin sulfate-tyramine conjugates. *Polym. Int.* **2013**, *62*, 353–361. [CrossRef]
37. Ni, Y.L.; Tang, Z.R.; Cao, W.X.; Lin, H.; Fan, Y.J.; Guo, L.K.; Zhang, X.D. Tough and elastic hydrogel of hyaluronic acid and chondroitin sulfate as potential cell scaffold materials. *Int. J. Biol. Macromol.* **2015**, *74*, 367–375. [CrossRef] [PubMed]
38. Jia, X.Q.; Yeo, Y.; Clifton, R.J.; Jiao, T.; Kohane, D.S.; Kobler, J.B.; Zeitels, S.M.; Langer, R. Hyaluronic acid-based microgels and microgel networks for vocal fold regeneration. *Biomacromolecules* **2006**, *7*, 3336–3344. [CrossRef]
39. Kirker, K.R.; Luo, Y.; Nielson, J.H.; Shelby, J.; Prestwich, G.D. Glycosaminoglycan hydrogel films as bio-interactive dressings for wound healing. *Biomaterials* **2002**, *23*, 3661–3671. [CrossRef]
40. Philandrianos, C.; Andrac-Meyer, L.; Mordon, S.; Feuerstein, J.-M.; Sabatier, F.; Veran, J.; Magalon, G.; Casanova, D. Comparison of five dermal substitutes in full-thickness skin wound healing in a porcine model. *Burns* **2012**, *38*, 820–829. [CrossRef] [PubMed]
41. Yan, S.; Zhang, Q.; Wang, J.; Liu, Y.; Lu, S.; Li, M.; Kaplan, D.L. Silk fibroin/chondroitin sulfate/hyaluronic acid ternary scaffolds for dermal tissue reconstruction. *Acta Biomater.* **2013**, *9*, 6771–6782. [CrossRef] [PubMed]
42. Kesti, M.; Mueller, M.; Becher, J.; Schnabelrauch, M.; D'Este, M.; Eglin, D.; Zenobi-Wong, M. A versatile bioink for three-dimensional printing of cellular scaffolds based on thermally and photo-triggered tandem gelation. *Acta Biomater.* **2015**, *11*, 162–172. [CrossRef]
43. Daamen, W.F.; van Moerkerk, H.T.B.; Hafmans, T.; Buttafoco, L.; Poot, A.A.; Veerkamp, J.H.; van Kuppevelt, T.H. Preparation and evaluation of molecularly-defined collagen-elastin-glycosaminoglycan scaffolds for tissue engineering. *Biomaterials* **2003**, *24*, 4001–4009. [CrossRef]
44. McFadden, T.M.; Duffy, G.P.; Allen, A.B.; Stevens, H.Y.; Schwarzmaier, S.M.; Plesnila, N.; Murphy, J.M.; Barry, F.P.; Guldberg, R.E.; O'Brien, F.J. The delayed addition of human mesenchymal stem cells to pre-formed endothelial cell networks results in functional vascularization of a collagen-glycosaminoglycan scaffold in vivo. *Acta Biomater.* **2013**, *9*, 9303–9316. [CrossRef]
45. Ko, C.-S.; Huang, J.-P.; Huang, C.-W.; Chu, I.M. Type II collagen-chondroitin sulfate-hyaluronan scaffold cross-linked by genipin for cartilage tissue engineering. *J. Biosci. Bioeng.* **2009**, *107*, 177–182. [CrossRef] [PubMed]
46. Chen, Y.-L.; Lee, H.-P.; Chan, H.-Y.; Sung, L.-Y.; Chen, H.-C.; Hu, Y.-C. Composite chondroitin-6-sulfate/dermatan sulfate/chitosan scaffolds for cartilage tissue engineering. *Biomaterials* **2007**, *28*, 2294–2305. [CrossRef] [PubMed]
47. Kuo, C.-Y.; Chen, C.-H.; Hsiao, C.-Y.; Chen, J.-P. Incorporation of chitosan in biomimetic gelatin/chondroitin-6-sulfate/hyaluronan cryogel for cartilage tissue engineering. *Carbohyd. Polym.* **2015**, *117*, 722–730. [CrossRef] [PubMed]
48. Fischer, R.L.; McCoy, M.G.; Grant, S.A. Electrospinning collagen and hyaluronic acid nanofiber meshes. *J. Mater. Sci. Mater. Med.* **2012**, *23*, 1645–1654. [CrossRef] [PubMed]

49. Ji, Y.; Ghosh, K.; Li, B.; Sokolov, J.C.; Clark, R.A.F.; Rafailovich, M.H. Dual-syringe reactive electrospinning of cross-linked hyaluronic acid hydrogel nanofibers for tissue engineering applications. *Macromol. Biosci.* **2006**, *6*, 811–817. [CrossRef]
50. Zhong, S.P.; Teo, W.E.; Zhu, X.; Beuertnan, R.; Ramakrishna, S.; Yung, L.Y.L. Development of a novel collagen-GAG nanofibrous scaffold via electrospinning. *Mater. Sci. Eng. C Biomim. Supramol. Syst.* **2007**, *27*, 262–266. [CrossRef]
51. Santo, V.E.; Gomes, M.E.; Mano, J.F.; Reis, R.L. Chitosan-chondroitin sulphate nanoparticles for controlled delivery of platelet lysates in bone regenerative medicine. *J. Tissue Eng. Regen. Med.* **2012**, *6*, s47–s59. [CrossRef] [PubMed]
52. Pomin, V.H.; Wang, X. Synthetic oligosaccharide libraries and microarray technology: A powerful combination for the success of current glycosaminoglycan interactomics. *ChemMedChem* **2018**, *13*, 648–661. [CrossRef] [PubMed]
53. Avci, F.Y.; DeAngelis, P.L.; Liu, J.; Linhardt, R.J. Enzymatic synthesis of glycosaminoglycans: improving on nature. In *Frontiers in Modern Carbohydrate Chemistry*; Demchenko, A.V., Ed.; ACS Symposium Series; American Chemical Society: Washington, DC, USA, 2007; Volume 960, pp. 253–284.
54. Linhardt, R.J.; Avci, F.Y.; Toida, T.; Kim, Y.S.; Cygler, M. CS lyases: Structure, activity, and applications in analysis and the treatment of diseases. *Adv. Pharmacol.* **2006**, *53*, 187–215. [PubMed]
55. Kasinathan, N.; Volety, S.M.; Josyula, V.R. Chondroitinase: A promising therapeutic enzyme. *Crit. Rev. Microbiol.* **2016**, *42*, 474–484. [CrossRef]
56. Orr, M.B.; Gensel, J.C. Spinal cord injury scarring and inflammation: Therapies targeting glial and inflammatory responses. *Neurotherapeutics* **2018**, *15*, 541–553. [CrossRef] [PubMed]
57. Denholm, E.M.; Lin, Y.Q.; Silver, P.J. Anti-tumor activities of chondroitinase AC and chondroitinase B: inhibition of angiogenesis, proliferation and invasion. *Eur. J. Pharmacol.* **2001**, *416*, 213–221. [CrossRef]
58. Yao, X.Y.; Hageman, G.S.; Marmor, M.F. Recovery of retinal adhesion after enzymatic perturbation of the interphotoreceptor matrix. *Invest. Ophthalmol. Vis. Sci.* **1992**, *33*, 498–503.
59. Kato, F.; Iwata, H.; Mimatsu, K.; Miura, T. Experimental chemonucleolysis with chondroitinase ABC. *Clin. Orthop. Relat. Res.* **1990**, *253*, 301–308. [CrossRef]
60. Buhren, B.A.; Schrumpf, H.; Hoff, N.P.; Bolke, E.; Hilton, S.; Gerber, P.A. Hyaluronidase: From clinical applications to molecular and cellular mechanisms. *Eur. J. Med. Res.* **2016**, *21*, 5. [CrossRef] [PubMed]
61. Rzany, B.; Becker-Wegerich, P.; Bachmann, F.; Erdmann, R.; Wollina, U. Hyaluronidase in the correction of hyaluronic acid-based fillers: A review and a recommendation for use. *J. Cosmet. Dermatol.* **2009**, *8*, 317–323. [CrossRef] [PubMed]
62. Khan, N.; Niazi, Z.R.; Rehman, F.U.; Akhtar, A.; Khan, M.M.; Khan, S.; Baloch, N.; Khan, S. Hyaluronidases: A Therapeutic Enzyme. *Protein Pept. Lett.* **2018**, *25*, 663–676. [CrossRef] [PubMed]
63. Zechel, D.L.; Withers, S.G. Glycosidase mechanisms: Anatomy of a finely tuned catalyst. *Acc. Chem. Res.* **2000**, *33*, 11–18. [PubMed]
64. Linhardt, R.J.; Galliher, P.M.; Cooney, C.L. Polysaccharide lyases. *Appl. Biochem. Biotechnol.* **1986**, *12*, 135–176. [CrossRef] [PubMed]
65. Lombard, V.; Bernard, T.; Rancurel, C.; Brumer, H.; Coutinho, P.M.; Henrissat, B. A hierarchical classification of polysaccharide lyases for glycogenomics. *Biochem. J.* **2010**, *432*, 437–444. [CrossRef] [PubMed]
66. Garron, M.L.; Cygler, M. Structural and mechanistic classification of uronic acid-containing polysaccharide lyases. *Glycobiology* **2010**, *20*, 1547–1573. [CrossRef] [PubMed]
67. Wang, W.; Wang, J.; Li, F. Hyaluronidase and chondroitinase. *Adv. Exp. Med. Biol.* **2017**, *925*, 75–87. [PubMed]
68. Stern, R.; Jedrzejas, M.J. Hyaluronidases: Their genomics, structures, and mechanisms of action. *Chem. Rev.* **2006**, *106*, 818–839. [CrossRef]
69. Hovingh, P.; Linker, A. Hyaluronidase activity in leeches (Hirudinea). *Comp. Biochem. Physiol. B Biochem. Mol. Biol.* **1999**, *124*, 319–326. [CrossRef]
70. Shaya, D.; Hahn, B.S.; Bjerkan, T.M.; Kim, W.S.; Park, N.Y.; Sim, J.S.; Kim, Y.S.; Cygler, M. Composite active site of chondroitin lyase ABC accepting both epimers of uronic acid. *Glycobiology* **2008**, *18*, 270–277. [CrossRef]
71. Linn, S.; Chan, T.; Lipeski, L.; Salyers, A.A. Isolation and characterization of two chondroitin lyases from Bacteroides thetaiotaomicron. *J. Bacteriol.* **1983**, *156*, 859–866.

72. Yamagata, T.; Saito, H.; Habuchi, O.; Suzuki, S. Purification and properties of bacterial chondroitinases and chondrosulfatases. *J. Biol. Chem.* **1968**, *243*, 1523–1535.
73. Huang, W.; Lunin, V.V.; Li, Y.; Suzuki, S.; Sugiura, N.; Miyazono, H.; Cygler, M. Crystal structure of Proteus vulgaris chondroitin sulfate ABC lyase I at 1.9A resolution. *J. Mol. Biol.* **2003**, *328*, 623–634. [CrossRef]
74. Prabhakar, V.; Raman, R.; Capila, I.; Bosques, C.J.; Pojasek, K.; Sasisekharan, R. Biochemical characterization of the chondroitinase ABC I active site. *Biochem. J.* **2005**, *390*, 395–405. [CrossRef] [PubMed]
75. Lunin, V.V.; Li, Y.; Linhardt, R.J.; Miyazono, H.; Kyogashima, M.; Kaneko, T.; Bell, A.W.; Cygler, M. High-resolution crystal structure of Arthrobacter aurescens chondroitin AC lyase: an enzyme-substrate complex defines the catalytic mechanism. *J. Mol. Biol.* **2004**, *337*, 367–386. [CrossRef] [PubMed]
76. Elmabrouk, Z.H.; Vincent, F.; Zhang, M.; Smith, N.L.; Turkenburg, J.P.; Charnock, S.J.; Black, G.W.; Taylor, E.J. Crystal structures of a family 8 polysaccharide lyase reveal open and highly occluded substrate-binding cleft conformations. *Proteins* **2011**, *79*, 965–974. [CrossRef] [PubMed]
77. Fethiere, J.; Eggimann, B.; Cygler, M. Crystal structure of chondroitin AC lyase, a representative of a family of glycosaminoglycan degrading enzymes. *J. Mol. Biol.* **1999**, *288*, 635–647. [CrossRef] [PubMed]
78. Huang, W.; Boju, L.; Tkalec, L.; Su, H.; Yang, H.O.; Gunay, N.S.; Linhardt, R.J.; Kim, Y.S.; Matte, A.; Cygler, M. Active site of chondroitin AC lyase revealed by the structure of enzyme-oligosaccharide complexes and mutagenesis. *Biochemistry* **2001**, *40*, 2359–2372. [CrossRef] [PubMed]
79. Pojasek, K.; Raman, R.; Kiley, P.; Venkataraman, G.; Sasisekharan, R. Biochemical characterization of the chondroitinase B active site. *J. Biol. Chem.* **2002**, *277*, 31179–31186. [CrossRef] [PubMed]
80. Huang, W.; Matte, A.; Li, Y.; Kim, Y.S.; Linhardt, R.J.; Su, H.; Cygler, M. Crystal structure of chondroitinase B from Flavobacterium heparinum and its complex with a disaccharide product at 1.7 A resolution. *J. Mol. Biol.* **1999**, *294*, 1257–1269. [CrossRef]
81. Rigden, D.J.; Jedrzejas, M.J. Structures of Streptococcus pneumoniae hyaluronate lyase in complex with chondroitin and chondroitin sulfate disaccharides. Insights into specificity and mechanism of action. *J. Biol. Chem.* **2003**, *278*, 50596–50606. [CrossRef]
82. Zhu, C.; Zhang, J.; Zhang, J.; Jiang, Y.; Shen, Z.; Guan, H.; Jiang, X. Purification and characterization of chondroitinase ABC from Acinetobacter sp. C26. *Int. J. Biol. Macromol.* **2017**, *95*, 80–86. [CrossRef]
83. Jandik, K.A.; Gu, K.; Linhardt, R.J. Action pattern of polysaccharide lyases on glycosaminoglycans. *Glycobiology* **1994**, *4*, 289–296. [CrossRef] [PubMed]
84. Michel, G.; Pojasek, K.; Li, Y.; Sulea, T.; Linhardt, R.J.; Raman, R.; Prabhakar, V.; Sasisekharan, R.; Cygler, M. The structure of chondroitin B lyase complexed with glycosaminoglycan oligosaccharides unravels a calcium-dependent catalytic machinery. *J. Biol. Chem.* **2004**, *279*, 32882–32896. [CrossRef] [PubMed]
85. Cordula, C.R.; Lima, M.A.; Shinjo, S.K.; Gesteira, T.F.; Pol-Fachin, L.; Coulson-Thomas, V.J.; Verli, H.; Yates, E.A.; Rudd, T.R.; Pinhal, M.A.; Toma, L.; Dietrich, C.P.; Nader, H.B.; Tersariol, I.L. On the catalytic mechanism of polysaccharide lyases: Evidence of His and Tyr involvement in heparin lysis by heparinase I and the role of Ca^{2+}. *Mol. BioSyst.* **2014**, *10*, 54–64. [CrossRef] [PubMed]
86. Li, L.; Li, Y.; Feng, D.; Xu, L.; Yin, F.; Zang, H.; Liu, C.; Wang, F. Preparation of low molecular weight chondroitin sulfates, screening of a high anti-complement capacity of low molecular weight chondroitin sulfate and its biological activity studies in attenuating osteoarthritis. *Int. J. Mol. Sci.* **2016**, *17*, 1685. [CrossRef] [PubMed]
87. Buyue, Y.; Sheehan, J.P. Fucosylated chondroitin sulfate inhibits plasma thrombin generation via targeting of the factor IXa heparin-binding exosite. *Blood* **2009**, *114*, 3092–3100. [CrossRef] [PubMed]
88. Tat, S.K.; Pelletier, J.P.; Mineau, F.; Duval, N.; Martel-Pelletier, J. Variable effects of 3 different chondroitin sulfate compounds on human osteoarthritic cartilage/chondrocytes: relevance of purity and Production Process. *J. Rheumatol.* **2010**, *37*, 656–664. [CrossRef] [PubMed]
89. Silva, L.C. Isolation and purification of chondroitin sulfate. *Adv. Pharmacol.* **2006**, *53*, 21–31. [PubMed]
90. Volpi, N. Analytical aspects of pharmaceutical grade chondroitin sulfates. *J. Pharm. Sci.* **2007**, *96*, 3168–3180. [CrossRef] [PubMed]
91. Leeb, B.F.; Schweitzer, H.; Montag, K.; Smolen, J.S. A metaanalysis of chondroitin sulfate in the treatment of osteoarthritis. *J. Rheumatol.* **2000**, *27*, 205–211.
92. Surapaneni, L.; Haley-Zitlin, V.; Bodine, A.; Jiang, X.P.; Brooks, J. Examination of chondroitin sulfate molecular weights on in vitro anti-inflammatory activity. *FASEB J.* **2013**, *27*, 1.

93. Cho, S.Y.; Sim, J.S.; Jeong, C.S.; Chang, S.Y.; Choi, D.W.; Toida, T.; Kim, Y.S. Effects of low molecular weight chondroitin sulfate on type II collagen-induced arthritis in DBA/1J mice. *Biol. Pharm. Bull.* **2004**, *27*, 47–51. [CrossRef]
94. Igarashi, N.; Takeguchi, A.; Sakai, S.; Akiyama, H.; Higashi, K.; Toida, T. Effect of molecular sizes of chondroitin sulfate on interaction with L-Selectin. *Int. J. Carbohydr. Chem.* **2013**, *2013*, 856142. [CrossRef]
95. Renukuntla, J.; Vadlapudi, A.D.; Patel, A.; Boddu, S.H.S.; Mitra, A.K. Approaches for enhancing oral bioavailability of peptides and proteins. *Int. J. Pharm.* **2013**, *447*, 75–93. [CrossRef] [PubMed]
96. Xiao, Y.; Li, P.; Cheng, Y.; Zhang, X.; Sheng, J.; Wang, D.; Li, J.; Zhang, Q.; Zhong, C.; Cao, R.; Wang, F. Enhancing the intestinal absorption of low molecular weight chondroitin sulfate by conjugation with alpha-linolenic acid and the transport mechanism of the conjugates. *Int. J. Pharm.* **2014**, *465*, 143–158. [CrossRef] [PubMed]
97. Martel-Pelletier, J.; Farran, A.; Montell, E.; Verges, J.; Pelletier, J.P. Discrepancies in composition and biological effects of different formulations of chondroitin sulfate. *Molecules* **2015**, *20*, 4277–4289. [CrossRef] [PubMed]
98. Pecly, I.M.D.; Melo, N.M.; Mourao, P.A.S. Effects of molecular size and chemical structure on renal and hepatic removal of exogenously administered chondroitin sulfate in rats. *Biochim. Biophys. Acta Gen. Subj.* **2006**, *1760*, 865–876. [CrossRef] [PubMed]
99. Volpi, N. About oral absorption and human pharmacokinetics of chondroitin sulfate. *Osteoarthritis Cartilage* **2010**, *18*, 1104–1105. [CrossRef]
100. Williams, A.; He, W.Q.; Cress, B.F.; Liu, X.Y.; Alexandria, J.; Yoshizawa, H.; Nishimura, K.; Toida, T.; Koffas, M.; Linhardt, R.J. Cloning and expression of recombinant chondroitinase acii and its comparison to the arthrobacter aurescens enzyme. *Biotechnol. J.* **2017**, *12*, 1700239. [CrossRef] [PubMed]
101. Chen, Z.; Li, Y.; Yuan, Q. Expression, purification and thermostability of MBP-chondroitinase ABC I from Proteus vulgaris. *Int. J. Biol. Macromol.* **2015**, *72*, 6–10. [CrossRef] [PubMed]
102. Li, Y.; Chen, Z.; Zhou, Z.; Yuan, Q. Expression, purification and characterization of GAPDH-ChSase ABC I from Proteus vulgaris in Escherichia coli. *Protein Expr. Purif.* **2016**, *128*, 36–41. [CrossRef]
103. Wang, J.P.; Zhang, L.; Jin, Z.Y. Separation and purification of low-molecular-weight chondroitin sulfates and their anti-oxidant properties. *Bangladesh J. Pharmacol.* **2016**, *11*, S61–S67. [CrossRef]
104. Valko, M.; Leibfritz, D.; Moncol, J.; Cronin, M.T.; Mazur, M.; Telser, J. Free radicals and antioxidants in normal physiological functions and human disease. *Int. J. Biochem. Cell Biol.* **2007**, *39*, 44–84. [CrossRef] [PubMed]
105. Fu, J.; Jiang, Z.; Chang, J.; Han, B.; Liu, W.; Peng, Y. Purification, characterization of Chondroitinase ABC from Sphingomonas paucimobilis and in vitro cardiocytoprotection of the enzymatically degraded CS-A. *Int. J. Biol. Macromol.* **2018**, *115*, 737–745. [CrossRef] [PubMed]
106. Vasconcelos, A.A.; Sucupira, I.D.; Guedes, A.L.; Queiroz, I.N.; Frattani, F.S.; Fonseca, R.J.; Pomin, V.H. Anticoagulant and antithrombotic properties of three structurally correlated sea urchin sulfated glycans and their low-molecular-weight derivatives. *Mar. Drugs* **2018**, *16*, 304. [CrossRef] [PubMed]
107. Zhao, L.; Wu, M.; Xiao, C.; Yang, L.; Zhou, L.; Gaoa, N.; Li, Z.; Chen, J.; Chen, J.; Liu, J.; et al. Discovery of an intrinsic tenase complex inhibitor: Pure nonasaccharide from fucosylated glycosaminoglycan. *Proc. Natl. Acad. Sci. USA* **2015**, *112*, 8284–8289. [CrossRef] [PubMed]
108. He, M.; Wang, J.; Hu, S.; Wang, Y.; Xue, C.; Li, H. The effects of fucosylated chondroitin sulfate isolated from Isostichopus badionotus on antimetastatic activity via down-regulation of Hif-1α and Hpa. *Food Technol. Biotechnol.* **2014**, *23*, 1643–1651. [CrossRef]
109. Zhao, Y.; Zhang, D.; Wang, S.; Tao, L.; Wang, A.; Chen, W.; Zhu, Z.; Zheng, S.; Gao, X.; Lu, Y. Holothurian glycosaminoglycan inhibits metastasis and thrombosis via targeting of nuclear factor-κB/tissue factor/Factor Xa pathway in melanoma B16F10 cells. *PLoS ONE* **2013**, *8*, e56557. [CrossRef]
110. Li, J.; Li, S.; Yan, L.; Ding, T.; Linhardt, R.J.; Yu, Y.; Liu, X.; Liu, D.; Ye, X.; Chen, S. Fucosylated chondroitin sulfate oligosaccharides exert anticoagulant activity by targeting at intrinsic tenase complex with low FXII activation: Importance of sulfation pattern and molecular size. *Eur. J. Med. Chem.* **2017**, *139*, 191–200. [CrossRef]
111. Wu, N.; Zhang, Y.; Ye, X.; Hu, Y.; Ding, T.; Chen, S. Sulfation pattern of fucose branches affects the anti-hyperlipidemic activities of fucosylated chondroitin sulfate. *Carbohyd. Polym.* **2016**, *147*, 1–7. [CrossRef] [PubMed]

112. Chen, S.; Li, G.; Wu, N.; Guo, X.; Liao, N.; Ye, X.; Liu, D.; Xue, C.; Chai, W. Sulfation pattern of the fucose branch is important for the anticoagulant and antithrombotic activities of fucosylated chondroitin sulfates. *BBA Gen. Subjects* **2013**, *1830*, 3054–3066. [CrossRef] [PubMed]
113. Sheehan, J.P.; Walke, E.N. Depolymerized holothurian glycosaminoglycan and heparin inhibit the intrinsic tenase complex by a common antithrombin-independent mechanism. *Blood* **2006**, *107*, 3876–3882. [CrossRef] [PubMed]
114. Solakyildirim, K.; Zhang, Z.; Linhardt, R.J. Ultraperformance liquid chromatography with electrospray ionization ion trap mass spectrometry for chondroitin disaccharide analysis. *Anal. Biochem.* **2010**, *397*, 24–28. [CrossRef] [PubMed]
115. Peng, C.N.; Wang, Q.B.; Wang, S.M.; Wang, W.S.; Jiao, R.M.; Han, W.J.; Li, F.C. A chondroitin sulfate and hyaluronic acid lyase with poor activity to glucuronyl 4,6-O-disulfated N-acetylgalactosamine (E-type)-containing structures. *J. Biol. Chem.* **2018**, *293*, 4230–4243. [CrossRef] [PubMed]
116. Mizumoto, S.; Murakoshi, S.; Kalayanamitra, K.; Deepa, S.S.; Fukui, S.; Kongtawelert, P.; Yamada, S.; Sugahara, K. Highly sulfated hexasaccharide sequences isolated from chondroitin sulfate of shark fin cartilage: Insights into the sugar sequences with bioactivities. *Glycobiology* **2013**, *23*, 155–168. [CrossRef] [PubMed]
117. Okamoto, Y.; Higashi, K.; Linhardt, R.J.; Toida, T. Comprehensive analysis of glycosaminoglycans from the edible shellfish. *Carbohyd. Polym.* **2018**, *184*, 269–276. [CrossRef]
118. Benito-Arenas, R.; Doncel-Pérez, E.; Fernández-Gutiérrez, M.; Garrido, L.; García-Junceda, E.; Revuelta, J.; Bastida, A.; Fernández-Mayoralas, A. A holistic approach to unravelling chondroitin sulfation: Correlations between surface charge, structure and binding to growth factors. *Carbohyd. Polym.* **2018**, *202*, 211–218. [CrossRef] [PubMed]
119. Doncel-Pérez, E.; Aranaz, I.; Bastida, A.; Revuelta, J.; Camacho, C.; Acosta, N.; Garrido, L.; Civera, C.; García-Junceda, E.; Heras, A.; et al. Synthesis, physicochemical characterization and biological evaluation of chitosan sulfate as heparan sulfate mimics. *Carbohyd. Polym.* **2018**, *191*, 225–233. [CrossRef] [PubMed]
120. Pudełko, A.; Wisowski, G.; Olczyk, K.; Koźma, E.M. The dual role of the glycosaminoglycan chondroitin-6-sulfate in the development, progression and metastasis of cancer. *FEBS J.* **2019**. [CrossRef]

Review

Biocatalyzed Synthesis of Statins: A Sustainable Strategy for the Preparation of Valuable Drugs

Pilar Hoyos [1], Vittorio Pace [2] and Andrés R. Alcántara [1,*]

1 Department of Chemistry in Pharmaceutical Sciences, Faculty of Pharmacy, Complutense University of Madrid, Campus de Moncloa, E-28040 Madrid, Spain; phoyosvi@ucm.es

2 Department of Pharmaceutical Chemistry, Faculty of Life Sciences, Althanstrasse 14, A-1090 Vienna, Austria; vittorio.pace@univie.ac.at

* Correspondence: andalcan@ucm.es; Tel.: +34-91-394-1823

Received: 25 February 2019; Accepted: 9 March 2019; Published: 14 March 2019

Abstract: Statins, inhibitors of 3-hydroxy-3-methylglutaryl coenzyme A (HMG-CoA) reductase, are the largest selling class of drugs prescribed for the pharmacological treatment of hypercholesterolemia and dyslipidaemia. Statins also possess other therapeutic effects, called pleiotropic, because the blockade of the conversion of HMG-CoA to (*R*)-mevalonate produces a concomitant inhibition of the biosynthesis of numerous isoprenoid metabolites (e.g., geranylgeranyl pyrophosphate (GGPP) or farnesyl pyrophosphate (FPP)). Thus, the prenylation of several cell signalling proteins (small GTPase family members: Ras, Rac, and Rho) is hampered, so that these molecular switches, controlling multiple pathways and cell functions (maintenance of cell shape, motility, factor secretion, differentiation, and proliferation) are regulated, leading to beneficial effects in cardiovascular health, regulation of the immune system, anti-inflammatory and immunosuppressive properties, prevention and treatment of sepsis, treatment of autoimmune diseases, osteoporosis, kidney and neurological disorders, or even in cancer therapy. Thus, there is a growing interest in developing more sustainable protocols for preparation of statins, and the introduction of biocatalyzed steps into the synthetic pathways is highly advantageous—synthetic routes are conducted under mild reaction conditions, at ambient temperature, and can use water as a reaction medium in many cases. Furthermore, their high selectivity avoids the need for functional group activation and protection/deprotection steps usually required in traditional organic synthesis. Therefore, biocatalysis provides shorter processes, produces less waste, and reduces manufacturing costs and environmental impact. In this review, we will comment on the pleiotropic effects of statins and will illustrate some biotransformations nowadays implemented for statin synthesis.

Keywords: biocatalysis; biotransformations; statins; pleiotropic effects

---

1. Introduction

It is very well known that raised cholesterol levels increase the risks of heart disease and stroke. Globally, a third of ischaemic heart disease is attributable to high cholesterol and, according to the World Health Organization, raised cholesterol is estimated to cause 2.6 million deaths (4.5% of total) and 29.7 million disability adjusted life years (DALYS) [1]. In this sense, inhibitors of 3-hydroxy-3-methylglutaryl coenzyme A (HMG-CoA) reductase, commonly known as statins (Figure 1), are the largest selling class of drugs prescribed for the pharmacological treatment of hypercholesterolemia and dyslipidaemia [2,3], and it has been also reported that since their introduction in 1987, the lives of millions of people have been extended through statin therapy and, more importantly, quality of life has been drastically improved [4].

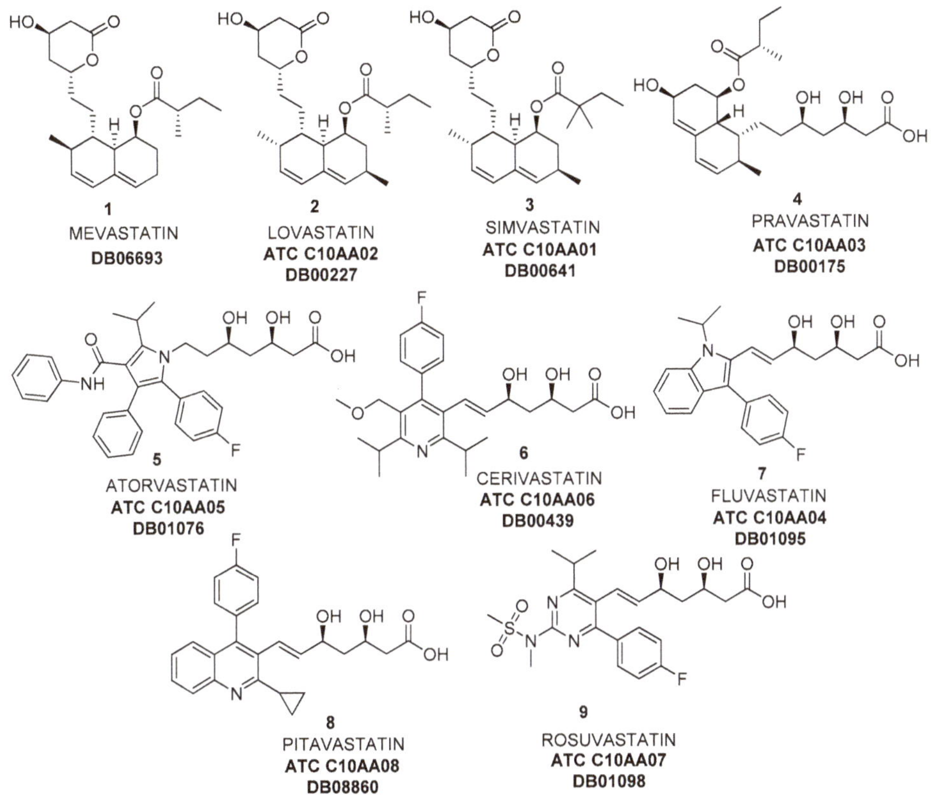

Figure 1. Some Statins, inhibitors of 3-hydroxy-3-methylglutaryl coenzyme A (HMG-CoA) reductase.

Since the discovery of the first statins from natural sources, mevastatin (Figure 1, **1**, also named compactin, from the fungi *Penicillium citrinum* [5] and *Penicillium brevicompactum* [6]), lovastatin (Figure 1, **2**, Mevinolin, found in *Aspergillus terreus* [7] and food such as oyster mushrooms [8] or red yeast rice [9]), simvastatin (Figure 1, **3**, Mevacor, also isolated from *Aspergillus terreus* [10]), and pravastatin (Figure 1, **4**, initially known as CS-514, originally identified in the bacterium *Nocardia autotrophica* [11]), synthetic more potent compounds (Figure 1, **5–9**), also known as superstatins, were introduced into the drug market [12,13]. As can be observed, the common structure of these compounds is formed by a central core of different heterocyclic aromatic rings containing nitrogen, and a lateral chain derived from (3*R*,5*R*)-3,5-dihydroxyheptanoic acid.

The economic impact of statins on the drug market is enormous. For instance, simvastatin was originally developed by Merck under the brand name Zocor™; in 2005, Zocor™ was Merck's best-selling drug and the second-largest selling statin in the world (more than US$3 million only in USA, according to different reports [14–21]). In 2006, Zocor™ went off patent, and the annual sales drastically dropped; anyhow, from that moment, generic simvastatin became the most-prescribed statin in the world between 2010 and 2015 [19,20,22]. On the other hand, atorvastatin (Figure 1, ATC (Anatomical Therapeutic Chemical) classification system, according to World Health Organization) Code C10AA05, DrugBank Code DB01076, **5**) is the greatest blockbuster drug in pharmaceutical history, and the best known representative of superstatins, receiving this name because of its pronounced ability to reduce low-density lipoprotein cholesterol levels and increase high-density lipoprotein cholesterol compared with other existing agents [13]. It was first synthesized in 1985 by Bruce Roth of Parke-Davis Warner-Lambert Company (now Pfizer), which commercialized it under the name of

Lipitor™. Since it was approved in 1996, sales have exceeded US$125 billion, and the drug has topped the list of best-selling branded pharmaceuticals in the world for nearly a decade. When Pfizer's patent on Lipitor™ expired in USA by the end of 2011 and in Europe in mid-2012, generic atorvastatin from other companies became available, and it is still being widely sold (US$2.16 billion in sales, standing as the year's fourth-best-selling cardiovascular drug, with analysts predicting sales of US$1.85 billion in 2024 [23]). Finally, rosuvastatin (Figure 1, **9**, ATC Code C10AA07, DrugBank Code DB01098) was marketed as calcium salt in 2003 by AstraZeneca under the name of Crestor™. Like atorvastatin, rosuvastatin is also a superstatin; the initial patent on rosuvastatin synthesis (purely chemical) was developed by Shionogi Research Laboratories [24] and later sold to AstraZeneca. This patent expired in June 2016, but anyhow, it still can be considered a blockbuster drug, by looking at the great volume of sales of Crestor™ (around U$2.7 billion in 2017, and US$727 million for the first half of 2018 [23]). A recent study [25] points toward global sales of statins of US$1 trillion by 2020, thus pharmaceutical companies are still interested in developing new synthetic strategies for putting these drugs on the market.

Thus, it is undoubtedly clear that the statin market involves a huge amount of money. Furthermore, the importance of this type of drug is even higher because of their new therapeutic uses that are recently becoming more and more recognized, which will be commented on in Section 2. Finally, as the absolute configuration of statins plays a crucial role in the activity of these compounds, the enormous potential of an enantioselective biocatalytic process for the sustainable synthesis of chiral building blocks involved in statin preparative procedures will further be commented on in Section 3.

2. New Therapeutic Effects of Statins

As mentioned before, these drugs act by reversibly and competitively inhibiting the bioreduction of *S*-3-hydroxy-3-methylglutaryl-coA (HMG-CoA), the rate-limiting step of the mevalonate pathway in cholesterol biosynthesis (Figure 2), because of the chemical similitude with mevalonyl-CoA, the intermediate obtained after the first reduction of HMG-CoA. Furthermore, there is extensive recent evidence suggesting that statins are more than simple lipid-lowering drugs [3,26]; in fact, a large amount of up-to-date experimental data have confirmed that statins may exert many different potentially beneficial therapeutic effects, by several mechanisms not essentially related to cholesterol metabolism. These so-called pleiotropic effects [27] could be attributed to their ability to prevent the conversion of HMG-CoA to *R*-mevalonate, which results in the concomitant inhibition of the downstream biosynthesis of cholesterol, as well as of numerous isoprenoid metabolites, such as geranylgeranyl pyrophosphate (GGPP) or farnesyl pyrophosphate (FPP), as shown in Figure 2.

These molecules are well-known key intermediates for prenylation of several cell signalling proteins (such as small GTPase family members: Ras, Rac, Rho, Rab), which act as molecular switches controlling multiple pathways and cell functions (maintenance of cell shape, motility, factor secretion, differentiation, and proliferation), so that they can be inhibited by statin treatment [28]. For instance, when Ras and Rho isoprenylation is inhibited, there is a concurrent accumulation of inactive forms of both proteins in cytoplasm and an inhibition of these signalling molecules [29]. Certainly, it has been reported that small G-proteins like Rho and Rac influence endothelial nitric oxide synthase (eNOS) expression and nitric oxide (NO) availability [30]. Rho negatively regulates eNOS expression, while Rac activates nicotinamide dinucleotide phosphate (NADPH)-oxidase and the correspondent superoxide production, which in turn inactivates NO. If statins block both Rho and Rac GTPase activity via inhibition of geranylgeranylation, this leads to eNOS upregulation [31,32]. Remarkably, some beneficial effects of statins were displayed before cholesterol levels were reduced [30], and it can be assumed that those effects, dependent on the enhancement of eNOS expression and/or activity, result in a decline of platelet activation, attenuation of adhesion molecules expression, decrease of inflammatory cytokine production, and increase of reactive oxygen species (ROS) [33]. Therefore, pleiotropic effects of statins include the reduction of haemostasis by reducing platelet activation and the pro-coagulation cascade; the increase of fibrinolysis and the anticoagulation cascade; the improvement of endothelial function; the increase of NO bioavailability; as well as antioxidant, immune modulatory, and anti-inflammatory activities and stabilization of atherosclerotic plaques [27,29,34–37]. Thus, the therapeutic effects of statins are nowadays present in areas such as cardiovascular health, regulation of the immune system, anti-inflammatory and immunosuppressive properties, prevention and treatment of sepsis, treatment of autoimmune diseases, osteoporosis, kidney and neurological disorders, and even in cancer therapy; some of these therapeutic areas will be commented on.

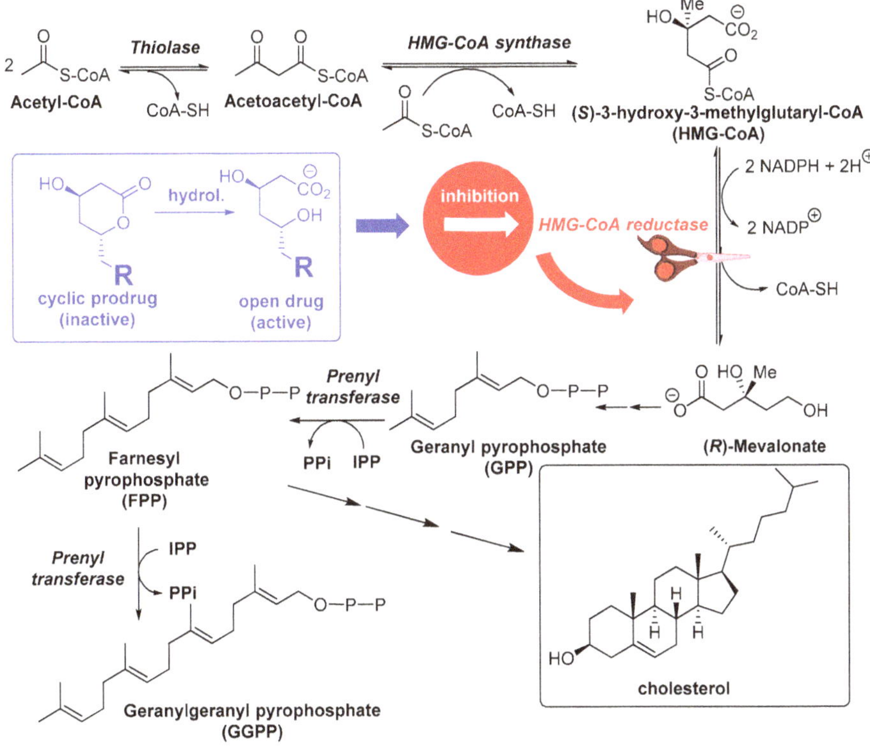

Figure 2. Mevalonate pathway.

2.1. Cardiovascular Effects

Aside from the main mechanism of action of lowering cholesterol levels, statins are also useful in the treatment of some other cardiovascular disorders, including acute coronary syndrome, heart failure, cardiac arrhythmias, aortic stenosis, peripheral arterial disease, cerebrovascular disease, and essential hypertension, as recently reviewed (see papers by Mihos et al. [38], Oesterle et al. [39], and references therein). In fact, chronic administration of statins is believed to produce what is known as PIC ("pre-ischemic conditioning"), protecting the myocardium during ischemic insult and injury [40], as a consequence of an increase in nitric oxide availability and immunomodulation; thus, statins increase the production of nitric oxide and blunt the formation of superoxide radicals via the upregulation of eNOS and stabilization of its mRNA, leading to an improved vascular function and a reduction in vascular inflammation [34]. In this sense, recent studies show the effectiveness of statins' cardiovascular primary prevention [41], also for elderly people [42], and point towards the special benefits of fluvastatin [43].

2.2. Immunomodulatory Effects

The main objectives of autoimmune therapies are to re-establish immunological homeostasis and reduce autoimmune damages. Different studies are increasingly suggesting that an imbalance between Th17 and Treg cells, as well as the incorrect release of potent pro-inflammatory mediators by Th17 cells, are crucial for the pathogenesis of a number of autoimmune disorders [44]. Thus, a new immunotherapeutic strategy could be based on increasing Treg or inhibiting Th17 differentiation/effector functions. In this respect, statins show an outstanding potential, especially considering the increasing evidence that they might inhibit Th17 differentiation/effector functions and conversely promote Treg differentiation/suppressive function selectively in the setting of autoimmune diseases [44]. Small GTPases have been centrally implicated in regulating the development and functions of T and B lymphocytes as well as of dendritic cells (DC) [45,46]. Thus, as a consequence of the inhibition of GTPases prenylation, statin-based therapy can be a potential alternative for the treatment of autoimmune diseases [44,47,48]. In fact, positive effects of statin treatment have been reported in numerous autoimmune diseases such as multiple sclerosis [49,50], systemic lupus erythematosus [51–53], autoimmune myocarditis [54–56], or rheumatoid arthritis [44,57–59].

2.3. Neurological Disorders

This is probably one of the most attractive therapeutic areas in which the use of statins introduces interesting advances. Pleiotropic effects of statins via GTPases inhibition might have potential therapeutic implications in many neurological disorders, as the current connection between neurodegenerative diseases and vascular risk factors is becoming more and more evident [30,60]; therefore, statin treatment could display beneficial effects in neurological disorders such as stroke, Alzheimer's disease (AD), Parkinson's disease (PD), multiple sclerosis (MS), primary brain tumours, or depression.

2.3.1. Stroke

The risk factors for cerebrovascular disease are well known and largely variable and, in this sense, reduction of serum cholesterol levels could be highly beneficial for reducing the hazard of suffering a cerebrovascular accident (CVA), also named stroke [30]. Anyhow, an indubitable link between high cholesterol level and stroke risk is difficult to establish, because controversial data from several clinical studies have been published in the literature, some of them finding no relationship between cholesterol and stroke [61,62], while in some other cases, the beneficial effects are indeed observed [63,64]. A possible explanation for these discrepancies could be based on the fact that stroke can be either ischemic or haemorrhagic, and there are evidences supporting an association between elevated cholesterol and increased risk of ischemic stroke, but also showing a relationship between low cholesterol levels and increased risk of haemorrhagic stroke [30]. So, while disagreements are still present on the usefulness of statins in the primary prevention of acute stroke, there is a wide consensus on the positive aspects of statins treatment in secondary prevention after stroke or transient ischemic attack for diminishing the menace of suffering a new stroke [65–68]. Even in haemorrhagic stroke, some data from recent studies suggest that statin therapy could improve the outcome after spontaneous intra-cerebral haemorrhage and statin therapy should be not discontinued [69–71]. In any case, the most feasible explanation for reduction in clinical events reported for patients treated with statins is the stabilization of atherosclerotic plaques, which are generated by lipids deposition and migration and proliferation of vascular smooth muscle cells (see report from Malfitano et al. [30] and references cited therein for a more detailed explanation).

2.3.2. Alzheimer's Disease (AD)

AD is a chronic neurodegenerative syndrome caused by the appearance of brain senile plaques composed of aggregated forms of β-amyloid peptide (Aβ), and it is the most common cause of dementia in elderly people, with a new case globally occurring every seven seconds [72]. Emerging evidence suggests a link between cholesterol and AD [37,72–76], and extensive studies have been published stressing the therapeutic utility of pleiotropic effects of statins, showing a dose-dependent beneficial effect on cognition, memory, and neuroprotection [72] by different mechanisms, such as altering the properties of plasma membrane by a reduction in cholesterol levels and a modulation of secretase activities, thus decreasing amyloid precursor protein (APP) processing [77], or by altering neuronal activity via modification of GTPases prenylation [28,74,78]. On the other hand, a possible effect of statins in cholinergic neurotransmission has been also described; in fact, simvastatin inhibits acetyl cholinesterase (AChE) activity in rats [79] and prevents the blockade caused by AChE inhibitors at α 7-nicotinic AChE receptors [80], thus increasing cholinergic neurotransmission. In this sense, Ghodke et al. [81] reported that statins treatment for 4 months, but not for 15 days, showed noteworthy enhancement in mice memory function, whereas a high cholesterol diet showed significant diminishing of memory. However, long-term statin treatment showed a significant decrease in serum cholesterol level as well as brain AChE level. Moreover, a high cholesterol diet showed a significant decrease in memory function with an increase in serum cholesterol level as well as brain AChE level. Thus, they concluded that there was no direct correlation between brain cholesterol level, as well as HMG-CoA activity with memory function regulation, although there is tangible link between plasma cholesterol level and AChE level, and long-standing plasma cholesterol alteration may be essential to regulate memory function through the AChE modulated pathway. Finally, a simvastatin-related rise of butyryl cholinesterase (BuChE) activity in mice brain, which may be a potential adverse effect in patients with AD, has been recently reported [82].

Another feasible mechanism for explaining statins' neuroprotective effect considers an activation of the heme oxygenase/biliverdin reductase (HO/BVR-A) system [37]. Statins can also be active in AD treatment because of their protecting effect against glutamate toxicity over primary cortical neurons [83,84]. Low-dose administration of statins avoids aberrant neuronal entry into mitosis [85], promotes anti-apoptotic pathways [86], and impairs inflammation [87], although higher doses of statins have been shown to induce toxic effects [88]. Recently, some studies point towards the utility of simvastatin administration in the improving of hippocampus-dependent spatial memory in mice, due to an activation of Akt (protein kinase B), via a depletion of FPP and inhibition of farnesylation [89,90]. This same group has recently shown how simvastatin administration potentiates the contribution of *N*-methyl *D*-aspartate receptor (NMDAR) to synaptic transmission, by increasing the surface distribution of the GluN2B subunit of the NMDAR without affecting cellular cholesterol content [91]. The influence of statins in these ionotropic glutamate-receptors, and the succeeding utility of these drugs on treatment of AD and other mental disorders, is undoubtedly a very attractive and innovative research field [91,92].

Lamentably, although most evidence consistently confirms how statins do afford neuroprotection and improve disease pathology in animal models [93,94], results are rather controversial or even disappointing in human trials [72,95–98], thus a very careful study design and analysis will be essential in the future [95].

2.3.3. Parkinson's Desease (PD)

PD, the second most common chronic neurodegenerative disorder in adults over the age of 65 years [99], is a progressive neurodegenerative disorder characterised by the presence of intracellular protein aggregates (Lewy bodies) and the loss of dopaminergic neurons from the *pars compacta* component of the *substantia nigra* in the midbrain; PD-related clinical manifestations of dopamine deficiency (gait, tremor, rigidity, and bradykinesia) are the most archetypical symptoms of this disease. There are several studies showing that some statins (simvastatin, but neither atorvastatin nor lovastatin) may reduce the incidence of PD in patients aged over 65 years [100]. Compared with discontinuation of statins, continuation of lipophilic statin use has been associated with a reduced risk of PD, particularly in the elderly [101]; nevertheless, in patients with existing PD, 10-day treatment of simvastatin

(40 mg/day) showed no significant effects on dyskinesia, functional impairment, or involuntary movement [102].

As inflammation is accepted to be a main contributor to the PD aetiology, the anti-inflammatory action of statins could be a rational explanation for their activity [30]; in fact, simvastatin has been reported useful for reversing the loss of striatal dopamine activity and the production of nitrosylated free radicals, thus inducing neuro-protection [103,104], by decreasing the release of inflammatory mediators from microglia. Also, some studies in rats have shown that simvastatin can protect against loss of NMDA receptors produced by 6-hydroxydopamine (6-OHDA) [105]; also using the 6-OHDA model in rats, Wang et al. [106] recently described the beneficial effect of simvastatin in reducing abnormal involuntary movements known as L-DOPA-induced dyskinesia, commonly observed in patients chronically treated with L-DOPA. Finally, simvastatin and pravastatin can decrease the dopaminergic neuronal loss induced by MPTP (1-methyl-4-phenyl-1,2,3,6-tetrahydropyridine) via inhibition of p21(Ras)-induced NF-κB (nuclear factor kappa-light-chain-enhancer of activated B cells) [107].

Anyhow, as mentioned in the previous paragraph, more definitive evidence from prospective and clinical studies is required before drawing any conclusions about statins efficacy for treatment of PD.

2.3.4. Depression

As well as for the previous neurological disorders discussed so far, there are reported discrepancies about statins' effect in depression, with some studies reporting positive effects of statins in reducing depression and depression-like symptoms in animals [108–115] or humans [116,117], while some others stated no relationships [118–120]. These divergences require more detailed studies, also for elucidating the possible mechanism of the positive effects, which, in some cases, have been associated with a modulation of NMDA receptors [121] or peroxisome proliferator-activated receptor gamma (PPAR-γ) receptors, by NO inhibition [122].

2.3.5. Epilepsy

In several studies, a reduced risk of developing epilepsy after age 50 has been reported [123–128], and the mechanism for this neuroprotective effect has been associated with a decrease in the association of subunit 1 of NMDA receptors to lipid rafts [129], as well as inhibition of calcium-dependent calpain activation, ROCK inhibition, the activation of the PI3K pathway, and increased APP cleavage [124], or the increased expression level of eNOS [130]; a recent publication by Scicchitano et al. [131] summarises the currently available data concerning statin effects in modulating epileptic seizure activity (sometimes adversely) and epileptogenesis in different experimental models, as well as in clinical studies [123,132].

2.4. *Cancer*

There are many studies dealing with the potential antitumor efficacy of statins, reporting effects in different cancer cell lines, as well as the possible risks of cancer development caused by statins treatment and the results of different clinical trials [133–138]. Once again, the molecular mechanisms explaining statins' effects are quite different, and clinical trials are not reporting conclusive results; in fact, although some large scale meta-analyses seem to indicate that statins do not have significant effects on cancer incidence [133,139,140], in some other cases, some beneficial effects associated with statins' administration in the treatment of different cancers have been described [141–144].

What is really clear is that there is not just one mechanism explaining the anticancer activity of statins, because depending on the type and dosages of statin used, the type of cancer cells, and the time of exposure of cells to statins, different effects leading to cell-cycle arrest, induction of apoptosis, or changes in molecular pathways are reported [138]. Concerning cell modifications, a common scheme is followed, starting with an arrest of cells in the G1 [145,146] or S-phase [147], and this inhibition of cell-cycle progression is mediated by cyclins (such as cyclin D1 [148]), cyclin-dependent kinases (CDKs, such as $p21^{WAF1/CIP1}$ [148], p27 [149], CD4 [148], or p53 [150,151]), and inhibitors of CDKs [145].

Simultaneously, inhibition of G-protein prenylation is produced, leading to the arrest of proliferation and/or induction of apoptosis in cancer cells [152,153], by an increase in caspases activity [147,154,155]; henceforward, inhibition of prenylation is a promising way to impede progression of cancer (see the recent review of Matusewicz et al. [138] and cites therein). On the other hand, it has been also reported that a substantial reduction in the amount of cholesterol leads to a reduction in the content of membrane lipid rafts in the cell membrane, altering cell signalling [156,157], and loosing membrane integrity; in this sense, it is known that the membrane of breast cancer and prostate cancer cells has higher level of cholesterol and lipid rafts, so these cells are more susceptible to apoptosis promoted by statins compared with normal cells [158,159]. In another feasible mechanism, statins are associated with inhibition of phosphorylation of caveolin-1 (Cav-1), the integral membrane protein that binds and transport cholesterol, which promotes tumour cell survival and resistance to chemotherapy by different mechanisms [160].

Anyhow, an exhaustive recompilation of all other mechanisms proposed to explain the action of statins in cancer treatment would be out of the scope of this manuscript and can be found in recent reviews [135,136,138,161]. However, once again, while the correlation between data obtained in vitro with those other ones reported in animal models is very high, clinical trials are not that irrefutable in their conclusions, and more detailed studies are demanded.

3. Biocatalyzed Synthesis of Statins

As previously indicated in the Introduction, the use of biocatalyzed steps for preparing homochiral synthons useful for the synthesis of statins is a smart strategy for gaining both efficiency and sustainability. In this section, we will present some examples.

3.1. *Simvastatin*

Lovastatin (Mevacor **2**, ATC Code C10AA02, DrugBank Code DB00227, Figure 1) is a naturally-occurring fungal polyketide produced by *Aspergillus terreus* [162], while simvastatin (**3**, ATC Code C10AA01, DrugBank Code DB00641, Figure 1) is a semisynthetic analogue of **2** and is more effective in treating hypercholesterolemia, because of the fact that the substitution of the α-methylbutyrate side chain with α-dimethylbutyrate significantly increases the inhibitory properties of **2**, while lowering undesirable side effects [10].

Because of the economic importance of simvastatin, as mentioned in the Introduction, various multistep syntheses of **3** starting from **2** have been described; thus, a widely used process (route #1) starts with the hydrolysis of the C8 ester in **2** to yield the triol Monacolin J **10**, followed by selective silylation of the C13 alcohol to yield **11**, esterification of C8 alcohol with dimethylbutyryl chloride to furnish **12**, and deprotection of C13 alcohol to finally yield **3** [163] (Figure 3).

In another option, namely route #2 [164], lovastatin **2** was treated with n-butylamine and TBSCl to obtain **13**, which was alkylated with another methyl group to furnish **14**, and finally transformed into **3** by hydrolysis and lactonization. Both multistep processes shown in Figure 3 were laborious, thus contributing to simvastatin being nearly five times more expensive than lovastatin [165].

Some enzymatic transformations using lipases and esterases were investigated as alternatives to chemical hydrolysis leading to Monacolin J **10** [166,167]. However, the requirement of regioselective esterification of the C8 alcohol invariably involves protection of other reactive alcohol groups in **10**, and generally leads to lowered overall yield. Therefore, a specific reagent that is able to selectively acylate C8 of **10** is important for the efficient synthesis of simvastatin **3** and additional statin analogues.

In this sense, Tang and co-workers [22] described an acyltransferase (LovD) able to catalyse the last step of lovastatin biosynthesis, as shown in Figure 4, by transferring a 2,2-dimethylbutyryl acyl group from dimethylbutyryl-S-methylmercaptopropionate (DMB-SMMP, **16**) regioselectively to the C8 hydroxyl of Monacolin J **10**, the immediate biosynthetic precursor of simvastatin. The reaction proceeds via a ping-pong mechanism, and LovD is inhibited by Monacolin J at moderate substrate concentrations. LovD displayed broad substrate specificity toward the acyl carrier, the acyl substrate, and the decalin core of the acyl acceptor. This same group developed a one-step, whole-cell biocatalyzed process for the synthesis of Simvastatin from Monacolin J using an *Escherichia coli* strain overexpressing LovD, leading to >99% conversion of monacolin J to simvastatin without the use of any chemical protection steps [165]. The process was scaled up for gram-scale synthesis of simvastatin, also showing that simvastatin synthesized via this method could be readily purified from the fermentation broth with >90% recovery and >98% purity, as determined by high-performance liquid chromatography.

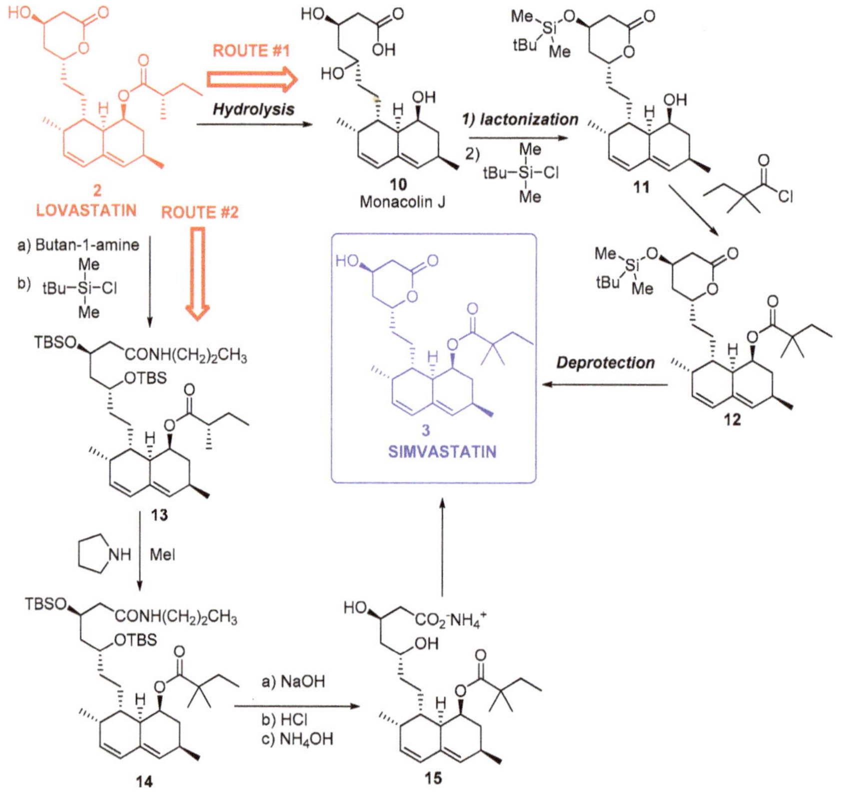

Figure 3. Chemical transformations of lovastatin **2** into simvastatin **3**.

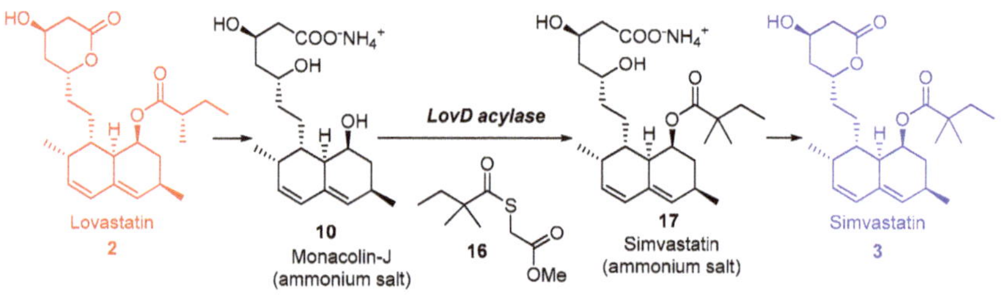

Figure 4. Biocatalyzed transformations of lovastatin **2** into simvastatin **3**.

Codexis improved not only the enzyme (previously modified used directed evolution at lab scale in an *E. coli*-based biocatalytic platform [168]) but also process chemistry to enable a large-scale simvastatin manufacturing process, by carrying out nine iterations of in vitro evolution, creating 216 libraries and screening 61,779 variants to develop a LovD variant with improved activity, in-process stability, and tolerance to product inhibition. The approximately 1000-fold improved enzyme and the new process pushed the reaction to completion at high substrate loading and minimized the amounts of acyl donor and of solvents for extraction and product separation. This process possesses many advantageous characteristics from a Green Chemistry point of view:

- Catalyst is produced efficiently from renewable feedstock.
- Reduced use of toxic and hazardous substances like tert-butyl dimethyl silane chloride, methyl iodide, and n-butyl lithium.
- Improved energy efficiency as the reaction is run at ambient temperature and at near atmospheric pressure.
- Reduction in solvent use because of the aqueous nature of the reaction conditions.
- The only by-product (methyl 3-mercaptopropionic acid) is recycled.
- The major waste streams generated are biodegraded in bio treatment facilities.
- Codexis' process can produce simvastatin with yields of 97%, significant when compared with <70% with other manufacturing routes.

For these reasons, Codexis and Prof. Tang obtained the U.S. Environmental Protection Agency's Green Chemistry Presidential Award in 2012 [169], inside the category of Greener Synthetic Pathway. Recently, identification of the complete biosynthetic pathway leading to monacolin J has been reported [170].

3.2. Biocatalyzed Synthesis of the Lateral Chain of Superstatins

Different biocatalytic routes have been proposed and implemented at industrial scale for the stereoselective preparation of the lateral chain (bearing the stereocentres) of superstatins. Thus, we would use the preparation of atorvastatin as a reference to illustrate how different biotransformations can be included in the overall protocol.

The chemical synthesis of atorvastatin originally described by researchers at Warner-Lambert Company [171], shown in Figure 5, started from a chiral building block, ethyl (*R*)-4-cyano-3-hydroxybutyrate **18**, also known as "hydroxynitrile" (**HN**), and the second stereogenic centre of **20** was obtained by diastereomeric induction, using cryogenic borohydride reduction of a boronate derivative of the 5-hydroxy-3-keto intermediate **19** derived from **HN**.

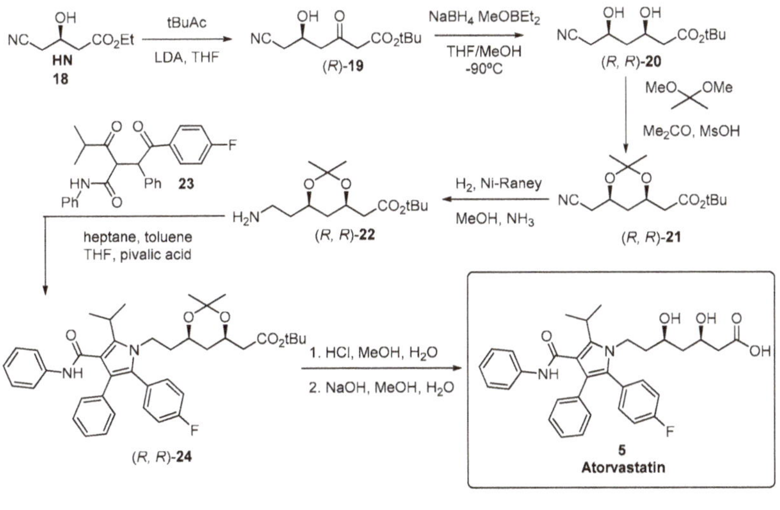

Figure 5. Chemical synthesis of atorvastatin **5**.

Taking this procedure as model, different strategies for generating the desired chirality can be envisaged from a biocatalytic retrosynthetic scheme [172], as depicted in Figure 6, in which purely chemical steps are denoted by a red **C**, while those syntheses feasible to be biocatalyzed are represented by a blue **BT** and a number, corresponding to the type of biocatalyst used. Thus, **route #1** creates the desired chirality by a stereoselective desymmetrization of dinitrile **25** using a *nitrilase* (**BT-1**), while **route #2** requires the preparation of **HN 18** via a bioreduction of ketoester **27**, so a *ketoreductase* (**BT-2**) is the biocatalyst required for that aim. Anyhow, in this synthetic path, another bioreduction should be used for avoiding the previously mentioned borohydride reduction of intermediate **19**, using another *ketoreductase* (**BT-3**), so this can be considered **route #3**. Finally, if an *aldolase* (**BT-4**) is the enzyme selected, it is possible to envisage **route #4** as an alternative through cyclic intermediate **28**. These different routes will be discussed in the following sections.

3.2.1. Hydrolases as Catalysts for the Preparation of the Lateral Chain of Atorvastatin and Other Superstatins

As shown in **route #1**, a nitrilase-catalyzed enzymatic desymmetrization of prochiral 3-hydroxyglutaronitrile **25** and subsequent esterification of the resulting (*R*)-3-hydroxy-4-cyanobutyric acid (*R*)**-26** can lead to **HN**. The use of enzymatic protocols for hydrolysing nitriles is a green alternative compared with chemical methodologies [173], because of the harsh reaction conditions required, demanding either strong mineral acids (e.g., hydrochloric or phosphoric acid) or bases (e.g., potassium or sodium hydroxide) and relatively high reaction temperatures. Moreover, chemical procedures sometimes give low yields because of both unwanted by-product formation and the generation of concentrated contaminating waste salt streams (e.g., 6 mol L^{-1}) when the acid or base is neutralized prior to disposal [174].

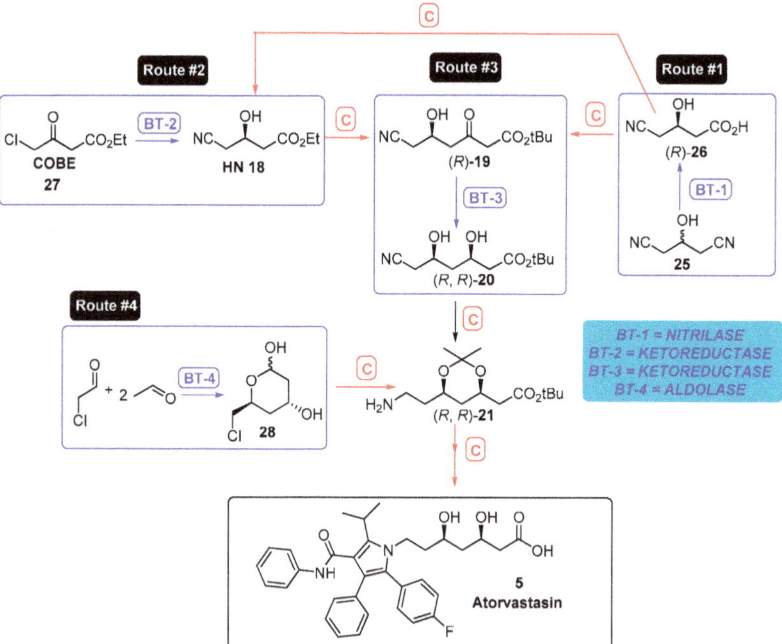

Figure 6. Biocatalytic retrosynthetic routes to atorvastatin. **BT** represents biotransformation step, while **C** stands for chemical processes.

Thus, researchers at Diversa described a wild type nitrilase enzyme that catalysed the desymmetrization of **25** at high substrate concentration (3M) at lab-scale reaction, with an enantiomeric excess (ee) of 88%. [175]. A mutant nitrilase, obtained by directed evolution using gene site saturation mutagenesis (GSSM), and showing Ala190His single mutation, resulted in an excellent biocatalyst; hence, after a 15 h reaction at 20 °C, (*R*)-**26** was isolated in 96% yield, with an excellent ee of 98.5% and a volumetric productivity of 619 g L^{-1} d^{-1} [176]. Subsequently, Dow Chirotech, a subsidiary of Dow Chemical Company, developed the Diversa nitrilase further into a biocatalysis process [177] and used the Pfenex expression system (a *Pseudomonas fluorescens*-based host expression system) to overproduce the enzyme. In this way, optimal reaction conditions for desymmetrization of **25** were as follows: 3 M (330 g L^{-1}), pH 7.5, 27 °C, under 16 h reaction time. A conversion of 100% and 99% product ee was obtained, and the so-formed (*R*)-**26** was consequently esterified to give **HN**. Overall, a highly efficient three-stage synthesis of **HN** starting from low-cost epichlorohydrin (required to produce **25**) was achieved with an overall yield of 23%, 98% ee, and 97% purity [177]. Recently, an enzymatic method has been described for the synthesis of ethyl (*R*)-3-hydroxyglutarate from **HN** using free and immobilized recombinant *Escherichia coli* BL21(DE3)pLysS harbouring a nitrilase gene from *Arabidopsis thaliana* (AtNIT2) [178]. The hydrolysis of **HN** proceeded with the freely suspended cells of the biocatalyst under the optimized conditions of 1.5 mol L^{-1} (235.5 g L^{-1}) substrate concentration and 6.0 wt % loading of wet cells at pH 8.0 and 25 °C, with 100% conversion obtained in 4.5 h. Furthermore, immobilization of the whole cells enhanced their substrate tolerance, stability, and reusability. Under the optimized conditions (100 mmol L^{-1} tris(hydroxymethyl)aminomethane hydrochloride buffer, pH 8.0, 25 °C), the immobilized biocatalyst could be reused for up to 16 batches, with a biocatalyst productivity of 55.6 g gwet $cells^{-1}$ and a space–time productivity of 625.5 g L^{-1} d^{-1}.

Hydrolases are also useful for preparing (*S*)-3-hydroxy butyrolactone (*S*)-**32**, another enantiopure intermediate to furnish **HN** (Figure 7). In fact, opening of (*S*)-**32** with HBr/EtOH will yield the corresponding ethyl (*S*)-4-bromo-3-hydroxybutanoate ((*S*)-BHBE, (*S*)-**33**) [179], later transformed into **HN** via S_N2 when treated with sodium (or potassium) bromide. Although (*S*)-**32** can be produced from chiral pool raw materials (lactose or malic acid), it can be conveniently obtained by enzymatic hydrolysis of the racemic ethyl 4-chloro-3-hydroxybutanoate (CHBE, *rac*-**29**) in the aqueous phase [180]. The lipase stereoselectively hydrolysed only the (*S*)-enantiomer; however, the resulting acid (*S*)-**30** is unstable, and it readily loses one HCl molecule to give the corresponding lactone of high enantiopurity (>99% ee). However, the enantiopurity of the lactone rapidly decreased when the process was operated at yields of more than 40%. The hydrolysis of the enantiopure benzoic ester of (*S*)-hydroxybutyrolactone (*S*)-**31** has also been described using lipase from *Candida rugosa* (CRL) immobilized on amberlite XAD-7 as polymeric support, with ee of 99% [181]. This enzymatic hydrolysis was observed to be non-stereoselective in nature, because the enzymatic hydrolysis of the racemic benzoic ester yielded the racemic lactone, so that a chiral pool precursor (L-malic acid) for this process was necessary. Anyhow, this method has been scaled up to a ton scale, with an overall yield of over 80%, and a reaction time of 14 h [182]. Recently, a platform pathway for the production of 3-hydroxyacids has been described as an alternative biosynthetic route to generate the enantiopure lactone [183].

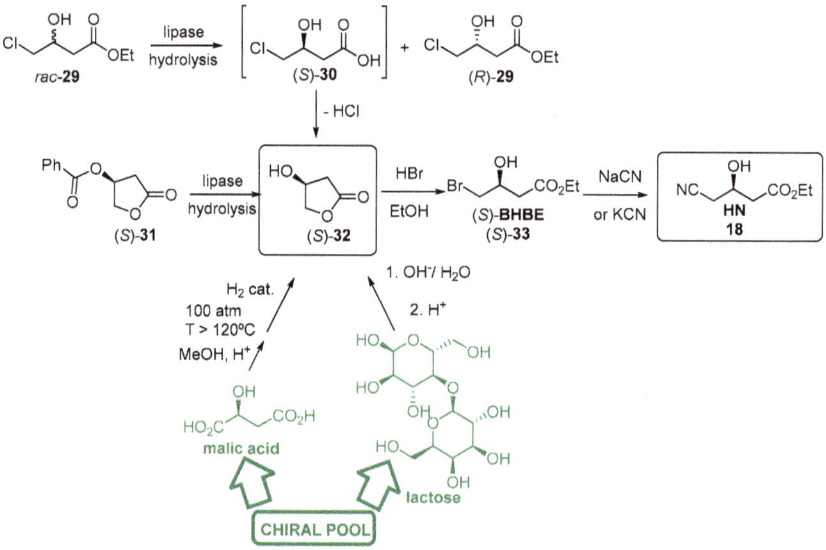

Figure 7. Preparation of **HN 18** starting from (S)-3-hydroxy butyrolactone (S)-**32**.

More recently, new biocatalytic approaches employing hydrolases have been described for furnishing the lateral chain of superstatins. Actually, Figure 8 shows a synthetic scheme for preparing rosuvastatin **9**. As can be seen, conjugated ketoester **36** is subsequently transformed into final calcium rosuvastatin **9** by different steps (silyl ether cleavage, diastereoselective Narasaka-Prasad [184,185] *syn*-reduction using diethylmethoxy borane leading to ester **37**, and finally ester hydrolysis and salt formation).

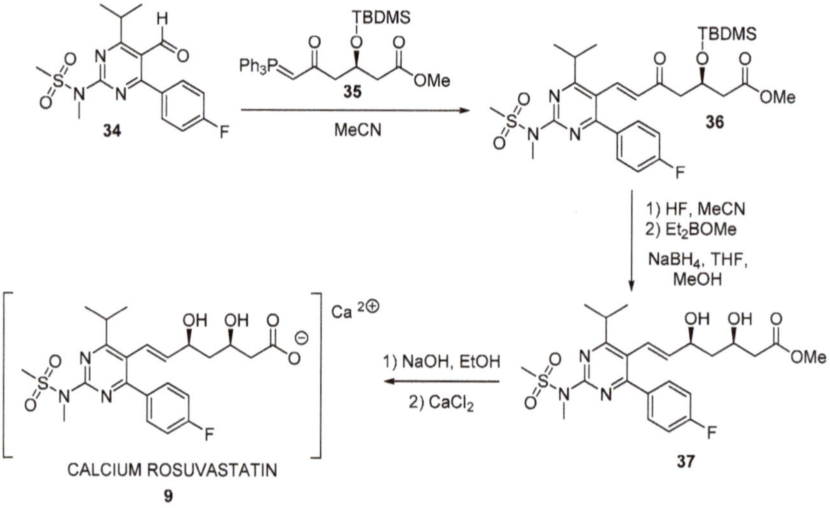

Figure 8. Final steps in the chemical synthesis of rosuvastatin **9**.

Aldehyde **34** can be easily obtained [186], while the preparation of enantiopure ylide **35** is much more complicated. Thus, several examples can be found in the literature starting from racemic diethyl 3-hydroxyglutarate, which had to be previously transformed in an activated derivative to react with the corresponding methyltriphenylphosphonium ylide to finally yield **35**; although this route has been described using an enzymatic desymmetrization step [187,188], different side reactions were observed to decrease either the final yield or the enantiomeric excess. Recently, a bi-enzymatic process has been described for obtaining enantiopure monoester (*R*)-**40** (Figure 9), combining a stereoselective hydrolysis of prochiral **38** to obtain (*R*)-**39** with high yield and enantiopurity, and a subsequent removal of the acetyl group with cephalosporin acetyl esterase [189].

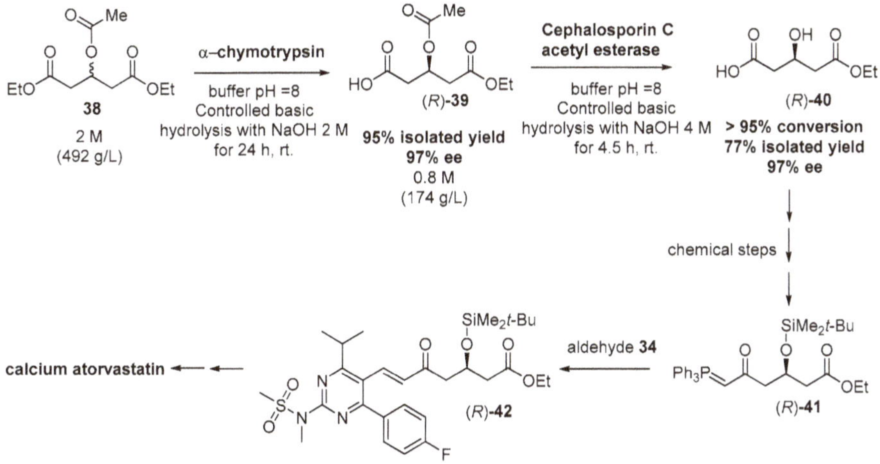

Figure 9. Two-enzymatic system for synthesizing chiral intermediates for Rosuvastatin, as described by Metzner et al. [189].

Furthermore, these same authors have optimized the overall procedure, using a smart engineering approach with an enzyme recycling of chymotrypsin and immobilized cephalosporin C acetyl esterase, with excellent volumetric productivity, transferring this technology to Sandoz for its industrial implementation [190].

3.2.2. Ketoreductases as Catalysts for the Preparation of the Lateral Chain of Atorvastatin and Other Superstatins

As commented before, **HN 18** was the starting point for the first synthesis of atorvastatin (Figure 5). For preparing **HN**, apart from the hydrolytic procedures described in Figure 7, some different purely chemical methodologies have been also described [191], and are depicted in Figure 10.

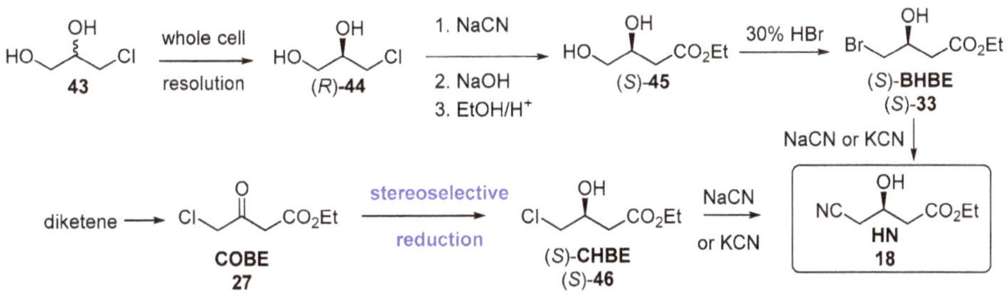

Figure 10. Different chemical methodologies for the preparation of **HN 18**.

The first synthetic protocols involved kinetic resolutions of prochiral 1,3-dichloropropan-2-ol **43** using microbes, and transformation to dihidroxyester (*S*)-**45** and subsequently to bromohydrine

(*S*)-**33** [179]. Later routes have involved asymmetric reduction of ethyl 4-chloroacetoacetate (COBE, **27**), produced from diketene, to furnish ethyl (*S*)-4-chloro-3-hydroxybutanoate ((*S*)-CHBE, (*S*)-**46**), using either chemical or biocatalytic reductions, as previously shown in **route #2**, Figure 6. Finally, the corresponding halohydrin ((*S*)-**33** or (*S*)-**46**) could be converted to **HN** by treatment with cyanide.

In this sense, the enzymatic asymmetric reduction of 4-bromo-3-oxobutyrate esters has hardly been investigated compared with the corresponding chlorine analogue, because of the lower reactivity and enantioselectivity of enzymes towards brominated compounds, although (*S*)-4-bromo-3-hydroxybutanoate esters would be better substrates for the ulterior cyanide treatment; anyhow, some examples can be found in the literature, starting from methyl 4-bromo-3-oxobutyrate (BAM), using *Escherichia coli* engineered cells containing a mutant β-keto ester reductase (KER-L54Q) from *Penicillium citrinum* and a cofactor-regeneration enzyme such as glucose dehydrogenase (GDH) or *Leifsonia sp.* alcohol dehydrogenase (LSADH) [192,193].

Regarding chlorine containing oxoesters, the seminal paper of Patel et al. using glucose-, acetate-, or glycerol-grown cell (10% w/v) suspensions of *Geotrichum candidum* SC 5469 [194] to produce (*S*)-**46** in reaction yield of 95% and optical purity of 96%, starting from 10 mg mL^{-1} of **27**, showed how the bio-reduction could be an interesting alternative to asymmetrical chemical reduction. Furthermore, the optical purity of (*S*)-**46** was increased to >99% by heat treatment of cell suspensions (55 °C for 30 min) prior to conducting bio-reduction at 28 °C.

Ye et al. [195] have reviewed a list of different yeast able to reduce **27** to furnish (*S*)-**46**, such as *Candida etchellsii* [196], *Candida parapsilosis* [197], *Candida magnoliae* [198], *Saccharomycopsis lipolytica* [196], or *Candida macedoniensis* [199], but in many cases, the stereoselectivity values obtained were not very high. Also, fungi as *Aureobasidium pullulans* CGMCC 1244 [200], *Cylindrocarpon sclerotigenum* IFO31855 [201], *Penicillium oxalicum* IFO 5748 [197], *Botrytis allii* IFO9430 [197], or *Pichia stipitis* CBS 6054 [202] can produce (*S*)-**46** with a higher enantiomeric excess compared with yeasts. This same group, through genome database mining of this yeast *Pichia stipitis*, found two carbonyl reductases (PsCRI and PsCRII) leading to (*S*)-**46** with >99% enantiomeric excess, which were subsequently characterized, cloned, and expressed in *E. coli* [195]. On the other hand, Cai et al. [203] described a substrate-coupled biocatalytic process based on the reactions catalyzed by an NADPH-dependent sorbose reductase (SOU1) from *Candida albicans* in which **27** was reduced to (*S*)-**46**, while NADPH was regenerated by the same enzyme via oxidation of sugar alcohols (sorbitol, mannitol, or xylitol). Optimization of COBE and sorbitol proportions yielded 2340 mM of (*S*)-**46** starting from 2500 mM **27** with an enantiomeric excess was 99%. This substrate-coupled system maintained a stable pH and a robust intracellular NADPH circulation, so that pH adjustment and the addition of extra coenzymes were unnecessary, thus making this system very attractive. The bio-reduction of **27** and the scaling up of the process using *Escherichia coli* cells expressing a reductase (ScCR) from *Streptomyces coelicolor* to afford enantiopure (*S*)-**46** has recently being described [204], at substrate loading of 100 g/L, while the concentration of coenzyme NAD^+ was limited to 0.1 mM based on cost considerations, other reaction parameters were optimized as 25 °C and pH 6.5, with a biocatalyst dose of 10 kU/L in the presence of isopropanol (1.5 equiv of **27**) as co-substrate for regenerating NADH. The reaction was performed in a toluene−aqueous biphasic system (1:1, v/v), with agitation at the maximal linear rate of 0.88 m/s. Finally, the bio-reaction was performed on a pilot scale using a 50 L thermostatised stirred-tank-reactor, affording (*S*)-**46** in 85.4% yield and 99.9% ee, and a total turnover number (TTN) of 6060 for the cofactor NAD^+. The specific production was calculated to be 36.8 g product/g dcw, which is the highest value reported to date among the whole-cell-mediated processes for producing (*S*)-**46**. Furthermore, from the point of view of sustainability, for this bio-reduction, the reaction and extraction solvent (toluene) was recycled with a loss of only 4.1%, so that the E factor (kg waste per kg product) for the process was determined as 1.8 if the process water was excluded, which was much lower than that value (2.3) obtained from the process using isolated ketoreductase, glucose dehydrogenase as the biocatalyst for cofactor regeneration, and glucose as the co-substrate [179]. The main contributors to the low E factor were the loss of the solvent toluene (46.1%), the use of excessive isopropanol, and the formation of coproduct acetone (combined ca. 35%). If water was also included, then the E factor would be 13.4.

Very recently, a recombinant *Escherichia coli* harbouring both the carbonyl reductase and glucose dehydrogenase has been described [205]. The recombinant *E. coli* was cultured in a 500-L fermenter, and the biocatalytic process for the synthesis of (*S*)-**46** in an aqueous-organic solvent system was constructed and optimized with a substrate fed-batch strategy. Concentration of **27** reached to 1.7 M, and (*S*)-**46** was obtained after a 4 h reaction in a 50-L reactor with yield of 97.2% and enantiomeric excess (ee) of 99%. Finally, (*S*)-**46** was extracted from the reaction mixture with 82% yield and 95% purity.

Nevertheless, because of the great overall demand of **HN** required for atorvastin synthesis (estimated to be in excess of 100 mT [179]), it is highly desirable to reduce the wastes and hazards involved in its manufacture, while reducing its cost and maintaining or, preferably, improving its quality. This has been successfully carried out on a multiton scale by Codexis by means of a three-enzyme two-step process, the detailed description of which is depicted in Figure 11.

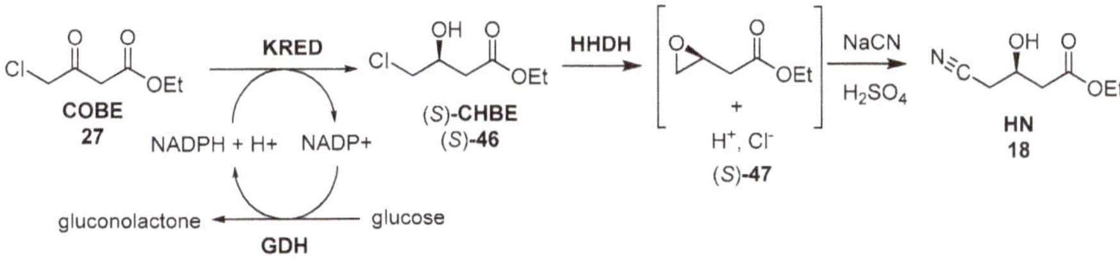

Figure 11. Codexis synthesis of **HN**.

Hence, the first step involves the biocatalytic reduction of **27**, using a ketoreductase (KRED) in combination with glucose and an NADP-dependent glucose dehydrogenase (GDH) for cofactor regeneration, leading to (*S*)-**46** in 96% isolated yield and >99.5% ee. In the second step, a halohydrin dehalogenase (HHDH), an enzyme capable of catalysing the elimination of halides from vicinal haloalcohols, resulting in epoxide ring formation [206], was employed to catalyse a nucleophilic substitution of chloride by cyanide, using HCN at neutral pH and ambient temperature. The efficiency and greenness of this protocol (Codexis was awarded the U.S. Environmental Protection Agency's Presidential Green Chemistry Challenge Award in 2006 for this work [207]) is based on the fact that all previous manufacturing routes to **HN** shown in Figure 10 involved, as the final step, a standard but troublesome S_N2 substitution of halide with cyanide ion in alkaline solution (pH = 10) at high temperatures (80 °C), being this reaction substituted in the Codexis protocol. In fact, in the S_N2 chlorine substitution, both (*S*)-**46** and **HN** are base-sensitive molecules, and extensive by-product formation is observed, leading to high E values [179]. Moreover, the product is a high-boiling oil, and a troublesome high-vacuum fractional distillation is required to recover **HN**, resulting in further yield losses and waste, and clearly contravening the first and sixth principles of Green Chemistry [208]. Thus, conducting the cyanation reaction under milder conditions at neutral pH, by employing the enzyme, HHDH, is the key step for increasing the greenness of the overall process.

Coming back to the Codexis protocol, awkwardly, both the wild-type KRED and GDH as well as HHDH displayed very low activities, so that in the first experiments, huge enzyme loadings were required to obtain an economically feasible reaction rate, thus leading to troublesome emulsions, which hampered the subsequent downstream processing. Additionally, severe product inhibition and poor stability under operating conditions were observed. To enable a practical large-scale process, the three enzymes were optimized by in vitro enzyme evolution using gene shuffling technologies according to predefined criteria and process parameters, resulting in an overall process in which the volumetric productivity per mass catalyst load of the cyanation process was improved ~2500-fold, comprising a 14-fold reduction in reaction time, a 7-fold increase in substrate loading, a 25-fold reduction in enzyme use, and a 50% improvement in isolated yield [179].

Also using bio-reductions, some other strategies have been developed for the preparation of chiral building blocks for statins synthesis. Thus, Figure 12 illustrates **route #3**, previously shown in Figure 6, depicting bioreduction of the corresponding 6-substitued-3,5-dioxohexanoates **48** to furnish (*R* or *S*)-**49** (similar to (*R*)-**19**, Figure 6). As depicted in Figure 5, the homochiral intermediate (3*R*,5*R*)-**20** was originally prepared by diastereoselective chemical reduction of (*R*)-**19**, using $NaBH_4$ and $MeOBEt_2$ and, so as to obtain a high diastereoselectivity (>99.5% de), an extremely low temperature (−90 °C) and pyrophoric triethyl borane were demanded [209], with a concomitant extensive energy consumption and substantial amount of waste formation. Another alternative chemical route using chlororuthenium(II) arene/β-amino alcohol as the catalyst for the reduction was described [210], although the diastereoselectivity was insufficient (80% de).

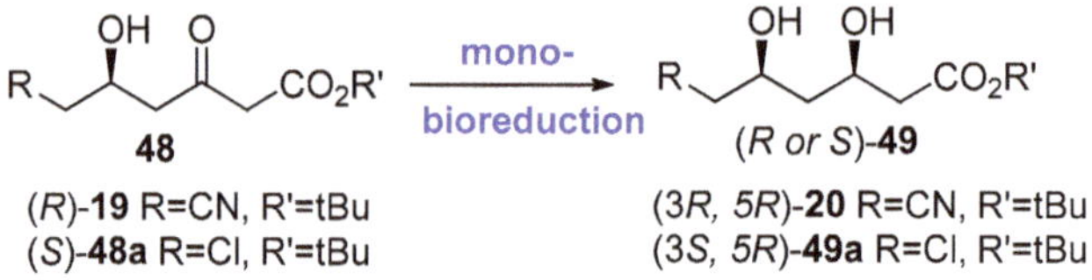

Figure 12. Bio-reductions to produce chiral building blocks for statins.

Therefore, the use of a ketoreductase is highly desirable to develop green and sustainable bioreduction. This process has been described [211,212] using NADP(H)-dependent alcohol dehydrogenase of *Lactobacillus brevis*. This enzyme was overexpressed in a recombinant *E. coli* and the cell extracts were then employed for carrying out the biocatalytic reactions on a gram scale, to reduce (*S*)-**48a** to give the corresponding (3*S*, 5*R*)-**49a** in >99.5% de and isolated yield of 72%, at 24 h. Alcohol dehydrogenase itself recycles its cofactor by a substrate coupled methodology, by oxidation of 2-propanol to acetone. This process was scaled up to 100 g [213] using a fed-batch reactor, with the conversion of more than 90% attained in a total reaction time of 24 h. For the same substrate, Liu and co-workers have reported the use of a ketoreductase from *Rhodosporidium toruloides*, wild-type and genetically evolved, under different reaction conditions [214–217], while Xu et al. used a ketoreductase from *Klebsiella oxytoca* [218]. On the other hand, for reducing (*R*)-**19** (up to 300 g L^{-1}), the ketoreductase from *L. brevis* overexpressed in *E. coli* cells has also been employed [219], coupled to glucose-GDH for cofactor recycling, yielding (*R*,*R*)-**20** in >99.5% de and 351 g L^{-1} d^{-1} space–time yield under the optimized conditions. Very recently, the same group has evolved the ketoreductase in order to improve the activity and thermostability of the enzyme [220]; thus, by coexpressing both the mutant ketoreductase and GDH, they describe the bioreduction of (*R*)-**19** to (*R*,*R*)-**20** at 40 °C in only 6 h, leading to values of >99.5% de and 1050 g L^{-1} d^{-1} space–time yield. Other similar bioreductions have also been reported using ketoreductases from other sources, such as *Rhodotorula glutinis* (whole cells [221]); engineered cells containing overexpressed NADPH-dependant ketoreductase from *Saccharomyces cerevisiae* and GDH [222,223]; a wild-type ketoreductase from *Kluyveromuces lactis* XP1461 (NADH-dependant) expressed in *E. coli* [224], subsequently improved by site-saturation mutagenesis [225]; or the ketoreductase from *Candida albicans* XP1463, also expressed in *E. coli* cells [226].

In a similar way, the double reduction of dioxoesters **50** (Figure 13) would directly lead to the target dihydroxyester **51**. For this purpose, whole cells of *Lactobacillus kefir*, which contain two different types of alcohol dehydrogenase, are able to convert **50b** into the dihydroxy ester (3*R*, 5*S*)-**51a** (99% ee in a total yield of 47.5% after 22 h, [227]) and the cofactor NADP(H) was regenerated by the usual glucose metabolism of the cell.

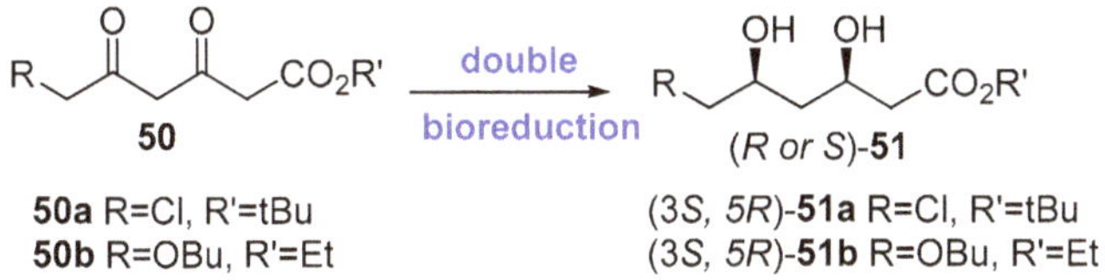

Figure 13. Bio-reductions of dioxoesters to produce chiral building blocks for statins.

The double bio-reduction has been also described using isolated enzymes, from *Acinetobacter* species; in fact, Patel et al. originally described the bio-reduction of **50b** using both whole cells and cell extracts from *Acinetobacter calcoaceticus* [228], and some years later, they also cloned and overexpressed [229] the diketoreductase responsible for the double reduction, which was efficiently carried out with the engineered enzyme [230]. Similarly, a diketoreductase from *Acinetobacter baylyi* ATCC 33305 was cloned and heterogeneously expressed in *Escherichia coli* by Wu et al. [231], showing an excellent biocatalytic performance at substrate concentration around 100 g L^{-1} [232] for the double reduction of **50a**. Interestingly, the 3D structure of this enzyme was reported, and the details of the catalytic mechanism were explained [233–235].

3.2.3. Aldolases for the Preparation of the Lateral Chain of Atorvastatin and Other Superstatins

Aldolases can also be used in the preparation of chiral building blocks for statin synthesis. This would correspond to **route #4** in Figure 6. In fact, Gijsen and Wong [236,237] first described the use of 2-deoxy-D-ribose 5-phosphate aldolase (DERA) from *E. coli* in the preparation of intermediate **28**, in a reaction mixture consisting of 133 mg of chloroacetaldehyde and 264 mg of acetaldehyde in a total reaction volume of 20 mL (Figure 14). The atorvastatin intermediate lactone (4*R*, 6*S*)-**54** can be easily formed by oxidation of lactol **28**. However, aldolase showed low affinity to chloroacetaldehyde and was promptly inactivated at required aldehyde concentrations, so that a huge amount of aldolase was required. Furthermore, a very long reaction time of 6 days was required because of the reversible nature of aldol reactions, making this process unpractical for scaling up.

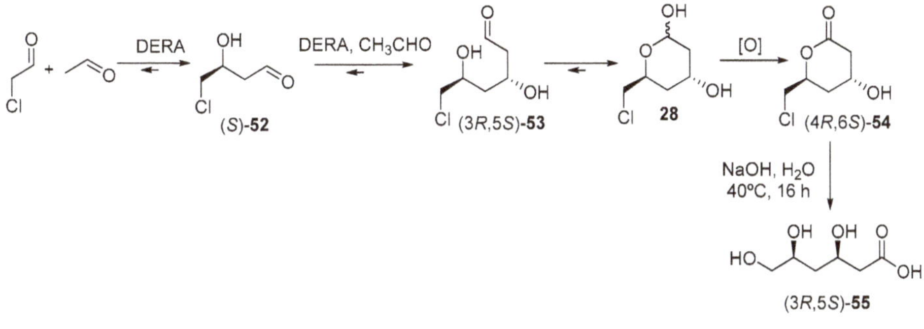

Figure 14. Aldolase-catalysed synthesis of chiral building blocks for statins.

Subsequent studies by Liu et al. [238] described a mutant aldolase, leading to an increased yield of (4*R*, 6*S*)-**54** to 43%, in comparison with 25% for the wild type aldolase, although the other reaction drawbacks were not overpassed. The process was markedly improved and scaled up by Greenberg et al. [239] of Diversa Corporation, by genetically modifying DERA by means of high throughput screenings of environmental DNA libraries, focussing on chloroacetaldehyde resistance and higher productivity; in a second step, the process was further improved by using a fed-batch bioreactor, in order to avoid significant substrate inhibition. Thus, the final synthesis of (4*R*, 6*S*)-**54** on a 100 g scale in a total reaction time of 3 h with an ee of >99.9% and a 10-fold reduction in catalyst load over the previous method [240]. More recently, the use of whole cells systems is being evaluated for this process [241,242], as well as new strategies for improving DERA by genetic engineering [243]. Finally, a simple basic hydrolysis of lactone (4*R*, 6*S*)-**54** leads to the trihydroxyacid (3*R*, 5*S*)-**55**, which is the precursor [244] of the lateral chain of superstatins. On the other hand, scientists from Lek Pharmaceutical (a Sandoz company) have described the use of whole cells of *Escherichia coli* BL21 (DE3) overexpressing the native *E. coli* deoC DERA gene for production of chiral lactols such as **28** [241], with excellent volumetric productivity (up to 50 g L^{-1} h^{-1}), >80% yield, and >80% chromatographic purity with titers reaching 100 g L^{-1}. This process is highly cost effective and environmentally friendly, and its sustainability is even improved if the oxidation of **28** to (4*R*, 6*S*)-**54** is also catalysed with an enzyme, as this same group has reported using PQQ-dependent glucose dehydrogenases [245]. Ohshima and co-workers described the sequential aldol reactions depicted in Figure 14 using DERA isolated from thermophilic organisms, describing a relatively lower activity compared with the enzyme from *E. coli*, although this fact was compensated by a better synthetic yield caused by the increased acetaldehyde resistance shown by the thermophilic enzyme [246]. Shen and co-workers reported higher conversions when chloroacetaldehyde was used as the acceptor substrate, as compared with acetaldehyde [243], and thus the development of new DERAs from different microorganisms is an open research area, as reported in recent revisions [247,248].

In any case, compared with other chemical protocols, most pharmaceutical processes are performed on a smaller scale, with the production volume of 1000 to 10,000 tons per year and product concentration ranging between 50 and 100 g/L; hence, the main drawback is the transfer of the biocatalytic process from laboratory to a larger scale, especially with respect to retention times, which are greater on a larger scale (compared with those in the laboratory). A good example of industrial scale-up has been described by Ručigaj and Krajnc [242], who used acetoxyacetaldehyde and acetaldehyde as substrates, which are presented in an aldol reaction catalyzed by a crude DERA expressing culture lysate. By optimizing addition regimes of both reactants into a reaction mixture, the corresponding lactol was produced at near 77 g/L. The complete process was designed in a practical and economical manner and could be used further on an industrial scale. Another industrial scale, low temperature process was developed by DSM, leading to a final product concentration of 100 g L^{-1} [249].

4. Prognosis and Conclusions

It is easily foreseen that because our diet habits are becoming progressively unhealthier, with an increased uptake of fats and abandoning the traditional "Mediterranean diet", hypercholesterolemia and dyslipidaemia will be typical maladies in Western society. Thus, statins would be gradually more present in our lives, being a very important piece of the global pharmaceutical market, either as branded or generic drugs. This fact, combined with the plethora of other pharmacological activities, called pleiotropic effects, that are being ascribed to statins, as revised in Section 2, makes us predict an ever-growing market for this type of drug. Anyway, more detailed and careful studies are demanded in order to be sure about the real efficiency of pleiotropic therapeutic effects of statins, by clearly identifying those patients who could be the best ones for responding to the desired effect of statins, and by establishing the most effective dose, duration of use, and statin drug entity required. Besides, more accurate clinical trials have to be conducted in order to evaluate the real effect upon the desired target, by designing more effective and truthful biomarkers.

For statins' preparation, new and more sustainable protocols would be demanded; in this context, the substitution of chemical by biocatalyzed processes will certainly help to gain sustainability, because of the well-known green features of biocatalysis—synthetic routes conducted under mild reaction conditions; at ambient temperature; using water as reaction medium in many cases; and, last but not least, avoiding functional group activation and protection/deprotection steps usually required in traditional organic synthesis. Thus, we also foresee a growing increase in the use of biocatalysis and biotransformations for the preparation of statins, mainly promoted by the enhancement of biocatalysts' performance through chemical modification and genetic engineering.

In another context, very recently, a new type of drug has emerged for dealing with those patients already using statins, but not reaching low-density lipoprotein cholesterol levels, rather by genetic and environmental factors or by pathological states—known as statin-resistance [250,251]. These drugs are the inhibitors of proprotein convertase subtilisin/kexin type 9 (PCSK9), a hepatic protease that becomes attached to low-density lipoproteins receptors (LDLRs), causing them to remain inside liposomes and get destroyed [252]. Nowadays, there are two PCSK9 inhibitors commercialized, both of them approved in 2015: alirocumab (Praluent®, from Sanofi) and evolucomab (Repatha®, from Amgen), both of them are used not as monotherapy, but are rather combined with a low cholesterol diet as well as with statins at maximally tolerated doses [253]. These two drugs are monoclonal antibodies, and their high price hampers their prior authorization practices and reduces their long-term adherence, so that the search for small molecules active as PCSK9 inhibitors is a "Holy Grail" in medicinal chemistry [254]. This situation leads us to think that (a) the statin market is not going to decrease, because they are going to be complemented (not substituted) with new drugs; and (b) as most of the new small molecules tested as PCSK9 inhibitors contain stereocenters in their structures [254], surely biocatalysis would become a very useful tool to facilitate more sustainable synthetic routes for their preparation.

Author Contributions: All authors (P.H., V.P., and A.R.A.) contributed equally in the preparation of this manuscript.

Funding: This research was partially funded by the Spanish Ministerio de Economia, Industria y Competitividad (MINECO), Project CTQ2015-66206-C2-1-R.

Acknowledgments: Authors thank Pablo Domínguez de María (CEO of Sustainable Momentum) for the critical reading of the manuscript.

Conflicts of Interest: The authors declare no conflict of interest.

References

1. World Health Organization. Raised Cholesterol. Situation and Trends. Available online: https://www.who.int/gho/ncd/risk_factors/cholesterol_text/en/ (accessed on 15 February 2019).
2. Stein, E.A. The power of statins: Aggressive lipid lowering. *Clin. Cardiol.* **2003**, *26*, 25–31. [CrossRef]
3. Bifulco, M.; Endo, A. Statin: New life for an old drug. *Pharmacol. Res.* **2014**, *88*, 1–2. [CrossRef]
4. Stossel, T.P. The discovery of statins. *Cell* **2008**, *134*, 903–905. [CrossRef]
5. Endo, A.; Kuroda, M.; Tsujita, Y. ML-236A, ML-236B, and ML-236C, new inhibitors of cholesterogenesis produced by *Penicillium citrinum*. *J. Antibiot.* **1976**, *29*, 1346–1348. [CrossRef]
6. Brown, A.G.; Smale, T.C.; King, T.J.; Hasenkamp, R.; Thompson, R.H. Crystal and molecular-structure of compactin, a new antifungal metabolite from *Penicillium brevicompactum*. *J. Chem. Soc. Perkin Trans. 1* **1976**, 1165–1173. [CrossRef]
7. Moore, R.N.; Bigam, G.; Chan, J.K.; Hogg, A.M.; Nakashima, T.T.; Vederas, J.C. Biosynthesis of the hypocholesterolemic agent mevinolin by *Aspergillus terreus*. Determination of the origin of carbon, hydrogen, and oxygen-atoms by ^{13}C NMR and Mass Spectrometry. *J. Am. Chem. Soc.* **1985**, *107*, 3694–3701. [CrossRef]
8. Gunde-Cimerman, N.; Cimerman, A. *Pleurotus* fruiting bodies contain the inhibitor of 3-hydroxy-3-methylglutaryl-coenzyme A reductase—Lovastatin. *Exp. Mycol.* **1995**, *19*, 1–6. [CrossRef]
9. Liu, J.; Zhang, J.; Shi, Y.; Grimsgaard, S.; Alraek, T.; Fonnebo, V. Chinese red yeast rice (*Monascus purpureus*) for primary hyperlipidemia: A meta-analysis of randomized controlled trials. *Chin. Med.* **2006**, *1*, 4. [CrossRef]
10. Mol, M.; Erkelens, D.W.; Leuven, J.A.G.; Schouten, J.A.; Stalenhoef, A.F.H. Simvastatin (MK-733)—A potent cholesterol-synthesis inhibitor in heterozygous familial hypercholesterolemia. *Atherosclerosis* **1988**, *69*, 131–137. [CrossRef]
11. Yoshino, G.; Kazumi, T.; Kasama, T.; Iwatani, I.; Iwai, M.; Inui, A.; Otsuki, M.; Baba, S. Effect of CS-514, an inhibitor of 3-hydroxy-3-methylglutaryl coenzyme A reductase, on lipoprotein and apolipoprotein in plasma of hypercholesterolemic diabetics. *Diabetes Res. Clin. Pract.* **1986**, *2*, 179–181. [CrossRef]
12. Li, J.J. *Triumph of the Heart: The Story of Statins*; Oxford University Press: New York, NY, USA, 2009.
13. Casar, Z. Historic Overview and Recent Advances in the Synthesis of Super-statins. *Curr. Org. Chem.* **2010**, *14*, 816–845. [CrossRef]

14. Lindsley, C.W. The top prescription drugs of 2010 in the United States: Antipsychotics show strong growth. *ACS Chem. Neurosci.* **2011**, *2*, 276–277. [CrossRef] [PubMed]
15. Lindsley, C.W. The top prescription drugs of 2011 in the United States: Antipsychotics and antidepressants once again lead CNS therapeutics. *ACS Chem. Neurosci.* **2012**, *3*, 630–631. [CrossRef]
16. Lindsley, C.W. 2012 Trends and statistics for prescription medications in the United States: CNS therapeutics continue to hold leading positions. *ACS Chem. Neurosci.* **2013**, *4*, 1133–1135. [CrossRef] [PubMed]
17. Lindsley, C.W. The top prescription drugs of 2012 globally: Biologics dominate, but small molecule CNS drugs hold on to top spots. *ACS Chem. Neurosci.* **2013**, *4*, 905–907. [CrossRef] [PubMed]
18. Lindsley, C.W. 2013 Trends and statistics for prescription medications in the United States: CNS highest ranked and record number of prescriptions dispensed. *ACS Chem. Neurosci.* **2015**, *6*, 356–357. [CrossRef] [PubMed]
19. Aitken, M.; Kleinrock, M.; Lyle, J.; Nass, D.; Caskey, L. *Medicines Use and Spending Shifts: A Review of the Use of Medicines in the U.S. in 2014*; IMS Institute for Healthcare Informatics: Parsippany, NJ, USA, 2015.
20. Lindsley, C.W. 2014 Prescription medications in the United States: Tremendous growth, specialty/orphan drug expansion, and dispensed prescriptions continue to increase. *ACS Chem. Neurosci.* **2015**, *6*, 811–812. [CrossRef] [PubMed]
21. Lindsley, C.W. 2014 Global prescription medication statistics: Strong growth and CNS well represented. *ACS Chem. Neurosci.* **2015**, *6*, 505–506. [CrossRef] [PubMed]
22. Xie, X.K.; Watanabe, K.; Wojcicki, W.A.; Wang, C.C.C.; Tang, Y. Biosynthesis of lovastatin analogs with a broadly specific acyltransferase. *Chem. Biol.* **2006**, *13*, 1161–1169. [CrossRef]
23. FiercePharma. The Cardiovascular Scene to Shift by 2024 as Next-Generation Drugs like Eliquis and Xarelto Eclipse Stalwarts. Available online: https://www.fiercepharma.com/pharma/cardiovascular-landscape-from-2017-to-2024-featuring-eliquis-xarelto-and-more (accessed on 14 February 2019).
24. Watanabe, M.; Koike, H.; Ishiba, T.; Okada, T.; Seo, S.; Hirai, K. Synthesis and biological activity of methanesulfonamide pyrimidine- and N-methanesulfonyl pyrrole-substituted 3,5-dihydroxy-6-heptenoates, a novel series of HMG-CoA reductase inhibitors. *Bioorg. Med. Chem.* **1997**, *5*, 437–444. [CrossRef]
25. Demasi, M. Statin wars: Have we been misled about the evidence? A narrative review. *Br. J. Sports Med.* **2018**, *52*, 905–909. [CrossRef] [PubMed]
26. Ferri, N.; Corsini, A. Clinical evidence of statin therapy in non-dyslipidemic disorders. *Pharmacol. Res.* **2014**, *88*, 20–30. [CrossRef] [PubMed]
27. Wang, C.Y.; Liu, P.Y.; Liao, J.K. Pleiotropic effects of statin therapy: Molecular mechanisms and clinical results. *Trends Mol. Med.* **2008**, *14*, 37–44. [CrossRef] [PubMed]
28. Cordle, A.; Koenigsknecht-Talboo, J.; Wilkinson, B.; Limpert, A.; Landreth, G. Mechanisms of statin-mediated inhibition of small G-protein function. *J. Biol. Chem.* **2005**, *280*, 34202–34209. [CrossRef] [PubMed]
29. Liao, J.K.; Laufs, U. Pleiotropic effects of statins. *Annu. Rev. Pharmacol. Toxicol.* **2005**, *45*, 89–118. [CrossRef] [PubMed]
30. Malfitano, A.M.; Marasco, G.; Proto, M.C.; Laezza, C.; Gazzerro, P.; Bifulco, M. Statins in neurological disorders: An overview and update. *Pharmacol. Res.* **2014**, *88*, 74–83. [CrossRef] [PubMed]
31. John, S.; Delles, C.; Jacobi, J.; Schlaich, M.P.; Schneider, M.; Schmitz, G.; Schmieder, R.E. Rapid improvement of nitric oxide bioavailability after lipid-lowering therapy with cerivastatin within two weeks. *J. Am. Coll. Cardiol.* **2001**, *37*, 1351–1358. [CrossRef]
32. Cheng, W.H.; Ho, W.Y.; Chang, C.F.; Lu, P.J.; Cheng, P.W.; Yeh, T.C.; Hong, L.Z.; Sun, G.C.; Hsiao, M.; Tseng, C.J. Simvastatin induces a central hypotensive effect via Ras-mediated signalling to cause eNOS up-regulation. *Br. J. Pharmacol.* **2013**, *170*, 847–858. [CrossRef] [PubMed]
33. Mason, R.P.; Walter, M.F.; Jacob, R.F. Effects of HMG-CoA reductase inhibitors on endothelial function—Role of microdomains and oxidative stress. *Circulation* **2004**, *109*, 34–41. [CrossRef] [PubMed]
34. Mihos, C.G.; Salas, M.J.; Santana, O. The pleiotropic effects of the hydroxy-methyl-glutaryl-CoA reductase inhibitors in cardiovascular disease a comprehensive review. *Cardiol. Rev.* **2010**, *18*, 298–304. [CrossRef] [PubMed]
35. Yanuck, D.; Mihos, C.G.; Santana, O. Mechanisms and clinical evidence of the pleiotropic effects of the Hydroxy-Methyl-Glutaryl-CoA Reductase inhibitors in Central Nervous System disorders: A comprehensive review. *Int. J. Neurosci.* **2012**, *122*, 619–629. [CrossRef]

36. Mihos, C.G.; Artola, R.T.; Santana, O. The pleiotropic effects of the hydroxy-methyl-glutaryl-CoA reductase inhibitors in rheumatologic disorders: A comprehensive review. *Rheumatol. Int.* **2012**, *32*, 287–294. [CrossRef]
37. Barone, E.; Di Domenico, F.; Butterfield, D.A. Statins more than cholesterol lowering agents in Alzheimer disease: Their pleiotropic functions as potential therapeutic targets. *Biochem. Pharmacol.* **2014**, *88*, 605–616. [CrossRef] [PubMed]
38. Mihos, C.G.; Pineda, A.M.; Santana, O. Cardiovascular effects of statins, beyond lipid-lowering properties. *Pharmacol. Res.* **2014**, *88*, 12–19. [CrossRef]
39. Oesterle, A.; Laufs, U.; Liao, J.K. Pleiotropic effects of statins on the cardiovascular system. *Circ. Res.* **2017**, *120*, 229–243. [CrossRef]
40. Cohen, M.V.; Yang, X.M.; Downey, J.M. Nitric oxide is a preconditioning mimetic and cardioprotectant and is the basis of many available infarct-sparing strategies. *Cardiovasc. Res.* **2006**, *70*, 231–239. [CrossRef]
41. Mills, E.J.; Wu, P.; Chong, G.; Ghement, I.; Singh, S.; Akl, E.A.; Eyawo, O.; Guyatt, G.; Berwanger, O.; Briel, M. Efficacy and safety of statin treatment for cardiovascular disease: A network meta-analysis of 170 255 patients from 76 randomized trials. *QJM-Int. J. Med.* **2011**, *104*, 109–124. [CrossRef] [PubMed]
42. Teng, M.; Lin, L.; Zhao, Y.J.; Khoo, A.L.; Davis, B.R.; Yong, Q.W.; Yeo, T.C.; Lim, B.P. Statins for primary prevention of cardiovascular disease in elderly patients: Systematic review and meta-analysis. *Drugs Aging* **2015**, *32*, 649–661. [CrossRef]
43. Tervonen, T.; Naci, H.; van Valkenhoef, G.; Ades, A.E.; Angelis, A.; Hillege, H.L.; Postmus, D. Applying multiple criteria decision analysis to comparative benefit-risk assessment: Choosing among statins in primary prevention. *Med. Decis. Mak.* **2015**, *35*, 859–871. [CrossRef]
44. Ulivieri, C.; Baldari, C.T. Statins: From cholesterol-lowering drugs to novel immunomodulators for the treatment of Th17-mediated autoimmune diseases. *Pharmacol. Res.* **2014**, *88*, 41–52. [CrossRef]
45. Scheele, J.S.; Marks, R.E.; Boss, G.R. Signaling by small GTPases in the immune system. *Immunol. Rev.* **2007**, *218*, 92–101. [CrossRef]
46. Greenwood, J.; Steinman, L.; Zamvil, S.S. Statin therapy and autoimmune disease: From protein prenylation to immunomodulation. *Nat. Rev. Immunol.* **2006**, *6*, 358–370. [CrossRef]
47. Chow, S.C. Immunomodulation by statins: Mechanisms and potential impact on autoimmune diseases. *Arch. Immunol. Ther. Exp. (Warsz.)* **2009**, *57*, 243–251. [CrossRef] [PubMed]
48. Khattri, S.; Zandman-Goddard, G. Statins and autoimmunity. *Immunol. Res.* **2013**, *56*, 348–357. [CrossRef]
49. Ciurleo, R.; Bramanti, P.; Marino, S. Role of statins in the treatment of multiple sclerosis. *Pharmacol. Res.* **2014**, *87*, 133–143. [CrossRef]
50. Pihl-Jensen, G.; Tsakiri, A.; Frederiksen, J.L. Statin treatment in multiple sclerosis: A systematic review and meta-analysis. *CNS Drugs* **2015**, *29*, 277–291. [CrossRef] [PubMed]
51. Soubrier, M.; Mathieu, S.; Hermet, M.; Makarawiez, C.; Bruckert, E. Do all lupus patients need statins? *Jt. Bone Spine* **2013**, *80*, 244–249. [CrossRef] [PubMed]
52. Yu, H.-H.; Chen, P.-C.; Yang, Y.-H.; Wang, L.-C.; Lee, J.-H.; Lin, Y.-T.; Chiang, B.-L. Statin reduces mortality and morbidity in systemic lupus erythematosus patients with hyperlipidemia: A nationwide population-based cohort study. *Atherosclerosis* **2015**, *243*, 11–18. [CrossRef]
53. Ruiz-Limon, P.; Barbarroja, N.; Perez-Sanchez, C.; Aguirre, M.A.; Bertolaccini, M.L.; Khamashta, M.A.; Rodriguez-Ariza, A.; Almaden, Y.; Segui, P.; Khraiwesh, H.; et al. Atherosclerosis and cardiovascular disease in systemic lupus erythematosus: Effects of in vivo statin treatment. *Ann. Rheum. Dis.* **2015**, *74*, 1450–1458. [CrossRef] [PubMed]
54. Liu, X.; Li, B.; Wang, W.; Zhang, C.; Zhang, M.; Zhang, Y.; Xia, Y.; Dong, Z.; Guo, Y.; An, F. Effects of HMG-CoA Reductase inhibitor on experimental autoimmune myocarditis. *Cardiovasc. Drugs Ther.* **2012**, *26*, 121–130. [CrossRef]
55. Lazzerini, P.E.; Capecchi, P.L.; Laghi-Pasini, F. Statins as a new therapeutic perspective in myocarditis and postmyocarditis dilated cardiomyopathy. *Cardiovasc. Drugs Ther.* **2013**, *27*, 365–369. [CrossRef]
56. Tajiri, K.; Shimojo, N.; Sakai, S.; Machino-Ohtsuka, T.; Imanaka-Yoshida, K.; Hiroe, M.; Tsujimura, Y.; Kimura, T.; Sato, A.; Yasutomi, Y.; et al. Pitavastatin regulates helper T-cell differentiation and ameliorates autoimmune myocarditis in mice. *Cardiovasc. Drugs Ther.* **2013**, *27*, 413–424. [CrossRef]
57. Lv, S.; Liu, Y.; Zou, Z.; Li, F.; Zhao, S.; Shi, R.; Bian, R.; Tian, H. The impact of statins therapy on disease activity and inflammatory factor in patients with rheumatoid arthritis: A meta-analysis. *Clin. Exp. Rheumatol.* **2015**, *33*, 69–76.

58. Tascilar, K.; Dell'Aniello, S.; Hudson, M.; Suissa, S. Statins and risk of rheumatoid arthritis: A nested case-control study. *Arthritis Rheumatol.* **2016**, *68*, 2603–2611. [CrossRef]
59. de Jong, H.J.I.; Tervaert, J.W.C.; Lalmohamed, A.; de Vries, F.; Vandebriel, R.J.; van Loveren, H.; Klungel, O.H.; van Staa, T.P. Pattern of risks of rheumatoid arthritis among patients using statins: A cohort study with the clinical practice research datalink. *PLoS ONE* **2018**, *13*, 1–17. [CrossRef]
60. McFarland, A.J.; Anoopkumar-Dukie, S.; Arora, D.S.; Grant, G.D.; McDermott, C.M.; Perkins, A.V.; Davey, A.K. Molecular mechanisms underlying the effects of statins in the central nervous system. *Int. J. Mol. Sci.* **2014**, *15*, 20607–20637. [CrossRef]
61. Qizilbash, N.; Lewington, S.; Duffy, S.; Peto, R.; Smith, T.; Spiegelhalter, D.; Iso, H.; Shimamoto, T.; Komachi, Y.; Iida, M.; et al. Cholesterol, diastolic blood pressure, and stroke: 13000 strokes in 450000 people in 45 prospective cohorts. *Lancet* **1995**, *346*, 1647–1653.
62. O'Brien, E.C.; Greiner, M.A.; Xian, Y.; Fonarow, G.C.; Olson, D.M.; Schwamm, L.H.; Bhatt, D.L.; Smith, E.E.; Maisch, L.; Hannah, D.; et al. Clinical effectiveness of statin therapy after ischemic stroke: Primary results from the statin therapeutic area of the Patient-centered Research into Outcomes Stroke Patients prefer and Effectiveness Research (PROSPER) study. *Circulation* **2015**, *132*, 1404–1413. [CrossRef]
63. Naci, H.; Brugts, J.J.; Fleurence, R.; Ades, A.E. Comparative effects of statins on major cerebrovascular events: A multiple-treatments meta-analysis of placebo-controlled and active-comparator trials. *QJM-Int. J. Med.* **2013**, *106*, 299–306. [CrossRef]
64. Markel, A. Statins and peripheral arterial disease. *Int. Angiol.* **2015**, *34*, 416–427.
65. Colivicchi, F.; Bassi, A.; Santini, M.; Caltagirone, C. Discontinuation of statin therapy and clinical outcome after ischemic stroke. *Stroke* **2007**, *38*, 2652–2657. [CrossRef]
66. Laloux, P. Risk and benefit of statins in stroke secondary prevention. *Curr. Vasc. Pharmacol.* **2013**, *11*, 812–816. [CrossRef]
67. Song, B.; Wang, Y.L.; Zhao, X.Q.; Liu, L.P.; Wang, C.X.; Wang, A.X.; Du, W.L.; Wang, Y.J. Association between statin use and short-term outcome based on severity of ischemic stroke: A cohort study. *PLoS ONE* **2014**, *9*, 7. [CrossRef]
68. Gutierrez-Vargas, J.; Cespedes-Rubio, A.; Cardona-Gomez, G. Perspective of synaptic protection after post-infarction treatment with statins. *J. Transl. Med.* **2015**, *13*, 118. [CrossRef]
69. Bustamante, A.; Montaner, J. Statin therapy should not be discontinued in patients with intracerebral hemorrhage. *Stroke* **2013**, *44*, 2060–2061. [CrossRef]
70. Molina, C.A.; Selim, M.H. Continued statin treatment after acute intracranial hemorrhage fighting fire with fire. *Stroke* **2013**, *44*, 2062–2063. [CrossRef]
71. Pan, Y.S.; Jing, J.; Wang, Y.L.; Zhao, X.Q.; Song, B.; Wang, W.J.; Wang, D.; Liu, G.F.; Liu, L.P.; Wang, C.X.; et al. Use of statin during hospitalization improves the outcome after intracerebral hemorrhage. *CNS Neurosci. Ther.* **2014**, *20*, 548–555. [CrossRef]
72. Anand, R.; Gill, K.D.; Mahdi, A.A. Therapeutics of Alzheimer's disease: Past, present and future. *Neuropharmacology* **2014**, *76*, 27–50. [CrossRef]
73. Silva, T.; Teixeira, J.; Remiao, F.; Borges, F. Alzheimer's disease, cholesterol, and statins: The junctions of important metabolic pathways. *Angew. Chem. Int. Ed.* **2013**, *52*, 1110–1121. [CrossRef]
74. Hottman, D.A.; Li, L. Protein prenylation and synaptic plasticity: Implications for Alzheimer's disease. *Mol. Neurobiol.* **2014**, *50*, 177–185. [CrossRef]
75. Mendoza-Oliva, A.; Zepeda, A.; Arias, C. The complex actions of statins in brain and their relevance for Alzheimer's disease treatment: An analytical review. *Curr. Alzheimer Res.* **2014**, *11*, 817–833. [CrossRef] [PubMed]
76. Wanamaker, B.L.; Swiger, K.J.; Blumenthal, R.S.; Martin, S.S. Cholesterol, statins, and dementia: What the cardiologist should know. *Clin. Cardiol.* **2015**, *38*, 243–250. [CrossRef] [PubMed]
77. Buxbaum, J.D.; Cullen, E.I.; Friedhoff, L.T. Pharmacological concentrations of the HMG-CoA reductase inhibitor lovastatin decrease the formation of the Alzheimer beta-amyloid peptide in vitro and in patients. *Front. Biosci.* **2002**, *7*, A50–A59. [CrossRef] [PubMed]
78. Li, L.; Zhang, W.; Cheng, S.W.; Cao, D.F.; Parent, M. Isoprenoids and related pharmacological interventions: Potential application in Alzheimer's disease. *Mol. Neurobiol.* **2012**, *46*, 64–77. [CrossRef] [PubMed]
79. Cibickova, L.; Palicka, V.; Cibicek, N.; Cermakova, E.; Micuda, S.; Bartosova, L.; Jun, D. Differential effects of statins and alendronate on cholinesterases in serum and brain of rats. *Physiol. Res.* **2007**, *56*, 765–770.

80. Mozayan, M.; Lee, T.J.F. Statins prevent cholinesterase inhibitor blockade of sympathetic alpha 7-nAChR-mediated currents in rat superior cervical ganglion neurons. *Am. J. Physiol.-Heart Circul. Physiol.* **2007**, *293*, H1737–H1744. [CrossRef] [PubMed]
81. Ghodke, R.M.; Tour, N.; Devi, K. Effects of statins and cholesterol on memory functions in mice. *Metab. Brain Dis.* **2012**, *27*, 443–451. [CrossRef] [PubMed]
82. Macan, M.; Vuksic, A.; Zunec, S.; Konjevoda, P.; Lovric, J.; Kelava, M.; Stambuk, N.; Vrkic, N.; Bradamante, V. Effects of simvastatin on malondialdehyde level and esterase activity in plasma and tissue of normolipidemic rats. *Pharmacol. Rep.* **2015**, *67*, 907–913. [CrossRef] [PubMed]
83. Bosel, J.; Gandor, F.; Harms, C.; Synowitz, M.; Harms, U.; Djoufack, P.C.; Megow, D.; Dirnagl, U.; Hortnagl, H.; Fink, K.B.; et al. Neuroprotective effects of atorvastatin against glutamate-induced excitotoxicity in primary cortical neurones. *J. Neurochem.* **2005**, *92*, 1386–1398. [CrossRef] [PubMed]
84. Krisanova, N.; Sivko, R.; Kasatkina, L.; Borisova, T. Neuroprotection by lowering cholesterol: A decrease in membrane cholesterol content reduces transporter-mediated glutamate release from brain nerve terminals. *Biochim. Biophys. Acta-Mol. Basis Dis.* **2012**, *1822*, 1553–1561. [CrossRef]
85. Sala, S.G.; Munoz, U.; Bartolome, F.; Bermejo, F.; Martin-Requero, A. HMG-CoA reductase inhibitor simvastatin inhibits cell cycle progression at the G(1)/S checkpoint in immortalized lymphocytes from Alzheimer's disease patients independently of cholesterol-lowering effects. *J. Pharmacol. Exp. Ther.* **2008**, *324*, 352–359. [CrossRef] [PubMed]
86. Merla, R.; Ye, Y.; Lin, Y.; Manickavasagam, S.; Huang, M.-H.; Perez-Polo, R.J.; Uretsky, B.F.; Birnbaum, Y. The central role of adenosine in statin-induced ERK1/2, Akt, and eNOS phosphorylation. *Am. J. Physiol.-Heart Circul. Physiol.* **2007**, *293*, H1918–H1928. [CrossRef] [PubMed]
87. Cordle, A.; Landreth, G. 3-Hydroxy-3-methylglutaryl-coenzyme A reductase inhibitors attenuate beta-amyloid-induced microglial inflammatory responses. *J. Neurosci.* **2005**, *25*, 299–307. [CrossRef]
88. Fonseca, A.C.R.G.; Resende, R.; Oliveira, C.R.; Pereira, C.M.F. Cholesterol and statins in Alzheimer's disease: Current controversies. *Exp. Neurol.* **2010**, *223*, 282–293. [CrossRef] [PubMed]
89. Mans, R.A.; McMahon, L.L.; Li, L. Simvastatin-mediated enhancement of long-term potentiation is driven by farnesyl-pyrophosphate depletion and inhibition of farnesylation. *Neuroscience* **2012**, *202*, 1–9. [CrossRef]
90. Mans, R.A.; Chowdhury, N.; Cao, D.; McMahon, L.L.; Li, L. Simvastatin enhances hippocampal long-term potentiation in C57BL/6 mice. *Neuroscience* **2010**, *166*, 435–444. [CrossRef]
91. Parent, M.; Hottman, D.A.; Cheng, S.W.; Zhang, W.; McMahon, L.L.; Yuan, L.L.; Li, L. Simvastatin treatment enhances NMDAR-mediated synaptic transmission by upregulating the surface distribution of the GluN2B subunit. *Cell. Mol. Neurobiol.* **2014**, *34*, 693–705. [CrossRef]
92. Wollmuth, L.P. Is cholesterol good or bad for your brain?—NMDARs have a say. *J. Physiol.-Lond.* **2015**, *593*, 2251–2252. [CrossRef] [PubMed]
93. Roy, A.; Jana, M.; Kundu, M.; Corbett, G.T.; Rangaswamy, S.B.; Mishra, R.K.; Luan, C.H.; Gonzalez, F.J.; Pahan, K. HMG-CoA Reductase Inhibitors bind to PPAR alpha to upregulate neurotrophin expression in the brain and improve memory in mice. *Cell Metab.* **2015**, *22*, 253–265. [CrossRef] [PubMed]
94. Jeong, J.H.; Yum, K.S.; Chang, J.Y.; Kim, M.; Ahn, J.Y.; Kim, S.; Lapchak, P.A.; Han, M.K. Dose-specific effect of simvastatin on hypoxia-induced HIF-1 alpha and BACE expression in Alzheimer's disease cybrid cells. *BMC Neurol.* **2015**, *15*, 7. [CrossRef]
95. Power, M.C.; Weuve, J.; Sharrett, A.R.; Blacker, D.; Gottesman, R.F. Statins, cognition, and dementia-systematic review and methodological commentary. *Nat. Rev. Neurol.* **2015**, *11*, 220–229. [CrossRef] [PubMed]
96. Strom, B.L.; Schinnar, R.; Karlawish, J.; Hennessy, S.; Teal, V.; Bilker, W.B. Statin therapy and risk of acute memory impairment. *JAMA Intern. Med.* **2015**, *175*, 1399–1405. [CrossRef] [PubMed]
97. Mendoza-Oliva, A.; Ferrera, P.; Fragoso-Medina, J.; Arias, C. Lovastatin differentially affects neuronal cholesterol and amyloid- production invivo and invitro. *CNS Neurosci. Ther.* **2015**, *21*, 631–641. [CrossRef] [PubMed]
98. Chuang, C.S.; Lin, C.L.; Lin, M.C.; Sung, F.C.; Kao, C.H. Decreased prevalence of dementia associated with statins: A national population-based study. *Eur. J. Neurol.* **2015**, *22*, 912–918. [CrossRef] [PubMed]
99. Phani, S.; Loike, J.D.; Przedborski, S. Neurodegeneration and inflammation in Parkinson's disease. *Parkinsonism Relat. Disord.* **2012**, *18* (Suppl. S1), S207–S209. [CrossRef]

100. Wolozin, B.; Kellman, W.; Ruosseau, P.; Celesia, G.G.; Siegel, G. Decreased prevalence of Alzheimer disease associated with 3-hydroxy-3-methyglutaryl coenzyme A reductase inhibitors. *Arch. Neurol.* **2000**, *57*, 1439–1443. [CrossRef]
101. Lee, Y.-C.; Lin, C.-H.; Wu, R.-M.; Lin, M.-S.; Lin, J.-W.; Chang, C.-H.; Lai, M.-S. Discontinuation of statin therapy associates with Parkinson disease. A population-based study. *Neurology* **2013**, *81*, 410–416. [CrossRef]
102. Tison, F.; Negre-Pages, L.; Meissner, W.G.; Dupouy, S.; Li, Q.; Thiolat, M.-L.; Thiollier, T.; Galitzky, M.; Ory-Magne, F.; Milhet, A.; et al. Simvastatin decreases levodopa-induced dyskinesia in monkeys, but not in a randomized, placebo-controlled, multiple cross-over ("n-of-1") exploratory trial of simvastatin against levodopa-induced dyskinesia in Parkinson's disease patients. *Parkinsonism Relat. Disord.* **2013**, *19*, 416–421. [CrossRef]
103. Selley, M.L. Simvastatin prevents 1-methyl-4-phenyl-1,2,3,6-tetrahydropyridine-induced striatal dopamine depletion and protein tyrosine nitration in mice. *Brain Res.* **2005**, *1037*, 1–6. [CrossRef]
104. Xu, Y.Q.; Long, L.; Yan, J.Q.; Wei, L.; Pan, M.Q.; Gao, H.M.; Zhou, P.; Liu, M.; Zhu, C.S.; Tang, B.S.; et al. Simvastatin induces neuroprotection in 6-OHDA-lesioned PC12 via the PI3K/AKT/caspase 3 pathway and anti-inflammatory responses. *CNS Neurosci. Ther.* **2013**, *19*, 170–177. [CrossRef]
105. Yan, J.; Xu, Y.; Zhu, C.; Zhang, L.; Wu, A.; Yang, Y.; Xiong, Z.; Deng, C.; Huang, X.-F.; Yenari, M.A.; et al. Simvastatin prevents dopaminergic neurodegeneration in experimental parkinsonian models: The association with anti-inflammatory responses. *PLoS ONE* **2011**, *6*, e20945. [CrossRef] [PubMed]
106. Wang, T.; Cao, X.B.; Zhang, T.; Shi, Q.Q.; Chen, Z.B.; Tang, B.S. Effect of simvastatin on *L*-DOPA-induced abnormal involuntary movements of hemiparkinsonian rats. *Neurol. Sci.* **2015**, *36*, 1397–1402. [CrossRef] [PubMed]
107. Ghosh, A.; Roy, A.; Matras, J.; Brahmachari, S.; Gendelman, H.E.; Pahan, K. Simvastatin inhibits the activation of p21(Ras) and prevents the loss of dopaminergic neurons in a mouse model of Parkinson's disease. *J. Neurosci.* **2009**, *29*, 13543–13556. [CrossRef] [PubMed]
108. Yang, C.C.; Jick, S.S.; Jick, H. Lipid-lowering drugs and the risk of depression and suicidal behavior. *Arch. Intern. Med.* **2003**, *163*, 1926–1932. [CrossRef]
109. Feng, L.; Tan, C.-H.; Merchant, R.A.; Ng, T.-P. Association between depressive symptoms and use of HMG-CoA reductase inhibitors (statins), corticosteroids and histamine H-2 receptor antagonists in community-dwelling older persons—Cross-sectional analysis of a population-based cohort. *Drugs Aging* **2008**, *25*, 795–805. [CrossRef] [PubMed]
110. Stafford, L.; Berk, M. The use of statins after a cardiac intervention is associated with reduced risk of subsequent depression: Proof of concept for the inflammatory and oxidative hypotheses of depression? *J. Clin. Psychiatry* **2011**, *72*, 1229–1235. [CrossRef] [PubMed]
111. Otte, C.; Zhao, S.; Whooley, M.A. Statin use and risk of depression in patients with coronary heart disease: Longitudinal data from the heart and soul study. *J. Clin. Psychiatry* **2012**, *73*, 610–615. [CrossRef] [PubMed]
112. Kim, J.-M.; Stewart, R.; Kang, H.-J.; Bae, K.-Y.; Kim, S.-W.; Shin, I.-S.; Kim, J.-T.; Park, M.-S.; Cho, K.-H.; Yoon, J.-S. A prospective study of statin use and poststroke depression. *J. Clin. Psychopharmacol.* **2014**, *34*, 72–79. [CrossRef]
113. Chuang, C.-S.; Yang, T.-Y.; Muo, C.-H.; Su, H.-L.; Sung, F.-C.; Kao, C.-H. Hyperlipidemia, statin use and the risk of developing depression: A nationwide retrospective cohort study. *Gen. Hosp. Psych.* **2014**, *36*, 497–501. [CrossRef] [PubMed]
114. Lin, P.-Y.; Chang, A.Y.W.; Lin, T.-K. Simvastatin treatment exerts antidepressant-like effect in rats exposed to chronic mild stress. *Pharmacol. Biochem. Behav.* **2014**, *124*, 174–179. [CrossRef]
115. ElBatsh, M.M. Antidepressant-like effect of simvastatin in diabetic rats. *Can. J. Physiol. Pharmacol.* **2015**, *93*, 649–656. [CrossRef]
116. Abbasi, S.H.; Mohammadinejad, P.; Shahmansouri, N.; Salehiomran, A.; Beglar, A.A.; Zeinoddini, A.; Forghani, S.; Akhondzadeh, S. Simvastatin versus atorvastatin for improving mild to moderate depression in post-coronary artery bypass graft patients: A double-blind, placebo-controlled, randomized trial. *J. Affect. Disord.* **2015**, *183*, 149–155. [CrossRef]
117. Gougol, A.; Zareh-Mohammadi, N.; Raheb, S.; Farokhnia, M.; Salimi, S.; Iranpour, N.; Yekehtaz, H.; Akhondzadeh, S. Simvastatin as an adjuvant therapy to fluoxetine in patients with moderate to severe major depression: A double-blind placebo-controlled trial. *J. Psychopharmacol.* **2015**, *29*, 575–581. [CrossRef] [PubMed]

118. Agostini, J.V.; Tinetti, M.E.; Han, L.; McAvay, G.; Foody, J.M.; Concato, J. Effects of statin use on muscle strength, cognition, and depressive symptoms in older adults. *J. Am. Geriatr. Soc.* **2007**, *55*, 420–425. [CrossRef] [PubMed]
119. Feng, L.; Yap, K.B.; Kua, E.H.; Ng, T.P. Statin use and depressive symptoms in a prospective study of community-living older persons. *Pharmacoepidemiol. Drug Saf.* **2010**, *19*, 942–948. [CrossRef]
120. Al Badarin, F.J.; Spertus, J.A.; Gosch, K.L.; Buchanan, D.M.; Chan, P.S. Initiation of statin therapy after acute myocardial infarction is not associated with worsening depressive symptoms: Insights from the Prospective Registry Evaluating Outcomes After Myocardial Infarctions: Events and Recovery (PREMIER) and Translational Research Investigating Underlying Disparities in Acute Myocardial Infarction Patients' Health Status (TRIUMPH) registries. *Am. Heart J.* **2013**, *166*, 879–886. [PubMed]
121. Ludka, F.K.; Zomkowski, A.D.E.; Cunha, M.P.; Dal-Cim, T.; Zeni, A.L.B.; Rodrigues, A.L.S.; Tasca, C.I. Acute atorvastatin treatment exerts antidepressant-like effect in mice via the *L*-arginine-nitric oxide-cyclic guanosine monophosphate pathway and increases BDNF levels. *Eur. Neuropsychopharmacol.* **2013**, *23*, 400–412. [CrossRef] [PubMed]
122. Shahsavarian, A.; Javadi, S.; Jahanabadi, S.; Khoshnoodi, M.; Shamsaee, J.; Shafaroodi, H.; Mehr, S.E.; Dehpour, A. Antidepressant-like effect of atorvastatin in the forced swimming test in mice: The role of PPAR-gamma receptor and nitric oxide pathway. *Eur. J. Pharmacol.* **2014**, *745*, 52–58. [CrossRef]
123. Citraro, R.; Chimirri, S.; Aiello, R.; Gallelli, L.; Trimboli, F.; Britti, D.; De Sarro, G.; Russo, E. Protective effects of some statins on epileptogenesis and depressive-like behavior in WAG/Rij rats, a genetic animal model of absence epilepsy. *Epilepsia* **2014**, *55*, 1284–1291. [CrossRef]
124. Etminan, M.; Samii, A.; Brophy, J.M. Statin use and risk of epilepsy A nested case-control study. *Neurology* **2010**, *75*, 1496–1500. [CrossRef]
125. Piermartiri, T.C.B.; Vandresen-Filho, S.; Herculano, B.d.A.; Martins, W.C.; Dal'Agnolo, D.; Stroeh, E.; Carqueja, C.L.; Boeck, C.R.; Tasca, C.I. Atorvastatin prevents hippocampal cell death due to quinolinic acid-induced seizures in mice by increasing Akt phosphorylation and glutamate uptake. *Neurotox. Res.* **2009**, *16*, 106–115. [CrossRef] [PubMed]
126. Ma, T.; Zhao, Y.; Kwak, Y.-D.; Yang, Z.; Thompson, R.; Luo, Z.; Xu, H.; Liao, F.-F. Statin's excitoprotection is mediated by sAPP and the subsequent attenuation of calpain-induced truncation events, likely via Rho-ROCK signaling. *J. Neurosci.* **2009**, *29*, 11226–11236. [CrossRef] [PubMed]
127. Lee, J.-K.; Won, J.-S.; Singh, A.K.; Singh, I. Statin inhibits kainic acid-induced seizure and associated inflammation and hippocampal cell death. *Neurosci. Lett.* **2008**, *440*, 260–264. [CrossRef] [PubMed]
128. Siniscalchi, A.; Mintzer, S. Statins for poststroke seizures The first antiepileptogenic agent? *Neurology* **2015**, *85*, 661–662. [CrossRef] [PubMed]
129. Ponce, J.; de la Ossa, N.P.; Hurtado, O.; Millan, M.; Arenillas, J.F.; Davalos, A.; Gasull, T. Simvastatin reduces the association of NMDA receptors to lipid rafts: A cholesterol-mediated effect in neuroprotection. *Stroke* **2008**, *39*, 1269–1275. [CrossRef] [PubMed]
130. Seker, F.B.; Kilic, U.; Caglayan, B.; Ethemoglu, M.S.; Caglayan, A.B.; Ekimci, N.; Demirci, S.; Dogan, A.; Oztezcan, S.; Sahin, F.; et al. HMG-CoA reductase inhibitor rosuvastatin improves abnormal brain electrical activity via mechanisms involving eNOs. *Neuroscience* **2015**, *284*, 349–359. [CrossRef] [PubMed]
131. Scicchitano, F.; Constanti, A.; Citraro, R.; De Sarro, G.; Russo, E. Statins and epilepsy: Preclinical studies, clinical trials and statin-anticonvulsant drug interactions. *Curr. Drug Targets* **2015**, *16*, 747–756. [CrossRef]
132. Sierra-Marcos, A.; Alvarez, V.; Faouzi, M.; Burnand, B.; Rossetti, A.O. Statins are associated with decreased mortality risk after status epilepticus. *Eur. J. Neurol.* **2015**, *22*, 402–405. [CrossRef] [PubMed]
133. Dale, K.M.; Coleman, C.I.; Henyan, N.N.; Kluger, J.; White, C.M. Statins and cancer risk—A meta-analysis. *JAMA-J. Am. Med. Assoc.* **2006**, *295*, 74–80. [CrossRef] [PubMed]
134. Caporaso, N.E. Statins and Cancer-Related Mortality—Let's Work Together. *N. Engl. J. Med.* **2012**, *367*, 1848–1850. [CrossRef] [PubMed]
135. Osmak, M. Statins and cancer: Current and future prospects. *Cancer Lett.* **2012**, *324*, 1–12. [CrossRef] [PubMed]
136. Pisanti, S.; Picardi, P.; Ciaglia, E.; D'Alessandro, A.; Bifulco, M. Novel prospects of statins as therapeutic agents in cancer. *Pharmacol. Res.* **2014**, *88*, 84–98. [CrossRef]
137. Kubatka, P.; Kruzliak, P.; Rotrekl, V.; Jelinkova, S.; Mladosievicova, B. Statins in oncological research: From experimental studies to clinical practice. *Crit. Rev. Oncol. Hematol.* **2014**, *92*, 296–311. [CrossRef]

138. Matusewicz, L.; Meissner, J.; Toporkiewicz, M.; Sikorski, A.F. The effect of statins on cancer cells-review. *Tumor Biol.* **2015**, *36*, 4889–4904. [CrossRef] [PubMed]
139. Baigent, C.; Blackwell, L.; Emberson, J.; Holland, L.E.; Reith, C.; Bhala, N.; Peto, R.; Barnes, E.H.; Keech, A.; Simes, J.; et al. Efficacy and safety of more intensive lowering of LDL cholesterol: A meta-analysis of data from 170000 participants in 26 randomised trials. *Lancet* **2010**, *376*, 1670–1681. [PubMed]
140. Mihaylova, B.; Emberson, J.; Blackwell, L.; Keech, A.; Simes, J.; Barnes, E.H.; Voysey, M.; Gray, A.; Collins, R.; Baigent, C.; et al. The effects of lowering LDL cholesterol with statin therapy in people at low risk of vascular disease: Meta-analysis of individual data from 27 randomised trials. *Lancet* **2012**, *380*, 581–590. [PubMed]
141. Shepherd, J.; Blauw, G.J.; Murphy, M.B.; Bollen, E.; Buckley, B.M.; Cobbe, S.M.; Ford, I.; Gaw, A.; Hyland, M.; Jukema, J.W.; et al. Pravastatin in elderly individuals at risk of vascular disease (PROSPER): A randomised controlled trial. *Lancet* **2002**, *360*, 1623–1630. [CrossRef]
142. Nielsen, S.F.; Nordestgaard, B.G.; Bojesen, S.E. Statin use and reduced cancer-related mortality. *N. Engl. J. Med.* **2012**, *367*, 1792–1802. [CrossRef] [PubMed]
143. Peng, Y.C.; Lin, C.L.; Hsu, W.Y.; Chang, C.S.; Yeh, H.Z.; Tung, C.F.; Wu, Y.L.; Sung, F.C.; Kao, C.H. Statins are associated with a reduced risk of cholangiocarcinoma: A population-based case-control study. *Br. J. Clin. Pharmacol.* **2015**, *80*, 755–761. [CrossRef] [PubMed]
144. Archibugi, L.; Maisonneuve, P.; Piciucchi, M.; Valente, R.; Delle Fave, G.; Capurso, G. Statins but not aspirin nor their combination have a chemopreventive effect on pancreatic cancer occurrence. *Pancreas* **2015**, *44*, 1359.
145. Sivaprasad, U.; Abbas, T.; Dutta, A. Differential efficacy of 3-hydroxy-3-methylglutaryl CoA reductase inhibitors on the cell cycle of prostate cancer cells. *Mol. Cancer Ther.* **2006**, *5*, 2310–2316. [CrossRef] [PubMed]
146. Song, X.; Liu, B.-C.; Lu, X.-Y.; Yang, L.-L.; Zhai, Y.-J.; Eaton, A.F.; Thai, T.L.; Eaton, D.C.; Ma, H.-P.; Shen, B.-Z. Lovastatin inhibits human B lymphoma cell proliferation by reducing intracellular ROS and TRPC6 expression. *Biochim. Biophys. Acta-Mol. Cell Res.* **2014**, *1843*, 894–901. [CrossRef] [PubMed]
147. Tu, Y.-S.; Kang, X.-L.; Zhou, J.-G.; Lv, X.-F.; Tang, Y.-B.; Guan, Y.-Y. Involvement of Chk1-Cdc25A-cyclin A/CDk2 pathway in simvastatin induced S-phase cell cycle arrest and apoptosis in multiple myeloma cells. *Eur. J. Pharmacol.* **2011**, *670*, 356–364. [CrossRef] [PubMed]
148. Yu, X.; Luo, Y.; Zhou, Y.; Zhang, Q.; Wang, J.; Wei, N.; Mi, M.; Zhu, J.; Wang, B.; Chang, H.; et al. BRCA1 overexpression sensitizes cancer cells to lovastatin via regulation of cyclin D1-CDK4-p21(WAF1/CIP1) pathway: Analyses using a breast cancer cell line and tumoral xenograft model. *Int. J. Oncol.* **2008**, *33*, 555–563. [PubMed]
149. Rao, S.; Porter, D.C.; Chen, X.M.; Herliczek, T.; Lowe, M.; Keyomarsi, K. Lovastatin-mediated G(1) arrest is through inhibition of the proteasome, independent of hydroxymethyl glutaryl-CoA reductase. *Proc. Natl. Acad. Sci. USA* **1999**, *96*, 7797–7802. [CrossRef] [PubMed]
150. Horiguchi, A.; Sumitomo, M.; Asakuma, J.; Asano, T.; Asano, T.; Hayakawa, M. 3-Hydroxy-3-methylglutaryl-coenzyme A reductase inhibitor, fluvastatin, as a novel agent for prophylaxis of renal cancer metastasis. *Clin. Cancer Res.* **2004**, *10*, 8648–8655. [CrossRef]
151. Denoyelle, C.; Vasse, M.; Korner, M.; Mishal, Z.; Ganne, F.; Vannier, J.P.; Soria, J.; Soria, C. Cerivastatin, an inhibitor of HMG-CoA reductase, inhibits the signaling pathways involved in the invasiveness and metastatic properties of highly invasive breast cancer cell lines: An in vitro study. *Carcinogenesis* **2001**, *22*, 1139–1148. [CrossRef]
152. Spampanato, C.; De Maria, S.; Sarnataro, M.; Giordano, E.; Zanfardino, M.; Baiano, S.; Carteni, M.; Morelli, F. Simvastatin inhibits cancer cell growth by inducing apoptosis correlated to activation of Bax and down-regulation of BCL-2 gene expression. *Int. J. Oncol.* **2012**, *40*, 935–941. [CrossRef]
153. Herrero-Martin, G.; Lopez-Rivas, A. Statins activate a mitochondria-operated pathway of apoptosis in breast tumor cells by a mechanism regulated by ErbB2 and dependent on the prenylation of proteins. *FEBS Lett.* **2008**, *582*, 2589–2594. [CrossRef]
154. Zhu, Y.; Casey, P.J.; Kumar, A.P.; Pervaiz, S. Deciphering the signaling networks underlying simvastatin-induced apoptosis in human cancer cells: Evidence for non-canonical activation of RhoA and Rac1 GTPases. *Cell Death Dis.* **2013**, *4*, e568. [CrossRef]
155. Miller, T.; Yang, F.; Wise, C.E.; Meng, F.; Priester, S.; Munshi, M.K.; Guerrier, M.; Dostal, D.E.; Glaser, S.S. Simvastatin stimulates apoptosis in cholangiocarcinoma by inhibition of Rac1 activity. *Dig. Liver Dis.* **2011**, *43*, 395–403. [CrossRef] [PubMed]

156. Simons, K.; Ehehalt, R. Cholesterol, lipid rafts, and disease. *J. Clin. Investig.* **2002**, *110*, 597–603. [CrossRef] [PubMed]
157. Gniadecki, R. Depletion of membrane cholesterol causes ligand-independent activation of Fas and apoptosis. *Biochem. Biophys. Res. Commun.* **2004**, *320*, 165–169. [CrossRef] [PubMed]
158. Zhuang, L.Y.; Kim, J.; Adam, R.M.; Solomon, K.R.; Freeman, M.R. Cholesterol targeting alters lipid raft composition and cell survival in prostate cancer cells and xenografts. *J. Clin. Investig.* **2005**, *115*, 959–968. [CrossRef] [PubMed]
159. Li, Y.C.; Park, M.J.; Ye, S.K.; Kim, C.W.; Kim, Y.N. Elevated levels of cholesterol-rich lipid rafts in cancer cells are correlated with apoptosis sensitivity induced by cholesterol-depleting agents. *Am. J. Pathol.* **2006**, *168*, 1107–1118. [CrossRef] [PubMed]
160. Boscher, C.; Nabi, I.R. CAVEOLIN-1: Role in Cell Signaling. In *Caveolins and Caveolae: Roles in Signaling and Disease Mechanisms*; Jasmin, J.F., Frank, P.G., Lisanti, M.P., Eds.; Springer: Berlin/Heidelberg, Germany, 2012; Volume 729, pp. 29–50.
161. Altwairgi, A.K. Statins are potential anticancerous agents (Review). *Oncol. Rep.* **2015**, *33*, 1019–1039. [CrossRef] [PubMed]
162. Barrios-Gonzalez, J.; Miranda, R.U. Biotechnological production and applications of statins. *Appl. Microbiol. Biotechnol.* **2010**, *85*, 869–883. [CrossRef] [PubMed]
163. Hoffman, W.F.; Alberts, A.W.; Anderson, P.S.; Chen, J.S.; Smith, R.L.; Willard, A.K. 3-hydroxy-3-methylglutaryl-coenzyme a reductase inhibitors. 4. Side-chain ester derivatives of mevinolin. *J. Med. Chem.* **1986**, *29*, 849–852. [CrossRef] [PubMed]
164. Askin, D.; Verhoeven, T.R.; Liu, T.M.H.; Shinkai, I. Synthesis of synvinolin—Extremely high conversion alkylation of an ester enolate. *J. Org. Chem.* **1991**, *56*, 4929–4932. [CrossRef]
165. Xie, X.K.; Tang, Y. Efficient synthesis of simvastatin by use of whole-cell biocatalysis. *Appl. Environ. Microbiol.* **2007**, *73*, 2054–2060. [CrossRef]
166. Schimmel, T.G.; Borneman, W.S.; Conder, M.J. Purification and characterization of a lovastatin esterase from *Clonostachys compactiuscula*. *Appl. Environ. Microbiol.* **1997**, *63*, 1307–1311. [PubMed]
167. Chen, L.C.; Lai, Y.K.; Wu, S.C.; Lin, C.C.; Guo, J.H. Production by *Clonostachys compactiuscula* of a lovastatin esterase that converts lovastatin to monacolin J. *Enzyme Microb. Technol.* **2006**, *39*, 1051–1059. [CrossRef]
168. Gao, X.; Xie, X.K.; Pashkov, I.; Sawaya, M.R.; Laidman, J.; Zhang, W.J.; Cacho, R.; Yeates, T.O.; Tang, Y. Directed Evolution and Structural Characterization of a Simvastatin Synthase. *Chem. Biol.* **2009**, *16*, 1064–1074. [CrossRef]
169. United States Environmental Protection Agency. Presidential Green Chemistry Challenge: 2012 Greener Synthetic Pathways Award. Available online: http://www2.epa.gov/green-chemistry/2012-greener-synthetic-pathways-award (accessed on 14 February 2019).
170. Xu, W.; Chooi, Y.H.; Choi, J.W.; Li, S.; Vederas, J.C.; Da Silva, N.A.; Tang, Y. LovG: The thioesterase required for dihydromonacolin L release and lovastatin nonaketide synthase turnover in lovastatin biosynthesis. *Angew. Chem. Int. Ed.* **2013**, *52*, 6472–6475. [CrossRef] [PubMed]
171. Butler, D.E.; Le, T.V.; Millar, A.; Nanninga, T.N. Process for the synthesis of (5*R*)-1,1-dimethylethyl-6-cyano-5-hydroxy-3-oxo-hexanoate. US Patent 5155251, 13 October 1992.
172. Turner, N.J.; O'Reilly, E. Biocatalytic retrosynthesis. *Nat. Chem. Biol.* **2013**, *9*, 285–288. [CrossRef]
173. Brady, D. Biocatalytic Hydrolysis of Nitriles. In *Handbook of Green Chemistry*; Wiley-VCH Verlag GmbH & Co. KGaA: Hoboken, NJ, USA, 2010; Volume 3, (Biocatalysis).
174. Faber, K. Biotransformations of non-natural compounds: State of the art and future development. *Pure Appl. Chem.* **1997**, *69*, 1613–1632. [CrossRef]
175. DeSantis, G.; Zhu, Z.L.; Greenberg, W.A.; Wong, K.V.; Chaplin, J.; Hanson, S.R.; Farwell, B.; Nicholson, L.W.; Rand, C.L.; Weiner, D.P.; et al. An enzyme library approach to biocatalysis: Development of nitrilases for enantioselective production of carboxylic acid derivatives. *J. Am. Chem. Soc.* **2002**, *124*, 9024–9025. [CrossRef] [PubMed]
176. DeSantis, G.; Wong, K.; Farwell, B.; Chatman, K.; Zhu, Z.L.; Tomlinson, G.; Huang, H.J.; Tan, X.Q.; Bibbs, L.; Chen, P.; et al. Creation of a productive, highly enantioselective nitrilase through gene site saturation mutagenesis (GSSM). *J. Am. Chem. Soc.* **2003**, *125*, 11476–11477. [CrossRef] [PubMed]

177. Bergeron, S.; Chaplin, D.A.; Edwards, J.H.; Ellis, B.S.W.; Hill, C.L.; Holt-Tiffin, K.; Knight, J.R.; Mahoney, T.; Osborne, A.P.; Ruecroft, G. Nitrilase-catalysed desymmetrisation of 3-hydroxyglutaronitrile: Preparation of a statin side-chain intermediate. *Org. Process Res. Dev.* **2006**, *10*, 661–665. [CrossRef]
178. Yao, P.; Li, J.; Yuan, J.; Han, C.; Liu, X.; Feng, J.; Wu, Q.; Zhu, D. Enzymatic synthesis of a key intermediate for rosuvastatin by nitrilase-catalyzed hydrolysis of ethyl (*R*)-4-cyano-3-hydroxybutyate at high substrate concentration. *ChemCatChem* **2015**, *7*, 271–275. [CrossRef]
179. Ma, S.K.; Gruber, J.; Davis, C.; Newman, L.; Gray, D.; Wang, A.; Grate, J.; Huisman, G.W.; Sheldon, R.A. A green-by-design biocatalytic process for atorvastatin intermediate. *Green Chem.* **2010**, *12*, 81–86. [CrossRef]
180. Chung, S.; Hwang, Y. Stereoselective hydrolysis of racemic ethyl 4-chloro-3-hydroxybutyrate by a lipase. *Biocatal. Biotransform.* **2008**, *26*, 327–330. [CrossRef]
181. Lee, S.H.; Park, O.J.; Uh, H.S. A chemoenzymatic approach to the synthesis of enantiomerically pure (*S*)-3-hydroxy-gamma-butyrolactone. *Appl. Microbiol. Biotechnol.* **2008**, *79*, 355–362. [CrossRef] [PubMed]
182. Patel, J.M. Biocatalytic synthesis of atorvastatin intermediates. *J. Mol. Catal. B-Enzym.* **2009**, *61*, 123–128. [CrossRef]
183. Martin, C.H.; Dhamankar, H.; Tseng, H.C.; Sheppard, M.J.; Reisch, C.R.; Prather, K.L.J. A platform pathway for production of 3-hydroxyacids provides a biosynthetic route to 3-hydroxy-gamma-butyrolactone. *Nat. Commun.* **2013**, *4*, 1414. [CrossRef] [PubMed]
184. Narasaka, K.; Pai, F.C. Stereoselective reduction of beta-hydroxyketones to 1,3-diols highly selective 1,3-asymmetric induction via boron chelates. *Tetrahedron* **1984**, *40*, 2233–2238. [CrossRef]
185. Chen, K.M.; Hardtmann, G.E.; Prasad, K.; Repic, O.; Shapiro, M.J. 1,3-*syn* diastereoselective reduction of beta-hydroxyketones utilizing alkoxydialkylboranes. *Tetrahedron Lett.* **1987**, *28*, 155–158. [CrossRef]
186. Sterk, D.; Casar, Z.; Jukic, M.; Kosmrlj, J. Concise and highly efficient approach to three key pyrimidine precursors for rosuvastatin synthesis. *Tetrahedron* **2012**, *68*, 2155–2160. [CrossRef]
187. Ohrlein, R.; Baisch, G. Chemo-enzymatic approach to statin side-chain building blocks. *Adv. Synth. Catal.* **2003**, *345*, 713–715. [CrossRef]
188. Santaniello, E.; Chiari, M.; Ferraboschi, P.; Trave, S. Enhanced and reversed enantioselectivity of enzymatic-hydrolysis by simple substrate modifications—The case of 3-hydroxyglutarate diesters. *J. Org. Chem.* **1988**, *53*, 1567–1569. [CrossRef]
189. Metzner, R.; Hummel, W.; Wetterich, F.; Konig, B.; Groger, H. Integrated biocatalysis in multistep drug synthesis without intermediate isolation: A de novo approach toward a rosuvastatin key building block. *Org. Process Res. Dev.* **2015**, *19*, 635–638. [CrossRef]
190. König, B.; Wetterich, F.; Gröger, H.; Metzner, R. Process for enantioselective synthesis of 3-hydroxy-glutaric acid monoesters and use thereof. WO 2014/140006, 18 September 2014.
191. You, Z.Y.; Liu, Z.Q.; Zheng, Y.G. Chemical and enzymatic approaches to the synthesis of optically pure ethyl (*R*)-4-cyano-3-hydroxybutanoate. *Appl. Microbiol. Biotechnol.* **2014**, *98*, 11–21. [CrossRef]
192. Asako, H.; Shimizu, M.; Itoh, N. Biocatalytic production of (*S*)-4-bromo-3-hydroxybutyrate and structurally related chemicals and their applications. *Appl. Microbiol. Biotechnol.* **2009**, *84*, 397–405. [CrossRef]
193. Asako, H.; Shimizu, M.; Makino, Y.; Itoh, N. Biocatalytic reduction system for the production of chiral methyl (*R*)/(*S*)-4-bromo-3-hydroxybutyrate. *Tetrahedron Lett.* **2010**, *51*, 2664–2666. [CrossRef]
194. Patel, R.N.; McNamee, C.G.; Banerjee, A.; Howell, J.M.; Robison, R.S.; Szarka, L.J. Stereoselective reduction of beta-keto-esters by *Geotrichum candidum*. *Enzyme Microb. Technol.* **1992**, *14*, 731–738. [CrossRef]
195. Ye, Q.; Ouyang, P.K.; Ying, H.J. A review-biosynthesis of optically pure ethyl (*S*)-4-chloro-3-hydroxybutanoate ester: Recent advances and future perspectives. *Appl. Microbiol. Biotechnol.* **2011**, *89*, 513–522. [CrossRef]
196. Yasohara, Y.; Kizaki, N.; Hasegawa, J.; Takahashi, S.; Wada, M.; Kataoka, M.; Shimizu, S. Synthesis of optically active ethyl 4-chloro-3-hydroxybutanoate by microbial reduction. *Appl. Microbiol. Biotechnol.* **1999**, *51*, 847–851. [CrossRef] [PubMed]
197. Kita, K.; Kataoka, M.; Shimizu, S. Diversity of 4-chloroacetoacetate ethyl ester-reducing enzymes in yeasts and their application to chiral alcohol synthesis. *J. Biosci. Bioeng.* **1999**, *88*, 591–598. [CrossRef]
198. Wada, M.; Kataoka, M.; Kawabata, H.; Yasohara, Y.; Kizaki, N.; Hasegawa, J.; Shimizu, S. Purification and characterization of NADPH-dependent carbonyl reductase, involved in stereoselective reduction of ethyl 4-chloro-3-oxobutanoate, from *Candida magnoliae*. *Biosci. Biotechnol. Biochem.* **1998**, *62*, 280–285. [CrossRef]

199. Kataoka, M.; Doi, Y.; Sim, T.S.; Shimizu, S.; Yamada, H. A novel NADPH-dependent carbonyl reductase of *Candida macedoniensis*. Purification and characterization. *Arch. Biochem. Biophys.* **1992**, *294*, 469–474. [CrossRef]
200. He, J.Y.; Sun, Z.H.; Ruan, W.Q.; Xu, Y. Biocatalytic synthesis of ethyl (*S*)-4-chloro-3-hydroxy-butanoate in an aqueous-organic solvent biphasic system using *Aureobasidium pullulans* CGMCC 1244. *Process Biochem.* **2006**, *41*, 244–249. [CrossRef]
201. Saratani, Y.; Uheda, E.; Yamamoto, H.; Nishimura, A.; Yoshizako, F. Stereoselective reduction of ethyl 4-chloro-3-oxobutanoate by fungi. *Biosci. Biochem. Biophys.* **2001**, *65*, 1676–1679. [CrossRef]
202. Ye, Q.; Yan, M.; Xu, L.; Cao, H.; Li, Z.; Chen, Y.; Li, S.; Ying, H. A novel carbonyl reductase from *Pichia stipitis* for the production of ethyl (*S*)-4-chloro-3-hydroxybutanoate. *Biotechnol. Lett.* **2009**, *31*, 537–542. [CrossRef]
203. Cai, P.; An, M.D.; Xu, L.; Xu, S.; Hao, N.; Li, Y.; Guo, K.; Yan, M. Development of a substrate-coupled biocatalytic process driven by an NADPH-dependent sorbose reductase from *Candida albicans* for the asymmetric reduction of ethyl 4-chloro-3-oxobutanoate. *Biotechnol. Lett.* **2012**, *34*, 2223–2227. [CrossRef]
204. Pan, J.; Zheng, G.-W.; Ye, Q.; Xu, J.-H. Optimization and scale-up of a bioreduction process for preparation of ethyl (*S*)-4-chloro-3-hydroxybutanoate. *Org. Process Res. Dev.* **2014**, *18*, 739–743. [CrossRef]
205. Liu, Z.-Q.; Ye, J.-J.; Shen, Z.-Y.; Hong, H.-B.; Yan, J.-B.; Lin, Y.; Chen, Z.-X.; Zheng, Y.-G.; Shen, Y.-C. Upscale production of ethyl (*S*)-4-chloro-3-hydroxybutanoate by using carbonyl reductase coupled with glucose dehydrogenase in aqueous-organic solvent system. *Appl. Microbiol. Biotechnol.* **2015**, *99*, 2119–2129. [CrossRef]
206. Schallmey, M.; Floor, R.J.; Hauer, B.; Breuer, M.; Jekel, P.A.; Wijma, H.J.; Dijkstra, B.W.; Janssen, D.B. Biocatalytic and structural properties of a highly engineered halohydrin dehalogenase. *ChemBioChem* **2013**, *14*, 870–881. [CrossRef]
207. Ritter, S.K. Going green keeps getting easier. *Chem. Eng. News* **2006**, *84*, 24–27. [CrossRef]
208. Li, C.-J.; Anastas, P.T. Green Chemistry: Present and future. *Chem. Soc. Rev.* **2012**, *41*, 1413–1414. [CrossRef]
209. Chen, X.F.; Xiong, F.J.; Chen, W.X.; He, Q.Q.; Chen, F.E. Asymmetric Synthesis of the HMG-CoA Reductase Inhibitor Atorvastatin Calcium: An Organocatalytic Anhydride Desymmetrization and Cyanide-Free Side Chain Elongation Approach. *J. Org. Chem.* **2014**, *79*, 2723–2728. [CrossRef]
210. Everaere, K.; Franceschini, N.; Mortreux, A.; Carpentier, J.F. Diastereoselective synthesis of syn-3,5-dihydroxyesters via ruthenium-catalyzed asymmetric transfer hydrogenation. *Tetrahedron Lett.* **2002**, *43*, 2569–2571. [CrossRef]
211. Wolberg, M.; Hummel, W.; Muller, M. Biocatalytic reduction of beta, delta-diketo esters: A highly stereoselective approach to all four stereoisomers of a chlorinated beta, delta-dihydroxy hexanoate. *Chem. Eur. J.* **2001**, *7*, 4562–4571. [CrossRef]
212. Wolberg, M.; Hummel, W.; Wandrey, C.; Muller, M. Highly regio- and enantioselective reduction of 3,5-dioxocarboxylates. *Angew. Chem. Int. Ed.* **2000**, *39*, 4306–4308. [CrossRef]
213. Wolberg, M.; Villela, M.; Bode, S.; Geilenkirchen, P.; Feldmann, R.; Liese, A.; Hummel, W.; Muller, M. Chemoenzymatic synthesis of the chiral side-chain of statins: Application of an alcohol dehydrogenase catalysed ketone reduction on a large scale. *Bioproc. Biosyst. Eng.* **2008**, *31*, 183–191. [CrossRef]
214. Liu, Z.Q.; Hu, Z.L.; Zhang, X.J.; Tang, X.L.; Cheng, F.; Xue, Y.P.; Wang, Y.J.; Wu, L.; Yao, D.K.; Zhou, Y.T.; et al. Large-scale synthesis of tert-butyl (3*R*,5*S*)-6-chloro-3,5-dihydroxyhexanoate by a stereoselective carbonyl reductase with high substrate concentration and product yield. *Biotechnol. Progr.* **2017**, *33*, 612–620. [CrossRef] [PubMed]
215. Liu, Z.Q.; Wu, L.; Zhang, X.J.; Xue, Y.P.; Zheng, Y.G. Directed evolution of carbonyl reductase from *Rhodosporidium toruloides* and its application in stereoselective synthesis of tert-butyl (3*R*,5*S*)-6-chloro-3,5-dihydroxyhexanoate. *J. Agric. Food Chem.* **2017**, *65*, 3721–3729. [CrossRef] [PubMed]
216. Liu, Z.Q.; Yin, H.H.; Zhang, X.J.; Zhou, R.; Wang, Y.M.; Zheng, Y.G. Improvement of carbonyl reductase activity for the bioproduction of tert-butyl (3*R*,5*S*)-6-chloro-3,5-dihydroxyhexanoate. *Bioorg. Chem.* **2018**, *80*, 733–740. [CrossRef] [PubMed]
217. Liu, Z.Q.; Wu, L.; Zheng, L.; Wang, W.Z.; Zhang, X.J.; Jin, L.Q.; Zheng, Y.G. Biosynthesis of tert-butyl (3*R*,5*S*)-6-chloro-3,5-dihydroxyhexanoate by carbonyl reductase from *Rhodosporidium toruloides* in mono and biphasic media. *Bioresour. Technol.* **2018**, *249*, 161–167. [CrossRef]

218. Xu, T.T.; Wang, C.; Zhu, S.Z.; Zheng, G.J. Enzymatic preparation of optically pure t-butyl 6-chloro-(3*R*,5*S*)-dihydroxyhexanoate by a novel alcohol dehydrogenase discovered from *Klebsiella oxytoca*. *Process Biochem.* **2017**, *57*, 72–79. [CrossRef]
219. Gong, X.M.; Zheng, G.W.; Liu, Y.Y.; Xu, J.H. Identification of a robust carbonyl reductase for diastereoselectively building *syn*-3,5-dihydroxy hexanoate: A bulky side chain of atorvastatin. *Org. Process Res. Dev.* **2017**, *21*, 1349–1354. [CrossRef]
220. Gong, X.M.; Qin, Z.; Li, F.L.; Zeng, B.B.; Zheng, G.W.; Xu, J.H. Development of an engineered ketoreductase with simultaneously improved thermostability and activity for making a bulky atorvastatin precursor. *ACS Catal.* **2019**, *9*, 147–153. [CrossRef]
221. Luo, X.; Wang, Y.-J.; Zheng, Y.-G. Improved stereoselective bioreduction of t-butyl 6-cyano-(5*R*)-hydroxy-3-oxohexanoate by *Rhodotorula glutinis* through heat treatment. *Biotechnol. Appl. Biochem.* **2016**, *63*, 795–804. [CrossRef]
222. Wu, X.; Gou, X.; Chen, Y. Enzymatic preparation of t-butyl-6-cyano-(3*R*,5*R*)-dihydroxyhexanoate by a whole-cell biocatalyst co-expressing carbonyl reductase and glucose dehydrogenase. *Process Biochem.* **2015**, *50*, 104–110. [CrossRef]
223. Wang, Y.J.; Chen, X.P.; Shen, W.; Liu, Z.Q.; Zheng, Y.G. Chiral diol t-butyl 6-cyano-(3*R*,5*R*)-dihydroxylhexanoate synthesis catalyzed by immobilized cells of carbonyl reductase and glucose dehydrogenase co-expression *E. coli*. *Biochem. Eng. J.* **2017**, *128*, 54–62. [CrossRef]
224. Luo, X.; Wang, Y.-J.; Zheng, Y.-G. Cloning and characterization of a NADH-dependent aldo-keto reductase from a newly isolated *Kluyveromyces lactis* XP1461. *Enzyme Microb. Technol.* **2015**, *77*, 68–77. [CrossRef]
225. Luo, X.; Wang, Y.-J.; Shen, W.; Zheng, Y.-G. Activity improvement of a *Kluyveromyces lactis* aldo-keto reductase KlAKR via rational design. *J. Biotechnol.* **2016**, *224*, 20–26. [CrossRef]
226. Wang, Y.-J.; Liu, X.-Q.; Luo, X.; Liu, Z.-Q.; Zheng, Y.-G. Cloning, expression and enzymatic characterization of an aldo-keto reductase from *Candida albicans* XP1463. *J. Mol. Catal. B-Enzym.* **2015**, *122*, 44–50. [CrossRef]
227. Pfruender, H.; Amidjojo, M.; Hang, F.; Weuster-Botz, D. Production of *Lactobacillus kefir* cells for asymmetric synthesis of a 3,5-dihydroxycarboxylate. *Appl. Microbiol. Biotechnol.* **2005**, *67*, 619–622. [CrossRef]
228. Patel, R.N.; Banerjee, A.; McNamee, C.G.; Brzozowski, D.; Hanson, R.L.; Szarka, L.J. Enantioselective microbial reduction of 3,5-dioxo-6-(benzyloxy) hexanoic acid, ethyl-ester. *Enzyme Microb. Technol.* **1993**, *15*, 1014–1021. [CrossRef]
229. Guo, Z.; Chen, Y.; Goswami, A.; Hanson, R.L.; Patel, R.N. Synthesis of ethyl and t-butyl (3*R*,5*S*)-dihydroxy-6-benzyloxy hexanoates via diastereo- and enantioselective microbial reduction. *Tetrahedron Asymm.* **2006**, *17*, 1589–1602. [CrossRef]
230. Goldberg, S.; Guo, Z.; Chen, S.; Goswami, A.; Patel, R.N. Synthesis of ethyl-(3*R*,5*S*)-dihydroxy-6-benzyloxyhexanoates via diastereo- and enantioselective microbial reduction: Cloning and expression of ketoreductase III from *Acinetobacter* sp. *SC 13874*. *Enzyme Microb. Technol.* **2008**, *43*, 544–549. [CrossRef]
231. Wu, X.; Liu, N.; He, Y.; Chen, Y. Cloning, expression, and characterization of a novel diketoreductase from *Acinetobacter baylyi*. *Acta Biochim. Biophys. Sin.* **2009**, *41*, 163–170. [CrossRef] [PubMed]
232. Chen, Y.; Chen, C.; Wu, X. Dicarbonyl reduction by single enzyme for the preparation of chiral diols. *Chem. Soc. Rev.* **2012**, *41*, 1742–1753. [CrossRef]
233. Lu, M.; Huang, Y.; White, M.A.; Wu, X.; Liu, N.; Cheng, X.; Chen, Y. Dual catalysis mode for the dicarbonyl reduction catalyzed by diketoreductase. *Chem. Commun.* **2012**, *48*, 11352–11354. [CrossRef] [PubMed]
234. Huang, Y.; Lu, Z.; Ma, M.; Liu, N.; Chen, Y. Functional roles of tryptophan residues in diketoreductase from *Acinetobacter baylyi*. *BMB Rep.* **2012**, *45*, 452–457. [CrossRef] [PubMed]
235. Huang, Y.; Lu, Z.; Liu, N.; Chen, Y. Identification of important residues in diketoreductase from *Acinetobacter baylyi* by molecular modeling and site-directed mutagenesis. *Biochimie* **2012**, *94*, 471–478. [CrossRef] [PubMed]
236. Gijsen, H.J.M.; Wong, C.H. Sequential 3-substrate and 4-substrate aldol reactions catalyzed by aldolases. *J. Am. Chem. Soc.* **1995**, *117*, 7585–7591. [CrossRef]
237. Gijsen, H.J.M.; Wong, C.H. Unprecedented asymmetric aldol reactions with 3 aldehyde substrates catalyzed by 2-deoxyribose-5-phosphate aldolase. *J. Am. Chem. Soc.* **1994**, *116*, 8422–8423. [CrossRef]
238. Liu, J.J.; Hsu, C.C.; Wong, C.H. Sequential aldol condensation catalyzed by DERA mutant Ser238Asp and a formal total synthesis of atorvastatin. *Tetrahedron Lett.* **2004**, *45*, 2439–2441. [CrossRef]

239. Greenberg, W.A.; Varvak, A.; Hanson, S.R.; Wong, K.; Huang, H.J.; Chen, P.; Burk, M.J. Development of an efficient, scalable, aldolase-catalyzed process for enantioselective synthesis of statin intermediates. *Proc. Natl. Acad. Sci. USA* **2004**, *101*, 5788–5793. [CrossRef]
240. Jennewein, S.; Schurmann, M.; Wolberg, M.; Hilker, I.; Luiten, R.; Wubbolts, M.; Mink, D. Directed evolution of an industrial biocatalyst: 2-deoxy-D-ribose 5-phosphate aldolase. *Biotechnol. J.* **2006**, *1*, 537–548. [CrossRef]
241. Oslaj, M.; Cluzeau, J.; Orkic, D.; Kopitar, G.; Mrak, P.; Casar, Z. A highly productive, whole-cell DERA chemoenzymatic process for production of key lactonized side-chain intermediates in statin synthesis. *PLoS ONE* **2013**, *8*, e62250. [CrossRef]
242. Ručigaj, A.; Krajnc, M. Optimization of a crude deoxyribose-5-phosphate aldolase lyzate-catalyzed process in synthesis of statin intermediates. *Org. Process Res. Dev.* **2013**, *17*, 854–862. [CrossRef]
243. You, Z.Y.; Liu, Z.Q.; Zheng, Y.G.; Shen, Y.C. Characterization and application of a newly synthesized 2-deoxyribose-5-phosphate aldolase. *J. Ind. Microbiol. Biotechnol.* **2013**, *40*, 29–39. [CrossRef]
244. Casar, Z.; Steinbucher, M.; Kosmrlj, J. Lactone pathway to statins utilizing the Wittig reaction. The synthesis of rosuvastatin. *J. Org. Chem.* **2010**, *75*, 6681–6684. [CrossRef]
245. Vajdic, T.; Oslaj, M.; Kopitar, G.; Mrak, P. Engineered, highly productive biosynthesis of artificial, lactonized statin side-chain building blocks: The hidden potential of *Escherichia coli* unleashed. *Metab. Eng.* **2014**, *24*, 160–172. [CrossRef]
246. Sakuraba, H.; Yoneda, K.; Yoshihara, K.; Satoh, K.; Kawakami, R.; Uto, Y.; Tsuge, H.; Takahashi, K.; Hori, H.; Ohshima, T. Sequential aldol condensation catalyzed by hyperthermophilic 2-deoxy-D-ribose-5-phosphate aldolase. *Appl. Environ. Microbiol.* **2007**, *73*, 7427–7434. [CrossRef]
247. Fei, H.; Zheng, C.C.; Liu, X.Y.; Li, Q. An industrially applied biocatalyst: 2-Deoxy-D-ribose-5-phosphate aldolase. *Process Biochem.* **2017**, *63*, 55–59. [CrossRef]
248. Haridas, M.; Abdelraheem, E.M.M.; Hanefeld, U. 2-Deoxy-D-ribose-5-phosphate aldolase (DERA): Applications and modifications. *Appl. Microbiol. Biotechnol.* **2018**, *102*, 9959–9971. [CrossRef]
249. Muller, M. Chemoenzymatic synthesis of building blocks for statin side chains. *Angew. Chem. Int. Ed.* **2005**, *44*, 362–365. [CrossRef]
250. Garattini, L.; Padula, A. Cholesterol-lowering drugs: Science and marketing. *J. R. Soc. Med.* **2017**, *110*, 57–64. [CrossRef]
251. Reiner, Z. Resistance and intolerance to statins. *Nutr. Metab. Carbiovasc. Dis.* **2014**, *24*, 1057–1066. [CrossRef] [PubMed]
252. Chaudhary, R.; Garg, J.; Shah, N.; Sumner, A. PCSK9 inhibitors: A new era of lipid lowering therapy. *World J. Cardiol.* **2017**, *9*, 76–91. [CrossRef] [PubMed]
253. Palmer, C. New directions in managing dyslipidemia. *J. Nurse Pract.* **2019**, *15*, 73–79. [CrossRef]
254. Xu, S.T.; Luo, S.S.; Zhu, Z.Y.; Xu, J.Y. Small molecules as inhibitors of PCSK9: Current status and future challenges. *Eur. J. Med. Chem.* **2019**, *162*, 212–233. [CrossRef] [PubMed]

Review

Microbial Reduction of Cholesterol to Coprostanol: An Old Concept and New Insights

Aicha Kriaa, Mélanie Bourgin, Héla Mkaouar, Amin Jablaoui, Nizar Akermi, Souha Soussou, Emmanuelle Maguin and Moez Rhimi *

UMR 1319 Micalis, INRA, Microbiota Interaction with Human and Animal Team (MIHA), AgroParisTech, Université Paris-Saclay, F-78350 Jouy-en-Josas, France; aicha.kriaa@inra.fr (A.K.); melanie.bourgin@inra.fr (M.B.); hela.mkaouar@inra.fr (H.M.); amin.jablaoui@gmail.com (A.J.); nizar.akermi@inra.fr (N.A.); souha.soussou@inra.fr (S.S.); emmanuelle.maguin@inra.fr (E.M.)

* Correspondence: moez.rhimi@inra.fr; Tel.: +33-1-3465-2294

Received: 23 January 2019; Accepted: 2 February 2019; Published: 8 February 2019

Abstract: The gut microbiota plays a key role in cholesterol metabolism, mainly through the reduction of cholesterol to coprostanol. The latter sterol exhibits distinct physicochemical properties linked to its limited absorption in the gut. Few bacteria were reported to reduce cholesterol into coprostanol. Three microbial pathways of coprostanol production were described based on the analysis of reaction intermediates. However, these metabolic pathways and their associated genes remain poorly studied. In this review, we shed light on the microbial metabolic pathways related to coprostanol synthesis. Moreover, we highlight current strategies and future directions to better characterize these microbial enzymes and pathways.

Keywords: cholesterol; coprostanol; reduction reaction; bacteria; metabolic pathways

1. Introduction

With more than 31% of all global deaths, atherosclerotic cardiovascular disease is now largely recognized as a major concern for global health [1]. A key risk factor for cardiovascular disease includes imbalanced blood cholesterol levels, notably a high serum concentration of low-density lipoprotein cholesterol [2]. Maintaining cholesterol homeostasis is achieved by balancing dietary cholesterol intake through its synthesis, absorption, and excretion [3].

Current therapeutic approaches aiming at reducing plasma cholesterol levels in humans rely on (i) the inhibition of cholesterol biosynthesis, (ii) an increased elimination of cholesterol from tissues into the intestinal lumen, and (iii) the reduction of cholesterol absorption from the gastrointestinal tract (GIT) [4–6]. While these three approaches seem attractive, they display several limitations including a significant proportion of non-responders among patients (20–30%), and the omission of the intestinal microbiota impact on cholesterol cycle [7]. On a daily basis, around 1 g of cholesterol from diet, bile, and desquamated intestinal cells reaches the human colon. This cholesterol amount is metabolized by commensal bacteria to coprostanol [8]. Unlike cholesterol, coprostanol is poorly absorbed by the human intestine [9]. For this reason, it was suggested to have an impact on cholesterol metabolism and modulating serum cholesterol levels [8]. This takes more importance if we consider the existence of an inverse relationship between plasma cholesterol levels and the ratio of cholesterol to coprostanol conversion in the feces [10]. Indeed, cholesterol displays chemical behaviors different from coprostanol, which confer to the latter a feeble intestinal absorption [9]. Cholesterol conversion to coprostanol could therefore constitute a new strategy, allowing a better management of cholesterol homeostasis in humans without influencing host physiology [8]. Notably, some studies reported that cholesterol metabolites could be associated with procarcinogenic effects and an increased risk of colon cancer [11,12]. The reduction of cholesterol to coprostanol and the properties of this reaction remain largely unknown. This concept is still poorly understood, as few studies are available regarding cholesterol-metabolizing bacteria and their associated genes. Therefore, achieving a molecular characterization of these bacterial pathways represents a challenge in this field of research in setting out new hypocholesterolemic strategies.

In this review, we provide an overview of cholesterol and coprostanol chemical behaviors, and outline the coprostanoligenic bacteria. Moreover, we analyze the current knowledge related to the reduction of cholesterol to coprostanol, and highlight available approaches and future opportunities to better characterize these metabolic pathways.

2. Cholesterol versus Coprostanol

2.1. Chemical Behaviors

Cholesterol, a major sterol in body tissues, was first discovered in solid form in gallstones and bile [13]. It is a complex lipid substance that plays a key role in many biochemical processes [14–16]. From a structural point of view, cholesterol ($C_{27}H_{46}O$) is mostly a hydrophobic molecule that has a rigid, steranic core coupled to a hydrophilic hydroxyl group at position 3. Such a unique structure is composed of four, fused hydrocarbon rings: A, B, C and D (Figure 1). Its physicochemical properties confer an amphiphilic nature and a specific orientation (i.e., A/B ring: *trans*) adapted to its integration into cell membranes [17,18]. Most plasma cholesterol exists in an esterified form (with a fatty acid attached at C3), making the molecule even more hydrophobic [19]. Cholesterol ester is the form of storage under which cholesterol is found in most tissues and plasma [19]. As a main structural component of cell membranes, cholesterol regulates membrane fluidity and cellular signaling, and serves as a precursor of important molecules including steroid hormones (e.g., progesterone, estrogen, etc.), vitamin D, and bile acids [20].

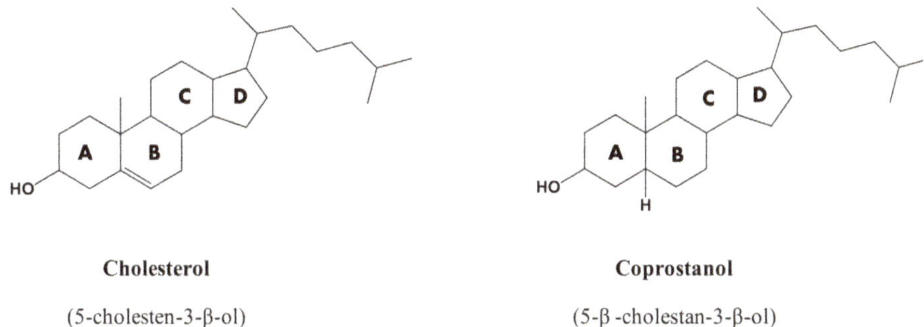

Figure 1. Chemical structures of cholesterol (Molecular weight (Mw): 386,654 g/mol) and coprostanol (Mw: 388,6756 g/mol).

Coprostanol, a saturated analog of cholesterol, was first isolated in the mid-1800s and characterized as an alcohol ($C_{27}H_{48}O$) [21]. This sterol shares the same basic structure as cholesterol, but it lacks the 5–6 double bond and it has an additional hydrogen atom at position 5. This gives different properties to the sterane nucleus, making it less rigid (Figure 1) [21]. At room temperature, coprostanol is a white crystalline solid with a melting point of 101 °C [22]. Like cholesterol, coprostanol is highly soluble in hexane, benzene, and chloroform. However, both sterols are insoluble in polar solutions like methanol [22]. The solubility of such sterols in alcoholic solvents increases with an increasing chain length of the alcohol–hydrocarbon moiety [23]. Accordingly, they tend to be more soluble in ethanol, but completely soluble in butan-1-ol [23]. Notably, cholesterol is insoluble in water at concentrations less than 1 μg/mL at 30 °C [24]; nevertheless, no report exists on coprostanol solubility.

Unlike most saturated sterols found in nature, which are A/B-*trans*, coprostanol seems to be the only exception with a *cis*-oriented A/B ring structure (the 3-hydroxyl group in coprostanol is in an equatorial configuration on ring A) [25]. The biological role of this compound remains to be elucidated.

2.2. Intestinal Uptake

The poor absorption of coprostanol in the intestine was associated with its very low uptake through the intestinal mucosa and its limited esterification in mucosal cells [9]. Indeed, the *cis*-A/-B ring structure in coprostanol, driving the difference in the 3-hydroxyl group from the axial to equatorial position, was suggested to affect coprostanol incorporation into mucosal cell membranes, thereby limiting its intestinal uptake [26]. Hence, a high-efficiency conversion of cholesterol to coprostanol was proposed to lower serum cholesterol levels [27,28]. Cholesterol absorption takes place mostly in the upper GIT [29,30], while coprostanol production occurs essentially in the large intestine [31]. This concept remains highly controversial, as earlier studies revealed the presence of significant amounts of coprostanol in the first half of the small intestine in rats [32,33] and humans [34].

3. Microbial Conversion of Cholesterol to Coprostanol

3.1. Coprostanoligenic Bacteria

Cholesterol conversion to coprostanol by intestinal bacteria was first reported in the 1930s [35]. Since coprostanol was not detected in tissues, but mainly in larger amounts in feces (more than 50% of total sterols in humans), this product was proposed to derive from bacterial conversion in the intestine [35,36]. A key role for the gut microbiota in this biotransformation was further investigated by comparing fecal sterols of conventional and germ-free rats. In fact, contrary to the conventional group, germ-free rats excreted only unmodified cholesterol [37]. These data provide evidence for the role of gut microbiota in cholesterol metabolism, including coprostanol synthesis. Numerous reports were aimed at identifying bacteria that were able to reduce cholesterol to coprostanol [38–43]. As anaerobic strains are difficult to culture, only few coprostanoligenic bacteria were isolated from rat cecal contents [38], baboons [40], and human feces [44,45]. Surprisingly, most strains exhibited similar properties and were assigned to the genus *Eubacterium* [40–44], except *Bacteroides* sp. strain D8 [45]. Bacteria belonging to *Bifidobacterium*, *Clostridium*, and *Lactobacillus* were also reported to reduce cholesterol to coprostanol in vitro [46–48], yet they were not explored in vivo. Based on metagenomic analysis, new bacterial phylotypes from Lachnospiraceae and Ruminococcaceae families were recently associated with high coprostanol levels in healthy humans [49]. Notably, associations of these bacterial taxa with coprostanol levels require further functional studies in order to elucidate the eventual causal relationship between these microbial communities and coprostanol synthesis.

3.2. Two Patterns Cepending on Gut Microbiota

In humans, microbial conversion of cholesterol to coprostanol is bimodal, with a majority of high producers (almost complete cholesterol conversion) that display low cholesterolemia and a minority of low, or inefficient, producers (coprostanol content representing less than one-third of the fecal neutral sterol content) [50,51]. Such conversion patterns were found to be equally distributed with respect to sex, and were independent of age [51–53]. Factors affecting cholesterol conversion were defined and proposed to be closely related to the abundance of cholesterol-metabolizing bacteria [39,46]. In fact, low converters harbored less than 10^6 cholesterol-metabolizing bacteria per gram of stool, while more than 10^8 cholesterol-metabolizing bacteria per gram of stool led to nearly complete cholesterol conversion [52]. Remarkably, Sekimoto et al. (1983) reported the existence of an inverse relationship between serum cholesterol levels and the coprostanol/cholesterol ratio in feces [9]. These findings denote that the conversion of cholesterol into coprostanol can significantly decrease the blood cholesterol level. A correlation between fecal microbial community structure and the conversion rate of cholesterol to coprostanol was also reported [46]. Interestingly, the high density of bacteria in the colon can explain the significant production levels of coprostanol seen in this part of the GIT. In fact, we previously reported the identification of the first human bacterium (*Bacteroides* sp. D8) that displayed a high coprostanol production level in human feces [45,54]. Furthermore, feeding the cholesterol-reducing bacterium *Eubacterium coprostanoligenes* significantly reduced plasma cholesterol levels in hypercholesterolemic rabbits [27]. Accordingly, greater coprostanol/cholesterol ratios were detected in the digestive contents of bacteria-fed rabbits [27]. Such effect was further ascribed to the reduction of cholesterol to coprostanol in the intestine, seeing that *E. coprostanoligenes* colonized and reduced cholesterol in the jejunum and ileum (sites for cholesterol absorption) [27]. The progress in functional knowledge related to the human gut microbiota, and the availability of new microbial culture facilities, offer a unique opportunity to isolate new coprostanoligenic bacteria belonging to different taxa.

3.3. Metabolic Pathways

The mechanism of cholesterol conversion to coprostanol was investigated and three major pathways were proposed [38,55–58]. The first one is a direct, stereospecific reduction of the 5,6-double bond of cholesterol, without an intermediate formation of a ketone at C3 (Figure 2) [59,60]. This statement was supported by the use of labeled cholesterol that led only to coprostanol synthesis [59,60]. The second one is an indirect pathway, which was described through the action of several bacteria on marked cholesterol (Figure 2). In fact, it was reported that cecal rat contents (*Eubacterium* strain, *E. coprostanoligenes,* and *Bacteroides* sp. D8) incubated with labeled cholesterol allowed for the production of cholestenone and coprostanone intermediates, demonstrating that this microbial pathway involved at least three steps [45,61,62]. The third pathway consisted of the isomerization of cholesterol to allocholesterol, which can be reduced to coprostanol by *Eubacterium* ATCC21, 408, and 403 species [38,57,58]. However, few data exist regarding this last bacterial metabolic pathway.

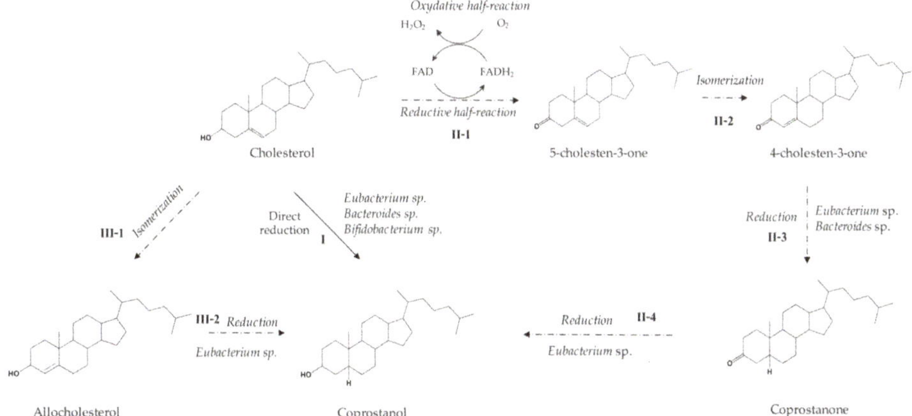

Figure 2. Metabolic pathways for cholesterol conversion to coprostanol. I: direct pathway; II: indirect pathway; II-1: Oxidation of the 3-β-hydroxyl group moiety catalyzed by cholesterol oxidase; II-2: Isomerization of Δ5 double bond to Δ4 by cholesterol oxidase; II-3: Reduction of the Δ4 double bond by 3-oxo-Δ4-steroid 5β-reductase; II-4: Reduction leading to coprostanol; III-1: Isomerization of cholesterol to allocholesterol; III-2: Reduction to coprostanol.

The first two steps of the indirect pathway are catalyzed by a single enzyme, cholesterol oxidase (EC 1.1.3.6), and result in the formation of 4-cholesten-3-one with a reduction of oxygen to hydrogen peroxide (Figure 2). Briefly, the 3-β-hydroxyl group of the steroid A ring is first oxidized to the corresponding ketone using the FAD (Flavin Adenine Dinucleotide) cofactor, which is reduced during this process (reductive first-reaction). Since the 5-cholesten-3-one intermediate is not stable and susceptible to radical oxygenation, the Δ5 double bond in the oxidized steroid ring is then isomerized to Δ4 to form 4-cholesten-3-one, the final steroid product. During the oxidative half-reaction, the FAD is finally re-oxidized by dioxygen to form hydrogen peroxide (Figure 2) [63,64]. To efficiently catalyze both reactions (oxidation and isomerization), distinct and essential features are required for cholesterol oxidase. As a starting point, the substrate (cholesterol) must be properly oriented to the cofactor to allow for hydride transfer from the steroid C3 site to the N5 site of the cofactor [64]. Then, during the dehydrogenation step a base is needed to: (i) deprotonate the steroid C3-OH, and (ii) transfer protons during the isomerization reaction [64].

Biochemical and structural studies were performed for cholesterol oxidases of several microorganisms, mostly deriving from soil [65–69]. Structural studies revealed the existence of two forms of cholesterol oxidases: one with the FAD linked covalently to the enzyme, and another with the FAD non-covalently bound [64]. Despite topological differences in these two forms, both enzymes exhibit a large buried hydrophobic pocket, able to accommodate the steroid substrate (Figure 3). In both forms, the binding site for cholesterol is sealed off by a number of amphipathic loops to prevent potential aggregation. Once opened, these loops allow the sterol to exit from the membrane so that the 8-carbon isoprenyl tail of cholesterol can pack and bind to the protein [64]. Following substrate binding, hydrophobic interactions between cholesterol and hydrophobic residues minimize energy loss. Two active-site residues, Histidine and Glutamate, were shown as important to catalyze the oxidation and isomerization reactions (Figure 3). The Histidine residue binds to the C3-OH of the steroid, allowing for proper positioning of cholesterol relative to the FAD cofactor. Glutamate acts as a base during the isomerization reaction [70,71].

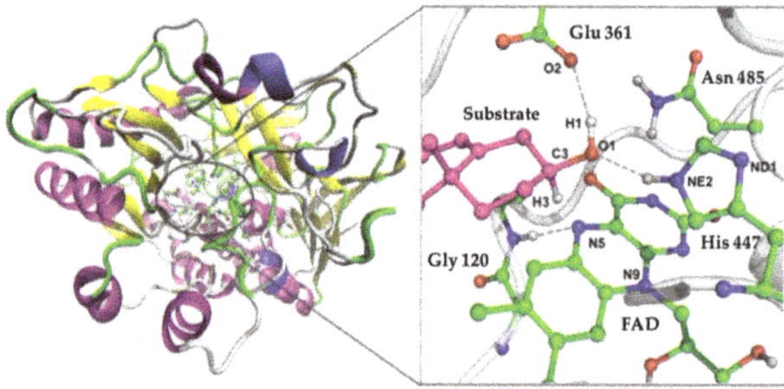

Figure 3. Structure of the cholesterol oxidase enzyme (pdb ID: 4U2T) with a close-up view of the active site showing the key catalytic residues. The substrate is shown in magenta and the dashed lines indicate the H-bond interactions. Atomic color designation: H, white; C, green; O, red; and N, blue (Adapted from Yu et al., 2017 [72]).

The third step in the conversion of 4-cholesten-3-one to coprostanone includes a reduction of the Δ4 double bond into 3-oxo-Δ4-steroid by microbial 3-oxo-Δ4-steroid 5β-reductase (EC 1.3.1.3). Purification and activity monitoring of this reductase was performed initially in cecal contents from rats [68]. This process was found to involve a partial transfer of hydrogen from the 4β-position of NADH (Nicotinamide adenine dinucleotide) to the 5β-position of the steroid [73].

Based on transcriptomic studies using *Bifidobacterium bifidum* PRL2010, Zanotti et al. (2015) reported the up-regulation of three genes encoding for a putative cholesterol reductase in the presence of cholesterol [74]. The in silico study of these genes demonstrate that they display significant similarities with human cholesterol reductase (34% similar), thus, highlighting their potential involvement in the conversion of cholesterol to coprostanol [74]. A comparative study showed that all *Bifidobacterium* genomes contained genes 26% to 100% similar to *Bifidobacterium bifidum* reductase PRL2010. [74]. Recently, we analyzed the *Eubacterium* genomes, and no significant similarities were found with reductase genes [75]. The same study demonstrated that the *Bacteroides* sp. D8 genome showed 37% to 46% similarity with a cholesterol reductase-encoding gene from *Bifidobacterium bifidum*

PRL2010 [75]. Overall, these results indicate that only few functional studies were focused on reductase, explaining the lack of mechanistic knowledge concerning this class of enzyme.

4. Towards a Better Understanding of the Coprostanol Production Pathway

Although cholesterol metabolism by gut microbiota has been known for almost a century, the genes and enzymes involved in the conversion of cholesterol to coprostanol are still largely unknown. Only a few coprostanoligenic bacteria have been isolated so far, and very few cholesterol-metabolizing strains are available [42,45]. Such strains are highly oxygen-sensitive, very difficult to grow, and seem to be non-dominant species within the human gut microbiota of high converters. Isolation of new cholesterol-metabolizing strains from the gut microbiota, and other ecological niches, will be of great interest to identify the genes involved in the reduction of cholesterol to coprostanol. Many approaches aiming at identifying bacterial functions are available, including activity-based screening and phenotypic assays [48]. With the availability of sequenced bacterial genomes and the diversity of coprostanol producing bacteria, comparative genomic approaches may also enable the identification of genes shared among coprostanoligenic microbial species [74]. The analysis of up-regulated genes in the presence of cholesterol using omic approaches can lead to the identification of candidate genes. Furthermore, functional metagenomics is a powerful way to characterize coprostanoligenic pathways from uncultured and metagenomic species. High throughput coprostanol analysis, coupled with targeted mutagenesis and genomic/metagenomic libraries, may provide additional mechanistic insights [76]. An in-depth, functional profiling of metagenomic data combined with biochemical insights may also help to achieve this goal.

5. Conclusions

Several reports stressed the key role of gut microbiota in cholesterol metabolism, essentially through the conversion of cholesterol to coprostanol. Microbial metabolic pathways remain poorly understood, as well as their relevance and distribution in the human gut microbiome. Significant efforts have been made to isolate new cholesterol-metabolizing bacteria and analyze their genomes. Unfortunately, these strategies did not allow the identification of genes responsible for coprostanol synthesis. Therefore, a critical need exists to set out new approaches and/or tools to analyze these microbial metabolic pathways and their associated genes. Deciphering the biochemical and structural features of the encoded proteins will be of interest to understand the properties of this reduction reaction and its importance for bacteria.

Author Contributions: A.K., M.B., H.M., A.J., N.A., S.S., E.M. and M.R. conceived the scientific ideas and designed the work; A.K., H.M., S.S., M.B. and M.R. performed the literature research and provided the first draft; A.K., A.J., N.A. and M.R. generated the figures; A.K., M.B., H.M., A.J., N.A., S.S., E.M. and M.R. discussed the work and edited the manuscript; E.M. and M.R. supervised the project. All authors reviewed the manuscript and provided critical feedbacks.

Funding: This work received a funding from the Microbiology and the Food Chain division (MICA) of the INRA institute through the metaprogramme MEM—Meta-omics and microbial ecosystems. AK, HM, NA, SS and AJ were supported by the project CMCU-PHC Utique (n°14G0816)-Campus France (n°30666QM).

Acknowledgments: The authors would like to express their gratitude to S. Bolotine for his helpful discussion concerning the analysis of genomic sequences of coprostanoligenic bacteria.

Conflicts of Interest: The authors declare no conflicts of interest.

References

1. Benjamin, E.J.; Virani, S.S.; Callaway, C.W.; Chamberlain, A.M.; Chang, A.R.; Rosamond, W.D.; Sampson, K.A.; Satou, G.M.; Shah, S.H.; Spartano, N.L.; et al. Heart Disease and Stroke Statistics—2018 Update: A Report from the American Heart Association. *Circulation* **2018**, *137*, e67–e492. [CrossRef] [PubMed]
2. Krobot, K.J.; Yin, D.D.; Alemao, E.; Steinhagen-Thiessen, E. Realworld effectiveness of lipid-lowering therapy in male and female outpatients with CHD: Relation to pre-treatment LDL-cholesterol, pre-treatment CHD risk and other factors. *Eur. J. Cardiovasc. Prev. Rehabil.* **2005**, *1*, 37–450. [CrossRef]
3. Groen, A.K.; Bloks, V.W.; Verkade, H.; Kuipers, F. Cross-talk between liver and intestine in control of cholesterol and energy homeostasis. *Mol. Aspects Med.* **2014**, *37*, 77–88. [CrossRef]
4. Lee, S.D.; Gershkovich, P.; Darlington, J.W.; Wasan, K.M. Inhibition of cholesterol absorption: Targeting the intestine. *Pharm. Res.* **2012**, *29*, 3235–3250. [CrossRef] [PubMed]
5. Chugh, A.; Ray, A.; Gupta, J.B. Squalene epoxidase as hypocholesterolemic drug target revisited. *Prog. Lipid Res.* **2003**, *42*, 37–50. [CrossRef]
6. Goldstein, J.L.; Brown, M.S. A century of cholesterol and coronaries: From plaques to genes to statins. *Cell* **2015**, *161*, 161–172. [CrossRef] [PubMed]
7. Hou, R.; Goldberg, A.C. Lowering low-density lipoprotein cholesterol: Statins, ezetimibe, bile acid sequestrants, and combinations: Comparative efficacy and safety. *Endocrinol. Metab. Clin. N. Am.* **2009**, *38*, 79–97. [CrossRef]
8. Gérard, P. Metabolism of cholesterol and bile acids by the gut microbiota. *Pathogens* **2014**, *3*, 14–24. [CrossRef]
9. Bhattacharyya, A.K. Differences in uptake and esterification of saturated analogues of cholesterol by rat small intestine. *Am. J. Physiol.* **1986**, *251*, 495–500. [CrossRef]
10. Sekimoto, H.; Shimada, O.; Makanishi, M.; Nakano, T.; Katayama, O. Interrelationship between serum and fecal sterols. *Jpn. J. Med.* **1983**, *22*, 14–20. [CrossRef]
11. Reddy, B.S.; Martin, C.W.; Wynder, E.L. Fecal bile acids and cholesterol metabolites of patients with ulcerative colitis, a high-risk group for development of colon cancer. *Cancer Res.* **1977**, *37*, 1697–1701. [PubMed]
12. Nomura, A.M.; Wilkins, T.D.; Kamiyama, S.; Heilbrun, L.K.; Shimada, A.; Stemmermann, G.N.; Mower, H.F. Fecal neutral steroids in two Japanese populations with different colon cancer risks. *Cancer Res.* **1983**, *43*, 1910–1913. [PubMed]
13. Dam, H. Historical introduction to cholesterol. In *Chemistry, Biochemistry and Pathology*; Cooked, R.P., Ed.; Academic Press: New York, NY, USA, 1958; pp. 1–14.
14. Liscum, L.; Underwood, K.W. Intracellular cholesterol transport and compartmentation. *J. Biol. Chem.* **1995**, *270*, 15443–15446. [CrossRef] [PubMed]
15. Simons, K.; Ikonene, E. How cells handle cholesterol. *Science* **2000**, *290*, 1721–1725. [CrossRef] [PubMed]
16. Sheng, R.; Chen, Y.; Yung Gee, H.; Stec, E.; Melowic, H.R.; Blatner, N.R.; Tun, M.P.; Kim, Y.; Källberg, M.; Fujiwara, T.K.; et al. Cholesterol modulates cell signaling and protein networking by specifically interacting with PDZ domain-containing scaffold proteins. *Nat. Commun.* **2012**, *3*, 1249. [CrossRef] [PubMed]
17. Ohvo-Rekilä, H.; Ramstedt, B.; Leppimäki, P.; Slotte, J.P. Cholesterol interactions with phospholipids in membranes. *Prog. Lipid Res.* **2002**, *41*, 66–97. [CrossRef]
18. Róg, T.; Pasenkiewicz-Gierula, M.; Vattulainen, I.; Karttunen, M. Ordering effects of cholesterol and its analogues. *Biochim. Biophys. Acta* **2009**, *1788*, 97–121. [CrossRef]
19. Do, T.Q.; Moshkani, S.; Castillo, P.; Anunta, S.; Pogosyan, A.; Cheung, A.; Marbois, B.; Faull, K.F.; Ernst, W.; Chiang, S.M.; et al. Lipids including cholesteryl linoleate and cholesteryl arachidonate contribute to the inherent antibacterial activity of human nasal fluid. *J. Immunol.* **2008**, *181*, 4177–4187. [CrossRef]
20. Tiangang, L.; Chiang, Y.L. Regulation of bile acid and cholesterol metabolism by PPARs. *PPAR Res.* **2009**, *2009*, 501739.
21. Walker, R.W.; Wun, C.; Litsky, W. Coprostanol as indicator of faecal pollution. *Crit. Rev. Env. Control* **1982**, *12*, 91–112. [CrossRef]
22. Singley, J.E.; Kirchmer, C.J.; Miura, R. *Analysis of Coprostanol, an Indicator of Fecal Contamination*; Environmental Protection Agency, Office of Research and Development: Washington, DC, USA, 1974; pp. 4–8.
23. Christie, W.W. *Lipid Analysis: Isolation, Separation, Identification and Structural Analysis of Lipids*, 2nd ed.; Pergamon Press: Oxford, UK, 1982.

24. Saad, H.Y.; Higuchi, W.I. Water solubility of cholesterol. *J. Pharm. Sci.* **1965**, *54*, 1205–1206. [CrossRef] [PubMed]
25. Kanazawa, A.; Teshima, S.I. The occurrence of coprostanol, an indicator of fecal pollution in seawater and sediments. *Oceanol. Acta* **1978**, *1*, 39–44.
26. Freier, T.A. Isolation and Characterization of Unique Cholesterol-Reducing Anaerobes. Ph.D. Thesis, Iowa State University, Ames, IA, USA, 1991.
27. Li, L.; Buhman, K.K.; Hartman, P.A.; Beitz, D.C. Hypocholesterolemic effect of *Eubacterium coprostanoligenes* ATCC 51222 in rabbits. *Lett. Appl. Microbiol.* **1995**, *20*, 137–140. [CrossRef] [PubMed]
28. Li, L.; Batt, S.M.; Wannemuehler, M.; Dispirito, A.; Beitz, D.C. Effect of feeding of a cholesterol-reducing bacterium, *Eubacterium coprostanoligenes*, to germ-free mice. *Lab. Anim. Sci.* **1998**, *48*, 253–255. [PubMed]
29. Iqbal, J.; Hussain, M.M. Intestinal lipid absorption. *Am. J. Physiol. Endocrinol. Metab.* **2009**, *296*, E1183–E11894. [CrossRef] [PubMed]
30. Swell, L.; Troutec, J.; Hopper, J.R.; Fieldh, J.; Treadwell, C.D. Mechanism of cholesterol absorption. II. Changes in free and esterified cholesterol pools of mucosa after feeding cholesterol-4-C14. *J. Biol. Chem.* **1958**, *233*, 49–53. [PubMed]
31. Kellogg, T.F. On the site of the microbiological reduction of cholesterol to coprostanol in the rat. *Lipids* **1973**, *8*, 658–659. [CrossRef]
32. Setty, C.S.; Ivy, A.C. Intestinal absorption of coprostanol (coprosterol) in the rat. *Am. J. Physiol.* **1960**, *199*, 1008–1010. [CrossRef]
33. Wells, W.; Anderson, S.A.; Ma, Q. Lactose diets and cholesterol metabolism. I. Cholesterol absorption, coprostanol formation and bile acid excretion in the rat. *J. Nutr.* **1960**, *71*, 405–410. [CrossRef]
34. Rosenfeld, R.S.; Zumoff, B.; Hellman, L. Metabolism of coprostanol-C14 and cholestanol-4-C14 in man. *J. Lipid Res.* **1963**, *4*, 337–340.
35. Schoenheimer, R. New contributions in sterol metabolism. *Science* **1931**, *74*, 579–584. [CrossRef]
36. Koppel, N.; Maini Rekdal, V.; Balskus, E.P. Chemical transformation of xenobiotics by the human gut microbiota. *Science* **2017**, *356*, eaag2770. [CrossRef] [PubMed]
37. Kellogg, T.F. Steroid balance and tissue cholesterol accumulation in germfree and conventional rats fed diets containing saturated and polyunsaturated fats. *J. Lipid Res.* **1974**, *15*, 574–579.
38. Eyssen, H.J.; Parmentier, G.G.; Compernolle, F.C.; De Pauw, G.; Piessens-Denef, M. Biohydrogenation of sterols by *Eubacterium* ATCC 21408 nova species. *Eur. J. Biochem.* **1973**, *36*, 411–421. [CrossRef] [PubMed]
39. Brinkley, A.W.; Gottesman, A.R.; Mott, G.E. Growth of cholesterol-reducing *Eubacterium* on cholesterol-brain agar. *Appl. Environ. Microbiol.* **1980**, *40*, 1130–1132. [PubMed]
40. Brinkley, A.W.; Gottesman, A.R.; Mott, G.E. Isolation and characterization of new strains of cholesterol-reducing bacteria from baboons. *Appl. Environ. Microbiol.* **1982**, *43*, 86–89. [PubMed]
41. Freier, T.A.; Beitz, D.C.; Li, L.; Hartman, P.A. Characterization of *Eubacterium coprostanoligenes* sp. nov., a cholesterol-reducing anaerobe. *Int. J. Syst. Bacteriol.* **1994**, *44*, 137–142. [CrossRef]
42. Li, L. Characterization and Application of a Novel Cholesterol-Reducing Anaerobe, Eubacterium coprostanoligenes ATCC 51222. Ph.D. Thesis, Iowa State University, Ames, IA, USA, 1995.
43. Mott, G.E.; Brinkley, A.W. Plasmenylethanolamine: Growth factor for cholesterol reducing *Eubacterium*. *J. Bacteriol.* **1979**, *139*, 755–760. [PubMed]
44. Sadzikowski, M.R.; Sperry, J.F.; Wilkins, T.D. Cholesterol-reducing bacterium from human feces. *Appl. Environ. Microbiol.* **1977**, *34*, 355–362.
45. Gérard, P.; Lepercq, P.; Leclerc, M.; Gavini, F.; Raibaud, P.; Juste, C. *Bacteroides* sp. strain D8, the first cholesterol-reducing bacterium isolated from human feces. *Appl. Environ. Microbiol.* **2007**, *73*, 5742–5749. [CrossRef]
46. Snog-Kjaer, A.; Prange, I.; Dam, H. Conversion of cholesterol into coprosterol by bacteria in vitro. *J. Gen. Microbiol.* **1956**, *14*, 256–260. [CrossRef] [PubMed]
47. Crowther, J.S.; Drasar, B.S.; Goddard, P.; Hill, M.J.; Johnson, K. The effect of a chemically defined diet on the faecal flora and faecal steroid concentration. *Gut* **1973**, *14*, 790–793. [CrossRef] [PubMed]
48. Lye, H.S.; Rusul, G.; Liong, M.T. Removal of cholesterol by lactobacilli via incorporation and conversion to coprostanol. *J. Dairy Sci.* **2010**, *93*, 1383–1392. [CrossRef] [PubMed]

49. Antharam, V.C.; McEwen, D.C.; Garrett, T.J.; Dossey, A.T.; Li, E.C.; Kozlov, A.N.; Mesbah, Z.; Wang, G.P. An Integrated metabolomic and microbiome analysis identified specific gut microbiota associated with fecal cholesterol and coprostanol in *Clostridium difficile* Infection. *PLoS ONE* **2016**, *11*, e0148824. [CrossRef] [PubMed]
50. Midtvedt, A.C.; Midtvedt, T. Conversion of cholesterol to coprostanol by the intestinal microflora during the first two years of human life. *J. Pediatr. Gastroenterol. Nutr.* **1993**, *17*, 161–168. [CrossRef] [PubMed]
51. Wilkins, T.D.; Hackman, A.S. Two patterns of neutral steroid conversion in the feces of normal North Americans. *Cancer Res.* **1974**, *34*, 2250–2254. [PubMed]
52. Veiga, P.; Juste, C.; Lepercq, P.; Saunier, K.; Beguet, F.; Gérard, P. Correlation between faecal microbial community structure and cholesterol-to-coprostanol conversion in the human gut. *FEMS Microbiol. Lett.* **2005**, *242*, 81–86. [CrossRef]
53. Macdonald, I.A.; Bokkenheuser, V.D.; Winter, J.; McLernon, A.M.; Mosbach, E.H. Degradation of steroids in the human gut. *J. Lipid Res.* **1983**, *24*, 675–700.
54. Gérard, P.; Béguet, F.; Lepercq, P.; Rigottier-Gois, L.; Rochet, V.; Andrieux, C.; Juste, C. Gnotobiotic rats harboring human intestinal microbiota as a model for studying cholesterol-to-coprostanol conversion. *FEMS Microbiol. Ecol.* **2004**, *47*, 337–343. [CrossRef]
55. Rosenfeld, R.S.; Fukushima, D.K.; Hellman, L.; Gallagher, T.F. The transformation of cholesterol to coprostanol. *J. Biol. Chem.* **1954**, *211*, 301–311.
56. Rosenfeld, R.S.; Hellman, L.; Gallagher, T.F. The transformation of cholesterol-3d to coprostanol-d. Location of deuterium in coprostanol. *J. Biol. Chem.* **1956**, *222*, 321–323.
57. Mott, G.E.; Brinkley, A.W.; Mersinger, C.L. Biochemical characterization of cholesterol-reducing Eubacterium. *Appl. Environ. Microbiol.* **1980**, *40*, 1017–1022. [PubMed]
58. Cuevas-Tena, M.; Alegría, A.; Lagarda, M.J. Relationship Between Dietary Sterols and Gut Microbiota: A Review. *Eur. J. Lipid Sci. Technol.* **2018**, *120*, 1800054. [CrossRef]
59. Rosenfeld, R.S.; Gallagher, T.F. Further studies of the biotransformation of cholesterol to coprostanol. *Steroids* **1964**, *4*, 515–520. [CrossRef]
60. Björkhem, I.; Gustafsson, J.A. Mechanism of microbial transformation of cholesterol into coprostanol. *Eur. J. Biochem.* **1971**, *21*, 428–432. [CrossRef] [PubMed]
61. Parmentier, G.; Eyssen, H. Mechanism of biohydrogenation of cholesterol to coprostanol by *Eubacterium* ATCC 21408. *Biochim. Biophys. Acta* **1974**, *348*, 279–284. [CrossRef]
62. Ren, D.; Li, L.; Schwabacher, A.W.; Young, J.W.; Beitz, D.C. Mechanism of cholesterol reduction to coprostanol by *Eubacterium coprostanoligenes* ATCC 51222. *Steroids* **1996**, *61*, 33–40. [CrossRef]
63. García, J.L.; Uhía, I.; Galán, B. Catabolism and biotechnological applications of cholesterol degrading bacteria. *Microb. Biotechnol.* **2012**, *5*, 679–699. [CrossRef]
64. Vrielink, A.; Ghisla, S. Cholesterol oxidase: Biochemistry and structural features. *FEBS J.* **2009**, *276*, 6826–6843. [CrossRef]
65. Cheetham, P.S.; Dunnill, P.; Lilly, M.D. The characterization and interconversion of three forms of cholesterol oxidase extracted from *Nocardia rhodochrous*. *Biochem. J.* **1982**, *201*, 515–521. [CrossRef]
66. Tomioka, H.; Kagawa, M.; Nakamura, S. Some enzymatic properties of 3beta-hydroxysteroid oxidase produced by *Streptomyces violascens*. *J. Biochem.* **1976**, *79*, 903–915. [CrossRef] [PubMed]
67. Fukuyama, M.; Miyake, Y. Purification and some properties of cholesterol oxidase from Schizophyllum commune with covalently bound flavin. *J. Biochem.* **1979**, *85*, 1183–1193. [PubMed]
68. Sojo, M.; Bru, R.; Lopez-Molina, D.; Garcia-Carmona, F.; Arguelles, J.C. Cell-linked and extracellular cholesterol oxidase activities from *Rhodococcus erythropolis* isolation and physiological characterization. *Appl. Microbiol. Biotechnol.* **1997**, *47*, 583–589. [CrossRef] [PubMed]
69. Vrielink, A.; Lloyd, L.F.; Blow, D.M. Crystal structure of cholesterol oxidase from *Brevibacterium sterolicum* refined at 1.8 Å resolution. *J. Mol. Biol.* **1991**, *219*, 533–554. [CrossRef]
70. Sampson, N.S.; Kwak, S. Catalysis at the membrane interface: Cholesterol oxidase as a case study. In Proceedings of the 3rd International Symposium on Experimental Standard Conditions of Enzyme Characterizations (ESCEC), Rudesheim Rhein, Germany, 23–26 September 2008; pp. 13–22.
71. Joseph, K.; Sampson, N.S. Cholesterol oxidase: Physiological functions. *FEBS J.* **2009**, *276*, 6844–6856.

72. Yu, L.J.; Golden, E.; Chen, N.; Zhao, Y.; Vrielink, A.; Karton, A. Computational insights for the hydride transfer and distinctive roles of key residues in cholesterol oxidase. *Sci. Rep.* **2017**, *7*, 17265. [CrossRef] [PubMed]
73. Björkhem, I.; Gustafsson, J.A.; Wrange, O. Microbial transformation of cholesterol into coprostanol. Properties of a 3-oxo-4-steroid-5 beta-reductase. *Eur. J. Biochem.* **1973**, *37*, 143–147. [CrossRef] [PubMed]
74. Zanotti, I.; Turroni, F.; Piemontese, A.; Mancabelli, L.; Milani, C.; Viappiani, A.; Prevedini, G.; Sanchez, B.; Margolles, A.; Elviri, L.; et al. Evidence for cholesterol-lowering activity by *Bifidobacterium bifidum* PRL2010 through gut microbiota modulation. *Appl. Microbiol. Biotechnol.* **2015**, *99*, 6813–6829. [CrossRef]
75. Bolotine, A.; (INRA-paris-Saclay University, Jouy-en-Josas, France). Personal communication, 2017.
76. Lynch, A.; Crowley, E.; Casey, E.; Cano, R.; Shanahan, R.; McGlacken, G.; Marchesi, J.R.; Clarke, D.J. The Bacteroidales produce an N-acylated derivative of glycine with both cholesterol-solubilising and hemolytic activity. *Sci. Rep.* **2017**, *7*, 13270. [CrossRef]

Article

Tailored Enzymatic Synthesis of Chitooligosaccharides with Different Deacetylation Degrees and Their Anti-Inflammatory Activity

P. Santos-Moriano [1,2], P. Kidibule [3], N. Míguez [1], L. Fernández-Arrojo [1], A.O. Ballesteros [1], M. Fernández-Lobato [3] and F.J. Plou [1,*]

1 Instituto de Catálisis y Petroleoquímica, CSIC, 28049 Madrid, Spain; palomacarmen.santos@universidadeuropea.es (P.S.-M.); noa.miguez@csic.es (N.M.); lucia@icp.csic.es (L.F.-A.); a.ballesteros@icp.csic.es (A.O.B.)

2 Applied Biotechnology Group, Faculty of Biomedical and Health Sciences, Universidad Europea de Madrid, Villaviciosa de Odón, 28670 Madrid, Spain

3 Centro de Biología Molecular Severo Ochoa, CSIC-UAM, 28049 Madrid, Spain; pkidibule@cbm.csic.es (P.K.); mfernandez@cbm.csic.es (M.F.-L.)

* Correspondence: fplou@icp.csic.es

Received: 5 April 2019; Accepted: 28 April 2019; Published: 30 April 2019

Abstract: By controlled hydrolysis of chitosan or chitin with different enzymes, three types of chitooligosaccharides (COS) with MW between 0.2 and 1.2 kDa were obtained: fully deacetylated (*fd*COS), partially acetylated (*pa*COS), and fully acetylated (*fa*COS). The chemical composition of the samples was analyzed by high-performance anion exchange chromatography with pulsed amperometric detection (HPAEC-PAD) and MALDI-TOF mass spectrometry. The synthesized *fd*COS was basically formed by GlcN, $(GlcN)_2$, $(GlcN)_3$, and $(GlcN)_4$. On the contrary, *fa*COS contained mostly GlcNAc, $(GlcNAc)_2$ and $(GlcNAc)_3$, while *pa*COS corresponded to a mixture of at least 11 oligosaccharides with different proportions of GlcNAc and GlcN. The anti-inflammatory activity of the three COS mixtures was studied by measuring their ability to reduce the level of TNF-α (tumor necrosis factor) in murine macrophages (RAW 264.7) after stimulation with a mixture of lipopolysaccharides (LPS). Only *fd*COS and *fa*COS were able to significantly reduce the production of tumor necrosis factor (TNF)-α at 6 h after stimulation with lipopolysaccharides.

Keywords: biocatalysis; glycosidases; chitinases; chitosanases; chitosan oligosaccharides; deacetylation degree; anti-inflammatory

1. Introduction

Chitin $[(C_8H_{13}O_5N)_n]$ is a linear biopolymer of N-acetyl-D-glucosamine (GlcNAc) moieties that gives toughness to the exoskeleton of arthropods (crustaceans, insects, etc.) and mollusks, as well as fungi cell walls [1,2]. The hydrolysis of chitin (and of its deacetylated product chitosan, more soluble than chitin) yields a series of chitooligosaccharides (COS) containing random GlcNAc and D-glucosamine (GlcN) units [3].

Three families of COS can be differentiated (Figure 1): fully acetylated chitooligosaccharides (*fa*COS) (formed exclusively by GlcNAc), partially acetylated chitooligosaccharides (*pa*COS) (composed of GlcN and GlcNAc), and fully deacetylated chitooligosaccharides (*fd*COS) (formed exclusively by GlcN) [4]. The bioactivity of COS is well reported [5–7], in particular their anti-inflammatory [8], neuroprotective [9], antibacterial [10], antiviral [11], antihypertensive [12] antiangiogenic [13], and antitumor [14] properties, among others. The size of COS (defined by the degree of polymerization, DP), degree of deacetylation (DD) and pattern of acetylation (PA) exert a notable influence on their properties [4,13,15,16].

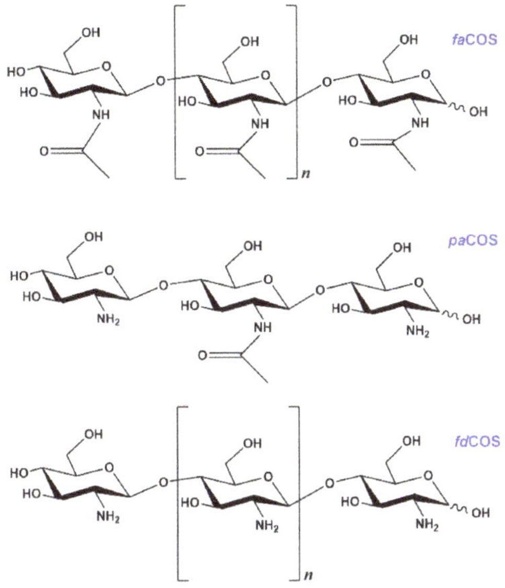

Figure 1. Structure of the three main types of chitooligosaccharides (COS): fully acetylated (*fa*COS), partially acetylated (*pa*COS), and fully deacetylated (*fd*COS).

Among the methodologies to perform the partial hydrolysis of chitin/chitosan into COS, the use of chitinolytic/chitosanolytic enzymes offer some advantages over physical, chemical, or electrochemical depolymerization [17]. Glycosidic enzymes require mild reaction conditions (moderate temperature and slightly acidic pH), display high efficiency, and allow for control of the composition of the final product on the basis of enzyme specificity [4,18–22], which can also be altered by protein engineering techniques [23,24]. The enzymatic strategies are more environmentally friendly and generate less waste than the chemical methods [3,25,26]. The physicochemical properties of the starting chitosan also influence the composition of the resulting oligosaccharides [27,28].

For the hydrolysis of chitosan, chitosanases (EC 3.2.1.132) catalyze specifically the cleavage of β (1→4) glycosidic linkages between GlcN moieties [29–32]. However, other enzymes such as pectinases [33], cellulases [34], and proteases [35] also display chitosanolytic activity yielding COS. We have recently reported that a proteolytic preparation from *Bacillus amyloliquefaciens* (Neutrase 0.8L) is able to produce a mixture of COS that is highly enriched in *fd*COS [18].

Chitinases (EC 3.2.1.14) are hydrolytic enzymes involved in chitin decomposition that play an important role as control agents against pathogenic fungi in plants [36–38]. In recent research, we cloned chitinase Chit42 from fungus *Trichoderma harzianum* in *Pichia pastoris* to produce 3 g/L using fed-batch fermentation, and its 3D structure was characterized [39]. This enzyme hydrolyzed chitin and chitosan with a low DD giving rise to mixtures enriched in *fa*COS and *pa*COS, respectively. In general, the binding site of chitinases of fungal origin is substantially long and interacts with a minimum of five sugar units. The glycosyl-binding subsites are designated as −3, − 2, −1, +1, and +2, and the split occurs between the −1 and +1 sugar. A detailed structural analysis of Chit42 indicated that this protein requires a GlcNAc residue in the substrate located at the −1 position for substrate hydrolysis.

In the present work, we have synthesized three COS mixtures enriched in *fd*COS, *pa*COS, and *fa*COS. The samples were chemically characterized by chromatography and mass spectrometry. The effect of COS composition on anti-inflammatory properties was studied using a murine macrophage cell line (RAW 264.7). Although the anti-inflammatory activity of chitosan oligosaccharides is well reported [40–43], most of the works have been performed with COS mixtures not fully characterized in terms of DP, DD, or PA, probably due to the difficulties in the controlled synthesis and analysis of COS mixtures.

2. Results and Discussion

2.1. Enzymatic Production and Characterization of fdCOS

Based on previous work [18], we selected the commercial proteolytic preparation Neutrase 0.8L and a chitosan (CHIT600) with a high DD (>90%) to scale up the production of fully deacetylated COS (*fd*COS). The reaction was carried out over 24 h with 1% (w/v) chitosan at pH 5.0 and 50 °C. After this time, the enzyme and the remaining chitosan were eliminated by ultrafiltration with a 10 kDa membrane. Then, the COS of high molecular weight were removed using a 1 kDa cut-off membrane. The resulting solution was dialyzed over 0.1–0.5 kDa cut-off tubing to eliminate the salts and other small contaminants, yielding a COS fraction with a molecular mass between 0.2 and 1.2 kDa. This fraction was further freeze-dried and characterized by high-performance anion exchange chromatography with pulsed amperometric detection (HPAEC-PAD) chromatography and MALDI-TOF mass spectrometry.

The HPAEC-PAD chromatogram of the COS obtained with Neutrase 0.8 L—purified as described above—is represented in Figure 2. The chromatogram shows five main peaks corresponding to *fd*COS, which was identified with the corresponding standards as GlcN (1), $(GlcN)_2$ (2), $(GlcN)_3$ (3), $(GlcN)_4$ (4), and $(GlcN)_5$ (5). In particular, chitobiose [$(GlcN)_2$] and chitotriose [$(GlcN)_3$] were the major products. Peaks marked with asterisks were not identified due to the lack of available standards.

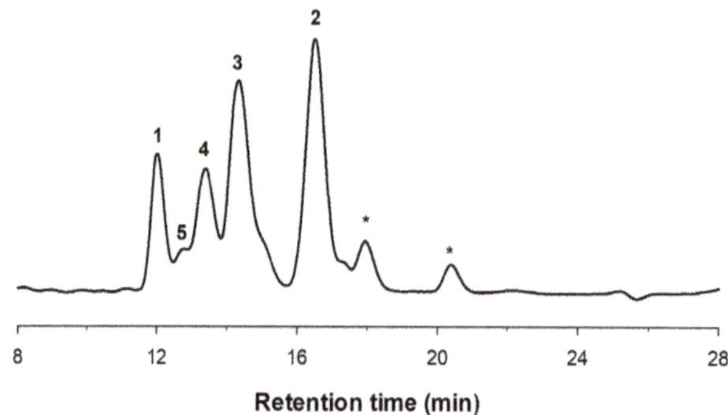

Figure 2. High-performance anion exchange chromatography with pulsed amperometric detection (HPAEC-PAD) chromatogram of the chitooligosaccharides produced by Neutrase 0.8 L using chitosan CHIT600. Reaction conditions: 1% (w/v) chitosan, 10% (v/v) Neutrase 0.8 L, 50 °C, 50 mM ammonium acetate buffer pH 5.0, 24 h. Identified peaks: (1) GlcN; (2) $(GlcN)_2$; (3) $(GlcN)_3$; (4) $(GlcN)_4$; (5) $(GlcN)_5$.

The MALDI-TOF spectrum of the COS mixture was in accordance with the chromatographic analysis since the main *m/z* peaks corresponded to the molecular weight of the *fd*COS. The main signals in the mass spectrum in positive mode belonged to the M + $[Na]^+$ and M + $[K]^+$ cations. Table 1 compiles the major *m/z* signals and the assigned composition. Several *m/z* values in agreement with partially acetylated COS (*pa*COS) appeared in the MS spectrum, with significantly lower intensity than the *fd*COS. These *pa*COS probably corresponded to the unidentified peaks in the HPAEC-PAD analysis (Figure 2). However, their chemical structure could not be unequivocally assigned from the obtained data. The deacetylation degree of this *fd*COS sample must be between 95 and 100%.

Table 1. Main identified signals in the MALDI-TOF mass spectrum of the reaction between chitosan CHIT600 and Neutrase 0.8 L. Reaction conditions were as described in Figure 2.

m/z	Assignation
180.0	GlcN + H^+
363.1	$(GlcN)_2$ + Na^+
524.2/540.2	$(GlcN)_3$ + Na^+/K^+
566.2/582.2	$(GlcN)_2$-GlcNAc + Na^+/K^+
685.3/701.3	$(GlcN)_4$ + Na^+/K^+
727.3/743.3	$(GlcN)_3$-GlcNAc + Na^+/K^+
846.3/862.2	$(GlcN)_5$ + Na^+/K^+
888.3/904.3	$(GlcN)_4$-GlcNAc + Na^+/K^+
1023.3	$(GlcN)_6$ + K^+
1049.4	$(GlcN)_5$-GlcNAc + Na^+
1210.4	$(GlcN)_6$-GlcNAc + Na^+

We calculated the efficiency of COS production with Neutrase 0.8 L. Starting from 1 g chitosan CHIT600, and after all the purification steps, approximately 210 mg of COS (mostly fully deacetylated) was obtained.

2.2. Enzymatic Production and Characterization of *fa*COS

For the production of *fa*COS, the first step was the transformation of chitin flakes into colloidal chitin as previously described [39]. Chitinase Chit42 was used for the hydrolysis of chitin into fully acetylated chitooligosaccharides. Figure 3 illustrates the HPAEC-PAD chromatogram of the reaction mixture obtained with Chit42, which was purified as described in the Experimental Section. The presence of the *fa*COS GlcNAc (1), $(GlcNAc)_2$ (2), and $(GlcNAc)_3$ (3) was verified by using the corresponding standards. The deacetylation degree of this *fa*COS sample was between 0 and 5%.

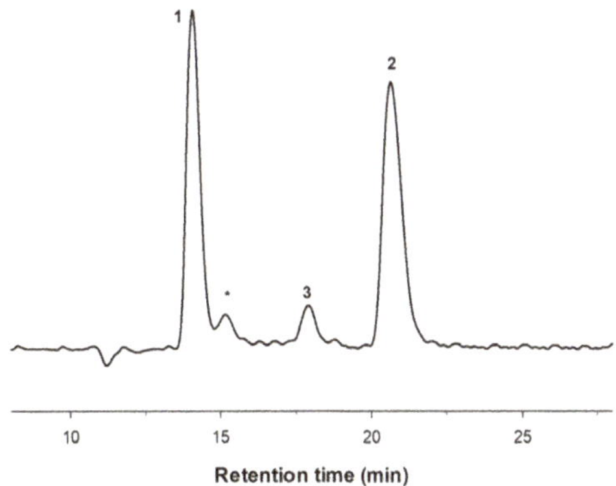

Figure 3. HPAEC-PAD chromatogram of the chitooligosaccharides obtained with chitinase Chit42 employing chitin (colloid). Reaction conditions: 1% (w/v) colloidal chitin, 10% (v/v) chitinase, 70 mM potassium phosphate pH 6.0. Identified peaks: (1) GlcNAc; (2) $(GlcNAc)_2$; (3) $(GlcNAc)_3$.

The MALDI-TOF spectrum of this mixture was simpler than that of *fd*COS. Table 2 summarizes the main m/z peaks and their assignations. The main signal in the mass spectrum corresponded to $(GlcNAc)_2$. It is worth noting that both the monomer GlcNAc and the trimer $(GlcNAc)_3$ did not appear in the MS spectrum, probably due to bad ionization or low stability of the formed ions. In contrast, several peaks containing a GlcN moiety were present, which probably corresponded to the minor peaks detected in the HPAEC-PAD analysis, indicating that GlcN units favor MALDI ionization. Starting from 1 g of colloidal chitin, 75.8 mg of the characterized *fa*COS was obtained.

Table 2. Main identified signals in the MALDI-TOF mass spectrum of the reaction between chitinase Chit42 and colloidal chitin. Reaction conditions were as described in Figure 3.

m/z	Assignation
405.2	GlcN-GlcNAc + Na^+
447.2/463.2	$(GlcNAc)_2$ + Na^+/K^+
608.3/624.2	GlcN-$(GlcNAc)_2$ + Na^+
769.3/785.2	$(GlcN)_2$-$(GlcNAc)_2$ + Na^+/K^+
811.3	GlcN-$(GlcNAc)_3$ + Na^+
853.3	$(GlcNAc)_4$ + Na^+

2.3. *Enzymatic Production and Characterization of paCOS*

Chitosan QS1—with a lower DD (81%) than CHIT600 (>90%)—and chitinase Chit42, which requires a GlcNAc residue at −1 position [39], were used for the preparation of partially acetylated chitooligosaccharides (*pa*COS). The analysis of this family of COS is quite difficult due to the lack of commercial standards. Figure 4 illustrates the HPAEC-PAD chromatogram of the resulting mixture. At least 11 unidentified peaks were detected.

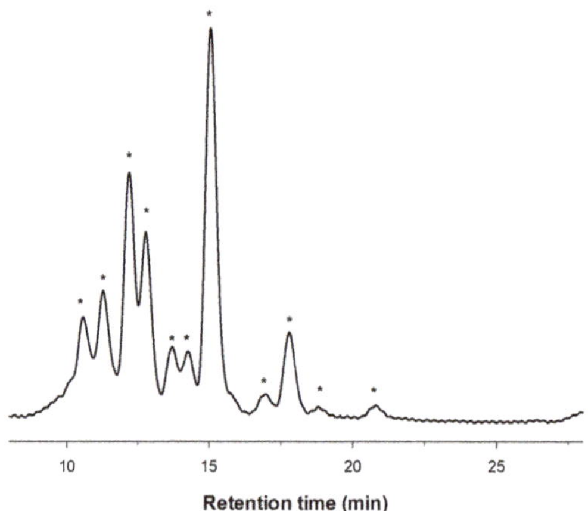

Figure 4. HPAEC-PAD chromatogram of the chitooligosaccharides produced by chitinase Chit42 using chitosan QS1 as substrate. Reaction conditions: 1% (w/v) chitosan, 10% (v/v) chitinase Chit42, 70 mM potassium phosphate pH 6.0.

Table 3 summarizes the main m/z peaks detected in the MALDI-TOF spectrum and the proposed composition. However, the chemical structure of these compounds cannot be inferred from the mass spectrometry data. Since chitinase Chit42 only cleaves chitosan when a GlcNAc residue is located at the −1 position, the synthesized COS should present a GlcNAc at the reducing end. Table 3 includes COS containing up to nine residues with GlcN as the main component, which correlates well with the degree of deacetylation of chitosan QS1. After the purification steps, starting from 500 mg of chitosan QS1, 18 mg of *pa*COS was isolated. The lower yield obtained in comparison with *fd*COS and *fa*COS was probably a consequence of the requirement of Chit42 for a GlcNAc at −1 position and the presence of a high proportion of GlcN (81%) in chitosan QS1.

Table 3. Main identified signals in the MALDI-TOF mass spectrum of the reaction between chitosan QS1 and chitinase Chit42. Reaction conditions were as described in Figure 4.

m/z	Assignation
405.2/421.2	GlcN-GlcNAc + Na^+/K^+
447.2/463.2	$(GlcNAc)_2$ + Na^+/K^+
566.3/582.2	$(GlcN)_2$-GlcNAc + Na^+/K^+
608.3/624.3	GlcN-$(GlcNAc)_2$ + Na^+/K^+
727.3/743.3	$(GlcN)_3$-GlcNAc + Na^+/K^+
769.3/785.3	$(GlcN)_2$-$(GlcNAc)_2$ + Na^+/K^+
811.3/827.3	GlcN-$(GlcNAc)_3$ + Na^+/K^+
888.4/904.3	$(GlcN)_4$-GlcNAc + Na^+/K^+
930.4/946.3	$(GlcN)_3$-$(GlcNAc)_2$ + Na^+/K^+
1049.4/1065.4	$(GlcN)_5$-GlcNAc + Na^+/K^+
1091.4/1107.4	$(GlcN)_4$-$(GlcNAc)_2$ + Na^+/K^+
1133.4/1149.4	$(GlcN)_3$-$(GlcNAc)_3$ + Na^+/K^+
1210.4/1226.4	$(GlcN)_6$-GlcNAc + Na^+/K^+
1252.5/1268.4	$(GlcN)_5$-$(GlcNAc)_2$ + Na^+/K^+
1294.5/1310.4	$(GlcN)_4$-$(GlcNAc)_3$ + Na^+/K^+
1413.5/1429.5	$(GlcN)_6$-$(GlcNAc)_2$ + Na^+/K^+
1532.6/1548.5	$(GlcN)_8$-GlcNAc + Na^+/K^+

2.4. Anti-Inflammatory Activity of *fd*COS, *fa*COS, and *pa*COS

Inflammation plays an important role in the development of a series of pathologies including autoimmune diseases and cancer [44]. The anti-inflammatory activity of the three samples of COS previously obtained was assessed by measuring their ability to reduce the level of TNF-α (tumor necrosis factor) in murine macrophages (RAW 264.7) after stimulation with a mixture of lipopolysaccharides (LPS). TNF-α is a cytokine involved in systemic inflammation and one of the cytokines released by activated macrophages during the acute phase reaction of inflammation [45].

Three concentrations of COS were tested in a multi-well plate: 100, 250, and 500 ng per well. The ELISA methodology for the detection of TNF-α was properly validated through the accuracy of the standard curve obtained, which allowed the quantification of samples with a TNF-α concentration between 30 and 1000 pg/mL. The amount of TNF-α was measured at 2 and 6 h after incubation with LPS (10 ng/well). To discard the possible inflammatory effect of the compounds, the cells were also exposed to the three COS samples in the absence of LPS. However, no significant effect was observed in samples supplemented only with *fd*COS, *fa*COS, or *pa*COS (data not shown). After the stimulations, both the culture supernatants and the cells were collected at 2 and 6 h.

Figure 5 shows the TNF-α concentration in the supernatants after 2 and 6 h stimulated with *fd*COS, *fa*COS, and *pa*COS (at the three concentrations) in combination with 10 ng LPS. The results of the control experiment of cells stimulated only with LPS are also included in the figure. As illustrated in Figure 5, the TNF-α concentration tended to increase over time for most samples. The highest TNF-α concentration (1575 pg/mL) was obtained after 6 h post-stimulation with 10 ng of LPS per well.

The three types of COS were able to decrease the production of TNF-α at 6 h after stimulation with LPS. The highest effect was obtained using 250 ng/well (Figure 5). It is worth noting that *fd*COS exhibited a negligible anti-inflammatory effect at concentrations of 100 and 500 ng/well after a 2-h incubation, but this effect increased significantly at 6 h. This relates well with our preliminary results with *fd*COS [18]. In contrast, *pa*COS (except for 100 ng/well) and *fa*COS displayed a more stable effect between 2 and 6 h. These results could be indicating a critical role of the acetamido group of COS in their properties. However, only the *fd*COS and *fa*COS at a concentration of 250 ng/well and after 6 h incubation displayed a statistically significant anti-inflammatory effect.

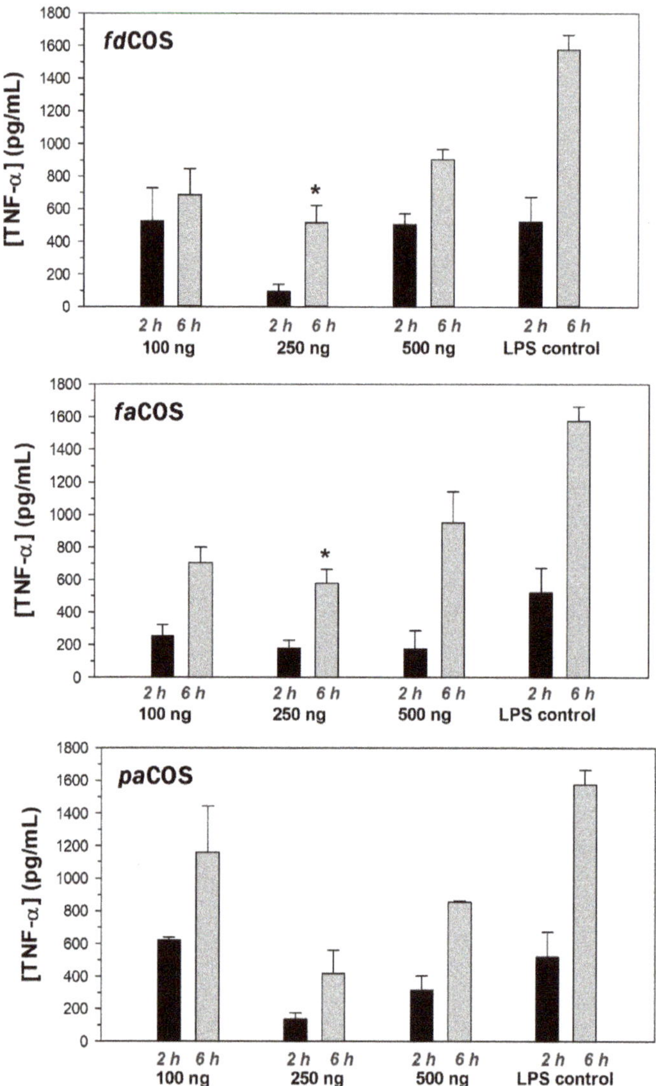

Figure 5. Tumor necrosis factor-alpha (TNF-α) concentration in the supernatants from cells stimulated with fdCOS, faCOS, and paCOS (at 100, 250, and 500 ng per well) in combination with 10 ng lipopolysaccharide (LPS) per well. Graphs show the amount of TNF-α at 2 and 6 h post-stimulation. The data is expressed as the mean ± SD (*$p < 0.05$ vs. LPS control).

These results correlate well with previous works that analyzed the anti-inflammatory activity of COS. Yoon et al. reported the attenuation of secretion of TNF-α and IL-6 induced by LPS upon incubation with COS, demonstrating that the expression of these cytokines was regulated by COS at the transcription level [41]. However, the authors employed a commercial COS that was not fully characterized (MW < 10000; 90–95% DD). The dependence of anti-inflammatory activity on the molecular weight of COS was studied by Fernandes et al. [8] and Pangestuti et al. [46], concluding that COS of a low molecular weight were the most efficient. Sánchez et al. reported that a mixture with a similar content of deacetylated and monoacetylated HMW COS (5–10 kDa) produced the best anti-inflammatory effects [47].

Our results correlate well with those of Lee et al. [48]. They demonstrated that COS with 90% N-deacetylation (90-COS) displayed a higher anti-inflammatory effect than COS with 50% N-deacetylation (50-COS, more related to *pa*COS); interestingly, the 90-COS with a molecular mass between 5 and 10 kDa showed the highest inhibition activity.

In conclusion, the secretion of TNF-α decreased during the first 6 h in macrophages treated with the three COS mixtures in comparison with macrophages stimulated with LPS only. These results confirm the inhibitory effect of COS against inflammation, which confirms their potential as ingredients in functional foods and nutraceutical and pharmaceutical preparations.

3. Materials and Methods

3.1. Enzymes and Reagents

Neutrase 0.8 L was kindly donated by Novozymes A/S (Bagsværd, Denmark). The expression and production of Chit42 (chitinase from *Trichoderma harzianum*) by *Pichia pastoris* was performed as previously described [39]. Chitosan CHIT600 from shrimp shells (600–800 kDa, DD > 90%) was purchased from Acros Organics (Geel, Belgium). Chitosan QS1 from *Paralomis granulosa* (98 kDa, 81% DD) was supplied by InFiQus (Madrid, Spain). Chitin (coarse flakes, DD ≤ 5%) from shrimp shells and N-acetyl-glucosamine (GlcNAc) were from Sigma-Aldrich (Madrid, Spain). Chitobiose [$(GlcN)_2$], chitotriose [$(GlcN)_3$], chitotetraose [$(GlcN)_4$], N,N′-di-N-acetyl-glucosamine [$(GlcNAc)_2$], and N,N′,N″-tri-N-acetyl-glucosamine [$(GlcNAc)_3$] were acquired from Carbosynth Ltd. (Compton, Berkshire, UK). All other reagents were of the highest purity grade available.

3.2. Preparation of Colloidal Chitin

Colloidal chitin was prepared following the method of Jeuniaux [49]. In particular, 10 g of chitin and 175 mL of 10 M HCl were stirred for 16 h at 4 °C. Then, the mixture was filtered using thick glass fibers and mixed with 1 L of ethanol. After 16 h at 4 °C, the precipitated chitin floccules were separated by centrifugation at 5000× g for 10 min and washed with distilled water. Finally, 200 mL of potassium phosphate buffer (70 mM, pH 6.0) was added to the pellet. To determine the concentration of colloidal chitin, 1 mL of solution was frozen at −70 °C, lyophilized, and weighed.

3.3. COS Production and Purification

Chitooligosaccharides with different deacetylation degrees were produced by different combinations of enzymes and substrates based on previous works [18,39]. The reaction conditions for the production of each type of COS are summarized in Table 4.

Table 4. Experimental conditions for the preparation of COS samples.

Enzyme	Substrate	Reaction Conditions	Main Products
Chit42	Colloidal chitin	35 °C, pH 6.0	*fa*COS
Chit42	Chitosan QS1	35 °C, pH 6.0	*pa*COS
Neutrase 0.8 L	Chitosan CHIT600	50 °C, pH 5.0	*fd*COS

Reactions were carried out in a final volume of 40 mL containing 4 mL of enzyme solution and 36 mL of 1% (w/w) substrate dissolved properly in ammonium acetate at the optimal pH for the reaction. The formation of COS was followed by HPAEC-PAD until the hydrolysis was complete. Samples were filtrated through a paper filter to remove any insoluble particles and further purified by a series of membranes. First, the reaction mixture was fractionated using a 50 mL Amicon system with a 10 kDa cut-off membrane. This step separated the enzyme and the unreacted high molecular weight chitosan from the chitooligosaccharides. Then, the fraction of COS (< 10 kDa) was further fractionated with a 1 kDa cut-off membrane. The fraction of COS whose MW was lower than 1 kDa was then dialyzed with a 0.1–0.5 kDa membrane (Biotech Cellulose Ester Dialysis Membrane, Spectra/Por, Fisher Scientific, Madrid, Spain) to remove the salts and small contaminants from the sample, yielding a COS fraction with a MW between 0.1 and 1 kDa. This fraction was lyophilized, analyzed by HPAEC-PAD and MALDI-TOF, and used for bioactivity assays.

3.4. *COS Characterization by HPAEC-PAD and MALDI-TOF*

COS samples were analyzed at 30 °C by HPAEC-PAD on a chromatograph ICS3000 (Dionex, Thermo Fischer Scientific Inc., Waltham, MA, USA) formed by a gradient pump (model SP), an electrochemical detector consisting of a working electrode (Au) and a reference electrode (Ag/AgCl), and a AS-HV autosampler. The column was an anion exchange Carbo-Pack PA-200 (Dionex, 4 × 250 mm) connected to a CarboPac PA-200 guard column (4 × 50 mm). A post-column delivery system (PC10) pumped 200 mM NaOH to enhance the detector response. The mobile phase was 1 mM NaOH at a flow rate of 0.3 mL/min for 20 min, followed by a gradient from 0 to 320 mM sodium acetate/100 mM NaOH in 10 min, that was kept for another 10 min. Equilibration of the column to the initial conditions was made for 40 min. The chromatograms were analyzed using Chromeleon software. The identification and quantification of the different carbohydrates were done based on commercial standards when available.

The molecular size of COS was analyzed by MALDI-TOF mass spectrometry using Ultraflex III TOF/TOF equipment (Bruker, Billerica, MA, USA) equipped with a NdYAG laser. The spectra were acquired in positive reflector mode in the mass interval 40–5000 Da, employing 20 mg/mL 2,5-dihydroxybenzoic acid (DHB) in acetonitrile: H_2O (3:7) (v/v) as matrix and external calibration. The samples were mixed with the DHB matrix in a 4:1 (v/v) ratio and 0.5 μL was injected.

3.5. *Anti-Inflammatory Activity of COS*

RAW 264.7 cells were cultured in DMEM (Dulbecco's Modified Eagle's medium) supplemented with 10% FBS (fetal bovine serum) and 1% penicillin/streptomycin. Cells were counted with a Neubauer chamber in order to seed a concentration of 2 million cells per well. Each of the compounds was tested in duplicate at three different concentrations: 500, 250, and 100 ng/well. A mixture of lipopolysaccharides (LPS, potent immune cell activator) at a concentration of 10 ng per well was used as a positive control, and PBS (phosphate buffered saline) was used as a negative control and added to the wells in the same volume as the rest of the compounds. The cells were also exposed to the three chitooligosaccharides (without LPS) as a control for the inflammatory effect of the compounds. After the stimulations, both the culture supernatants and the cells were collected at 2 and 6 h. Samples were frozen and kept at −80 °C until analysis.

The quantification of anti-TNF-α antibodies in culture supernatants was assessed by ELISA using a murine TNF-α ELISA kit (Diaclone, Besançon, France) following the protocol provided by the manufacturer. Samples were diluted 1:2 (*v/v*) in an appropriate buffer (provided by the kit) and 100 μL was added in duplicate to the plate. Serial dilutions of the standard were made to provide a concentration range from 1000 to 31.25 pg/mL, and 100 μL was added per well in duplicate to the ELISA plate. Biotinylated anti-murine TNF-α antibody was properly diluted according to the manufacturer's protocol and 50 μL was added to each well. Plates were sealed and incubated at room temperature for 3 h. After incubation, plates were washed three times with the wash buffer provided with the kit. Streptavidin-Horseradish Peroxidase (HRP) solution (100 μL) was added for the detection of the biotinylated detection antibody and plates were incubated at room temperature for 30 min. Then, 100 μL of ready-to-use 3,3′,5,5′-Tetramethylbenzidine (TMB) substrate solution was transferred into each well. Plates were incubated in the dark at room temperature (10 min) for color development. To stop the color reaction, 50 μL of 2 N H_2S0_4 was added. The optical density (OD) for each well was measured with a microplate reader set to 450 nm.

Data were expressed as the mean ± standard deviation (SD) with n = 2. Brown–Forsythe test and post-hoc Games–Howell method were used to find differences with respect to the LPS control. Statistical analysis was performed with IBM® SPSS® Statistics v25 and differences were considered significant when $p < 0.05$.

Author Contributions: F.J.P., M.F.-L. and A.O.B. conceived and designed the experiments; P.S.-M. and N.M. performed most of the experiments; P.K. and M.F.-L. contributed with Chit42 production; L.F.-A. contributed with chromatographic methods for COS analysis; F.J.P. and P.S.-M. wrote the paper, which was improved by the rest of the authors.

Funding: This work was supported by the Fundación Ramón Areces (XIX Call of Research Grants in Life and Materials Sciences) and the Spanish Ministry of Economy and Competitiveness (Grants BIO2016-76601-C3-1-R and BIO2016-76601-C3-3-R). We also acknowledge the institutional grant from Ramon Areces Foundation to the Centre of Molecular Biology "Severo Ochoa". The European Union's Horizon 2020 program also financed this work (Blue Growth: Unlocking the potential of Seas and Oceans; grant agreement No. 634486; INMARE). We thank the support of the EU COST-Action CM1303 on Systems Biocatalysis.

Acknowledgments: We thank Laura Córdoba and Miguel Angel Llamas (Diomune S.L.) for their help in the analysis of the anti-inflammatory properties. We thank Ramiro Martinez (Novozymes A/S) for supplying Neutrase 0.8 L and for relevant suggestions. We are also grateful for the support of the publication fee by the CSIC Open Access Publication Support Initiative through its Unit of Information Resources for Research (URICI).

Conflicts of Interest: The authors declare no conflict of interest.

References

1. Hamed, I.; Özogul, F.; Regenstein, J.M. Industrial applications of crustacean by-products (chitin, chitosan, and chitooligosaccharides): A review. *Trends Food Sci. Technol.* **2016**, *48*, 40–50. [CrossRef]
2. Gortari, M.C.; Hours, R.A. Biotechnological processes for chitin recovery out of crustacean waste: A mini-review. *Electron. J. Biotechnol.* **2013**, *16*, 1–14.
3. Kumar, M.; Brar, A.; Vivekanand, V.; Pareek, N. Bioconversion of chitin to bioactive chitooligosaccharides: Amelioration and coastal pollution reduction by microbial resources. *Mar. Biotechnol.* **2018**, *20*, 269–281. [CrossRef] [PubMed]
4. Hamer, S.N.; Cord-Landwehr, S.; Biarnés, X.; Planas, A.; Waegeman, H.; Moerschbacher, B.M.; Kolkenbrock, S. Enzymatic production of defined chitosan oligomers with a specific pattern of acetylation using a combination of chitin oligosaccharide deacetylases. *Sci. Rep.* **2015**, *5*, 8716. [CrossRef] [PubMed]
5. Liaqat, F.; Eltem, R. Chitooligosaccharides and their biological activities: A comprehensive review. *Carbohydr. Polym.* **2018**, *184*, 243–259. [CrossRef]
6. Je, J.Y.; Kim, S.K. Chitooligosaccharides as potential nutraceuticals: production and bioactivities. *Adv. Food Nutr. Res.* **2012**, *65*, 321–336.
7. Xia, W.; Liu, P.; Zhang, J.; Chen, J. Biological activities of chitosan and chitooligosaccharides. *Food Hydrocolloids* **2011**, *25*, 170–179. [CrossRef]
8. Fernandes, J.C.; Spindola, H.; de Sousa, V.; Santos-Silva, A.; Pintado, M.E.; Malcata, F.X.; Carvalho, J.E. Anti-inflammatory activity of chitooligosaccharides in vivo. *Mar. Drugs* **2010**, *8*, 1763–1768. [CrossRef]
9. Jiang, M.; Guo, Z.; Wang, C.; Yang, Y.; Liang, X.; Ding, F. Neural activity analysis of pure chito-oligomer components separated from a mixture of chitooligosaccharides. *Neurosci. Lett.* **2014**, *581*, 32–36. [CrossRef]
10. Wu, S.-J.; Pan, S.-K.; Wang, H.-B.; Wu, J.-H. Preparation of chitooligosaccharides from cicada slough and their antibacterial activity. *Int. J. Biol. Macromol.* **2013**, *62*, 348–351. [CrossRef]
11. Artan, M.; Karadeniz, F.; Karagozlu, M.Z.; Kim, M.M.; Kim, S.K. Anti-HIV-1 activity of low molecular weight sulfated chitooligosaccharides. *Carbohydr. Res.* **2010**, *345*, 656–662. [CrossRef]
12. Huang, R.; Mendis, E.; Kim, S.K. Improvement of ACE inhibitory activity of chitooligosaccharides (COS) by carboxyl modification. *Bioorg. Med. Chem.* **2005**, *13*, 3649–3655. [CrossRef] [PubMed]
13. Wu, H.; Aam, B.B.; Wang, W.; Norberg, A.L.; Sørlie, M.; Eijsink, V.G.H.; Du, Y. Inhibition of angiogenesis by chitooligosaccharides with specific degrees of acetylation and polymerization. *Carbohydr. Polym.* **2012**, *89*, 511–518. [CrossRef]
14. Kim, E.K.; Je, J.Y.; Lee, S.J.; Kim, Y.S.; Hwang, J.W.; Sung, S.H.; Moon, S.H.; Jeon, B.T.; Kim, S.K.; Jeon, Y.J.; et al. Chitooligosaccharides induce apoptosis in human myeloid leukemia HL-60 cells. *Bioorg. Med. Chem. Lett.* **2012**, *22*, 6136–6138. [CrossRef] [PubMed]

15. Mengíbar, M.; Mateos-Aparicio, I.; Miralles, B.; Heras, A. Influence of the physico-chemical characteristics of chito-oligosaccharides (COS) on antioxidant activity. *Carbohydr. Polym.* **2013**, *97*, 776–782. [CrossRef]
16. Yarullina, L.G.; Sorokan, A.V.; Burkhanova, G.F.; Cherepanova, E.A.; Maksimov, I.V. Influence of chitooligosaccharides with different acetylation degrees on the H_2O_2 content and the activity of pathogenesis-related proteins in potato plants infected with *Phytophthora infestans*. *Appl. Biochem. Microbiol.* **2018**, *54*, 528–534. [CrossRef]
17. Liang, S.; Sun, Y.X.; Dai, X.L. A review of the preparation, analysis and biological functions of chitooligosaccharide. *Int. J. Mol. Sci.* **2018**, *19*, 2197. [CrossRef] [PubMed]
18. Santos-Moriano, P.; Fernandez-Arrojo, L.; Mengibar, M.; Belmonte-Reche, E.; Peñalver, P.; Acosta, F.N.; Ballesteros, A.O.; Morales, J.C.; Kidibule, P.; Fernandez-Lobato, M.; et al. Enzymatic production of fully deacetylated chitooligosaccharides and their neuroprotective and anti-inflammatory properties. *Biocatal. Biotransform.* **2018**, *36*, 57–67. [CrossRef]
19. de Araujo, N.K.; de Assis, C.F.; Dos Santos, E.S.; de Macedo, G.R.; de Farias, L.F.; Arimateia, H., Jr.; de Freitas Fernandes Pedrosa, M.; Pagnoncelli, M.G. Production of enzymes by *Paenibacillus chitinolyticus* and *Paenibacillus ehimensis* to obtain chitooligosaccharides. *Appl. Biochem. Biotechnol.* **2013**, *170*, 292–300. [CrossRef] [PubMed]
20. Plou, F.J.; Gómez de Segura, A.; Ballesteros, A. Application of glycosidases and transglycosidases for the synthesis of oligosaccharides. In *Industrial enzymes: Structure, Function and Application*; Polaina, J., MacCabe, A.P., Eds.; Springer: New York, NY, USA, 2007; pp. 141–157.
21. Fernandez-Arrojo, L.; Marin, D.; Gomez de Segura, A.; Linde, D.; Alcalde, M.; Gutierrez-Alonso, P.; Ghazi, I.; Plou, F.J.; Fernandez-Lobato, M.; Ballesteros, A. Transformation of maltose into prebiotic isomaltooligosaccharides by a novel alpha-glucosidase from *Xantophyllomyces dendrorhous*. *Process Biochem.* **2007**, *42*, 1530–1536. [CrossRef]
22. Linde, D.; Rodriguez-Colinas, B.; Estevez, M.; Poveda, A.; Plou, F.J.; Fernandez-Lobato, M. Analysis of neofructooligosaccharides production mediated by the extracellular beta-fructofuranosidase from *Xanthophyllomyces dendrorhous*. *Bioresour. Technol.* **2012**, *109*, 123–130. [CrossRef]
23. Liu, S.; Shao, S.; Li, L.; Cheng, Z.; Tian, L.; Gao, P.; Wang, L. Substrate-binding specificity of chitinase and chitosanase as revealed by active-site architecture analysis. *Carbohydr. Res.* **2015**, *418*, 50–56. [CrossRef]
24. de Abreu, M.; Alvaro-Benito, M.; Sanz-Aparicio, J.; Plou, F.J.; Fernandez-Lobato, M.; Alcalde, M. Synthesis of 6-kestose using an efficient beta-fructofuranosidase engineered by directed evolution. *Adv. Synth. Catal.* **2013**, *355*, 1698–1702. [CrossRef]
25. Song, J.Y.; Alnaeeli, M.; Park, J.K. Efficient digestion of chitosan using chitosanase immobilized on silica-gel for the production of multisize chitooligosaccharides. *Process Biochem.* **2014**, *49*, 2107–2113. [CrossRef]
26. Alcalde, M.; Ferrer, M.; Plou, F.J. Environmental biocatalysis: From remediation with enzymes to novel green processes. *Biocatal. Biotransform.* **2007**, *25*, 113. [CrossRef]
27. Olicón-Hernández, D.R.; Vázquez-Landaverde, P.A.; Cruz-Camarillo, R.; Rojas-Avelizapa, L.I. Comparison of chito-oligosaccharide production from three different colloidal chitosans using the endochitonsanolytic system of *Bacillus thuringiensis*. *Prep. Biochem. Biotechnol.* **2017**, *47*, 116–122. [CrossRef] [PubMed]
28. Santos-Moriano, P.; Woodley, J.M.; Plou, F.J. Continuous production of chitooligosaccharides by an immobilized enzyme in a dual-reactor system. *J. Mol. Catal. B Enzym.* **2016**, *133*, 211–217. [CrossRef]
29. Li, H.; Fei, Z.; Gong, J.; Yang, T.; Xu, Z.; Shi, J. Screening and characterization of a highly active chitosanase based on metagenomic technology. *J. Mol. Catal. B Enzym.* **2015**, *111*, 29–35. [CrossRef]
30. Thadathil, N.; Velappan, S.P. Recent developments in chitosanase research and its biotechnological applications: A review. *Food Chem.* **2014**, *150*, 392–399. [CrossRef]
31. Sinha, S.; Chand, S.; Tripathi, P. Production, purification and characterization of a new chitosanase enzyme and improvement of chitosan pentamer and hexamer yield in an enzyme membrane reactor. *Biocatal. Biotransform.* **2014**, *32*, 208–213. [CrossRef]
32. Santos-Moriano, P.; Kidibule, P.E.; Alleyne, E.; Ballesteros, A.O.; Heras, A.; Fernandez-Lobato, M.; Plou, F.J. Efficient conversion of chitosan into chitooligosaccharides by a chitosanolytic activity from *Bacillus thuringiensis*. *Process Biochem.* **2018**, *73*, 102–108. [CrossRef]
33. Kittur, F.S.; Vishu Kumar, A.B.; Varadaraj, M.C.; Tharanathan, R.N. Chitooligosaccharides—Preparation with the aid of pectinase isozyme from *Aspergillus niger* and their antibacterial activity. *Carbohydr. Res.* **2005**, *340*, 1239–1245. [CrossRef]

34. Xia, W.; Liu, P.; Liu, J. Advance in chitosan hydrolysis by non-specific cellulases. *Bioresour. Technol.* **2008**, *99*, 6751–6762. [CrossRef] [PubMed]
35. Kumar, A.B.V.; Tharanathan, R.N. A comparative study on depolymerization of chitosan by proteolytic enzymes. *Carbohydr. Polym.* **2004**, *58*, 275–283.
36. Oyeleye, A.; Normi, Y.M. Chitinase: Diversity, limitations, and trends in Engineering for suitable applications. *Biosci. Rep.* **2018**, *38*, 4. [CrossRef]
37. Moon, C.; Seo, D.J.; Song, Y.S.; Hong, S.H.; Choi, S.H.; Jung, W.J. Antifungal activity and patterns of N-acetyl-chitooligosaccharide degradation via chitinase produced from *Serratia marcescens* PRNK-1. *Microb. Pathog.* **2017**, *113*, 218–224. [CrossRef] [PubMed]
38. Sha, L.; Shao, E.; Guan, X.; Huang, Z. Purification and partial characterization of intact and truncated chitinase from *Bacillus thuringiensis* HZP7 expressed in *Escherichia coli*. *Biotechnol. Lett.* **2016**, *38*, 279–284. [CrossRef]
39. Kidibule, P.E.; Santos-Moriano, P.; Jiménez-Ortega, E.; Ramírez-Escudero, M.; Limón, M.C.; Remacha, M.; Plou, F.J.; Sanz-Aparicio, J.; Fernández-Lobato, M. Use of chitin and chitosan to produce new chitooligosaccharides by chitinase Chit42: Enzymatic activity and structural basis of protein specificity. *Microb. Cell Fact.* **2018**, *17*, 47. [CrossRef] [PubMed]
40. Wei, P.; Ma, P.; Xu, Q.S.; Bai, Q.H.; Gu, J.G.; Xi, H.; Du, Y.G.; Yu, C. Chitosan oligosaccharides suppress production of nitric oxide in lipopolysaccharide-induced N9 murine microglial cells in vitro. *Glycoconjugate J.* **2012**, *29*, 285–295. [CrossRef]
41. Yoon, H.J.; Moon, M.E.; Park, H.S.; Im, S.Y.; Kim, Y.H. Chitosan oligosaccharide (COS) inhibits LPS-induced inflammatory effects in RAW 264.7 macrophage cells. *Biochem. Biophys. Res. Commun.* **2007**, *358*, 954–959. [CrossRef] [PubMed]
42. Li, Y.; Liu, H.T.; Xu, Q.S.; Du, Y.G.; Xu, J. Chitosan oligosaccharides block LPS-induced O-GlcNAcylation of NF-kappa B and endothelial inflammatory response. *Carbohydr. Polym.* **2014**, *99*, 568–578. [CrossRef] [PubMed]
43. Yousef, M.; Pichyangkura, R.; Soodvilai, S.; Chatsudthipong, V.; Muanprasat, C. Chitosan oligosaccharide as potential therapy of inflammatory bowel disease: Therapeutic efficacy and possible mechanisms of action. *Pharmacol. Res.* **2012**, *66*, 66–79. [CrossRef]
44. Lee, H.J.; Park, J.M.; Han, Y.M.; Gil, H.K.; Kim, J.; Chang, J.Y.; Jeong, M.; Go, E.J.; Hahm, K.B. The role of chronic inflammation in the development of gastrointestinal cancers: reviewing cancer prevention with natural anti-inflammatory intervention. *Expert Rev. Gastroenterol. Hepatol.* **2016**, *10*, 129–139. [CrossRef] [PubMed]
45. Zhou, Z.J.; Wang, L.; Feng, P.P.; Yin, L.H.; Wang, C.; Zhi, S.X.; Dong, J.Y.; Wang, J.Y.; Lin, Y.; Chen, D.P.; et al. Inhibition of epithelial TNF-alpha receptors by purified fruit bromelain ameliorates intestinal inflammation and barrier dysfunction in colitis. *Front. Immunol.* **2017**, *8*, 1468. [CrossRef] [PubMed]
46. Pangestuti, R.; Bak, S.S.; Kim, S.K. Attenuation of pro-inflammatory mediators in LPS-stimulated BV2 microglia by chitooligosaccharides via the MAPK signaling pathway. *Int. J. Biol. Macromol.* **2011**, *49*, 599–606. [CrossRef] [PubMed]
47. Sánchez, Á.; Mengíbar, M.; Fernández, M.; Alemany, S.; Heras, A.; Acosta, N. Influence of preparation methods of chitooligosaccharides on their physicochemical properties and their anti-inflammatory effects in mice and in RAW 264.7 macrophages. *Mar. Drugs* **2018**, *16*, 430. [CrossRef] [PubMed]
48. Lee, S.-H.; Senevirathne, M.; Ahn, C.-B.; Kim, S.-K.; Je, J.-Y. Factors affecting anti-inflammatory effect of chitooligosaccharides in lipopolysaccharides-induced RAW264.7 macrophage cells. *Bioorg. Med. Chem.* **2009**, *19*, 6655–6658.
49. Jeuniaux, C. Chitinases. *Methods Enzymol.* **1966**, *8*, 644–650.

Article

OcUGT1-Catalyzed Glucosylation of Sulfuretin Yields Ten Glucosides

Shuai Yuan [1], Yan-Li Xu [1], Yan Yang [1] and Jian-Qiang Kong [1,2,*]

[1] State Key Laboratory of Bioactive Substance and Function of Natural Medicines & Ministry of Health Key Laboratory of Biosynthesis of Natural Products, Institute of Materia Medica, Chinese Academy of Medical Sciences & Peking Union Medical College, Beijing 100050, China; yuanshuai@imm.ac.cn (S.Y.); yanlixu@imm.ac.cn (Y.-L.X.); yangyan@imm.ac.cn (Y.Y.)

[2] Hebei LANSEN Biotech. Co. Ltd., Jinzhou 052263, China

* Correspondence: jianqiangk@imm.ac.cn; Tel.: +86-106-303-3559

Received: 2 September 2018; Accepted: 20 September 2018; Published: 25 September 2018

Abstract: Sulfuretin glucosides are important sources of innovative drugs. However, few glucosides of sulfuretin have been observed in nature. Therefore, it is urgent to diversify sulfuretin glycosides. Herein, glycosyltransferase (GT)-catalyzed glycodiversification of sulfuretin was achieved. Specifically, a flavonoid GT designated as OcUGT1 was used as a biocatalyst for the glucosylation of sulfuretin with UDP-Glc. The OcUGT1-assisted glucosylation of sulfuretin yielded ten glycosylated products, including three monoglucosides, five diglucosides and two triglucosides. The three monoglucosides were thus identified to be sulfuretin 3′-, 4′- and 6-glucoside according to HR-ESI-TOFMS data and their coelution with respective standards. A major diglucoside was assigned as sulfuretin 4′,6-diglucoside by HR-ESI-TOFMS in conjunction with NMR analysis. The exact structure of the other four diglucosides was not well characterized due to their trace amount. However, they were reasonably inferred as sulfuretin 3′,6-diglucoside, sulfuretin 3′,4′-diglucoside and two disaccharide glucosides. In addition, the structural identification of the remaining two triglucosides was not performed because of their small amount. However, one of the triglucosides was deduced to be sulfuretin 3′,4′,6-triglucoside based on the catalytic behavior of OcUGT1. Of the ten sulfuretin glucosides, at least six were new compounds. This is the first time to obtain monoglucosides, diglucosides and triglucosides of sulfuretin simultaneously by a single glycosyltransferase.

Keywords: glycosyltransferase; glycodiverfication; sulfuretin; OcUGT1

---

1. Introduction

Glycodiversification is a collective strategy of natural product glycosylation, in which varied activated sugars are attached to natural-product acceptors by enzymatic or chemical means, thereby providing diverse carbohydrate structures and functions [1,2]. The resultant glycosylated bioactive compounds have been shown to exert various biological and pharmacological activities with improved physicochemical characters, such as solubility and stability [3,4]. Many glycosides are thus developed to clinical drugs, e.g., rutin [5–7], puerarin [8] and scutellarin [9]. Hence, glycodiversification of natural products is deemed an effective strategy to broaden the scope of new compounds [2].

Owing to the structural complexity of many glycosylated compounds, glycodiversification of natural products by chemical synthesis may be a formidable task [2]. Conversely, enzymatic glycodiversification is becoming a main strategy for diversifying glycosylated natural products due to the great strides made in the generation of glycosyltransferase with catalytic promiscuity [10–13].

Sulfuretin (1, also designated as sulphuretin, Figure 1 and Figure S1), a naturally occurring aurone [14–16], is found to display a remarkable spectrum of biological activities such as therapeutic activity against acquired lymphedema [14], anti-Parkinson's disease activity [15], antioxidant action [16], therapeutic benefits in bone disease and regeneration [17,18] and neuroprotective effect [19], suggesting sulfuretin is a promising molecule for drug development. Accordingly, the interest for the discovery or synthesis of sulfuretin derivatives is increasing. Many sulfuretin derivatives featuring varied functional groups were thus observed to display a wide range of biological activities [20–23]. Of these derivatives, glycosides of sulfuretin, e.g., sulfuretin 6-glucoside (sulfurein) [24–26], sulfuretin 3′-glucoside [25] and palasitrin (sulfuretin 3′,6-diglucoside) [27], have been determined to exhibit diverse activities such as antioxidant activity [26,28] and influenza A neuraminidase inhibitory activity [25], suggesting sulfuretin glycosides are a potent source of drug discovery. Thus far, however, few sulfuretin glycosides have been obtained through direct extraction or enzymatic synthesis [29,30], which limited their druggability study. Therefore, it is urgent to diversify sulfuretin glycosides for drug screening.

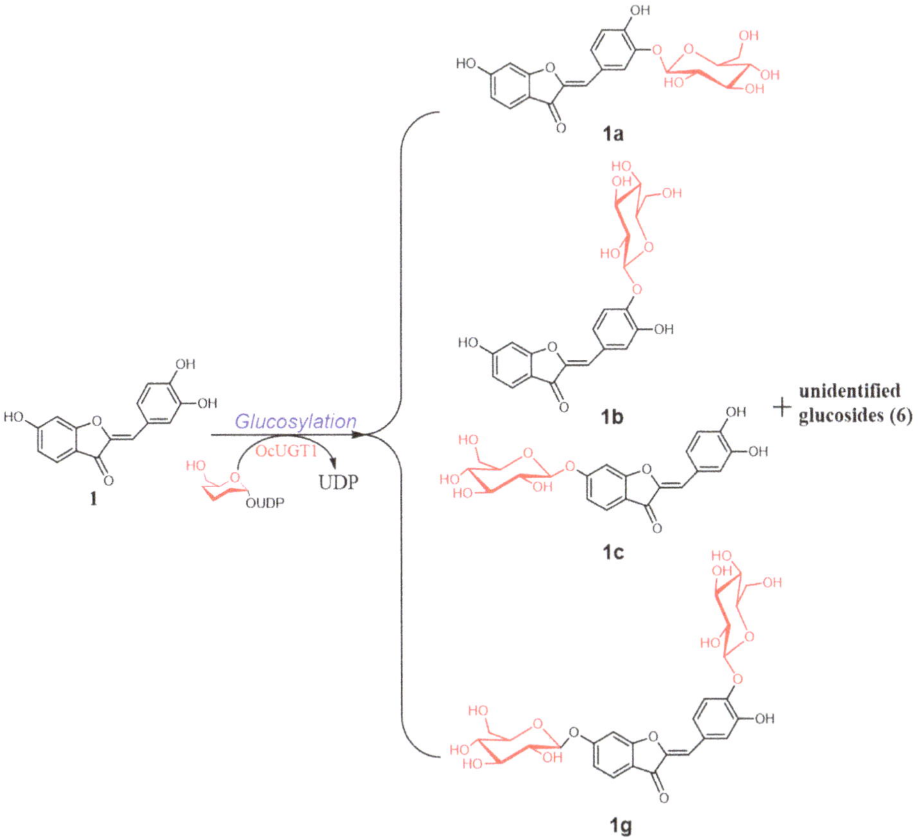

Figure 1. OcUGT1-catalyzed glucosylation of sulfuretin (**1**) resulted in the generation of ten glucosides (**1a**, **1b**, **1c**, **1g** and six unidentified compounds).

OcUGT1 (*Ornithogalum caudatum* UDP-glycosyltransferase), isolated from *O. caudatum* previously [13], is a flavonoid glycosyltransferase (GT) with catalyzing promiscuity. OcUGT1 can glucosylate diverse sugar acceptors including flavonoids. Moreover, OcUGT1 has been observed to function on multiple sites of flavonoids, yielding a number of flavonoid glycosides [13]. Both indicate OcUGT1 is an ideal tool for glycodiversification of small molecules. OcUGT1 was used as a biocatalyst for the glucosylation of sulfuretin with UDP-D-glucose (UDP-Glc). OcUGT1-assisted glucosylation of sulfuretin resulted in the formation of ten glucosides including three monoglucosides, five diglucosides and two triglucosides. Of these ten newly formed glycosides, at least six glucosides were new compounds (Figure 1). Thus, the use of single glycosyltransferases capable of forming multiple glycosides is an effective way to achieve glycosidic diversification, and can significantly increase the probability of drug discovery.

2. Results and Discussion

2.1. Protein Expression and Purification

After induction by IPTG, total proteins of *Escherichia coli* strain BL21(DE3) [pET28a-OcUGT1 + pKJE7] were subject to sodium dodecyl sulfate polyacrylamide gel electrophoresis (SDS-PAGE) analysis [13]. As shown in Figure 2A, an intense band with 53 kDa was detected in the sample. No corresponding band was present in the control strain, suggesting a soluble OcUGT1 was expressed in *E. coli* (Figure 2). The expressed OcUGT1 was thus purified to near homogeneity and its concentration was determined for glucosylation reaction.

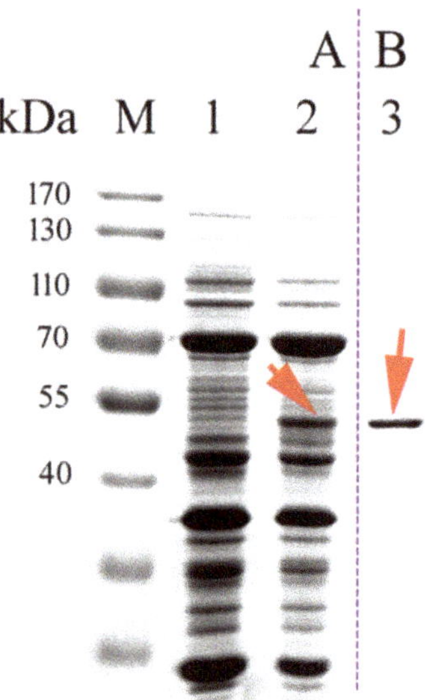

Figure 2. Heterologous expression (**A**) and affinity purification (**B**) of OcUGT1. 1, total protein of the control strain BL21(DE3) [pET-28a (+)+pKJE7]; 2, total protein of BL21(DE3) [pET28a-OcUGT1 + pKJE7]; and 3, the purified OcUGT1 protein. Values at the left margin indicate the position and molecular mass of protein standards. Red arrows show the recombinant OcUGT1.

2.2. OcUGT1-Catalyzed Glycosylation towards Sulfuretin

After incubated at 50 °C for 2 h, the reaction mixture containing purified OcUGT1, sulfuretin and UDP-Glc was analyzed by reverse phase high performance liquid chromatography (RP-HPLC). As shown in Figure 3, ten new peaks 1a–j were present in the reaction mixture (Figure 3), while there were no new peaks in the control reaction harboring no purified OcUGT1 (Figure 3) suggesting the ten peaks might be glucosylated metabolites of sulfuretin.

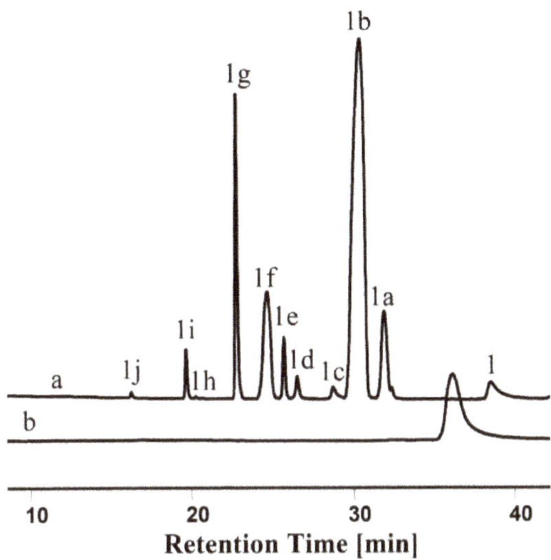

Figure 3. HPLC chromatogram of the glucosylated metabolites of sulfuretin with (**a**) or without (**b**) purified OcUGT1.

2.3. Structural Identification of Sulfuretin Monoglucosides

The ten metabolites were then subjected to high-resolution electrospray ionization mass spectrometry (HR-ESI-MS) analyses. The positive ion HR-ESI-MS spectrum of 1a displayed a molecular ion peak at m/z 455.0928 $[M + Na]^+$ corresponding to $C_{21}H_{20}O_{10}Na$ (Figure S2).The major metabolite 1b exhibited a pseudomolecular ion peak $[M + Na]^+$ at m/z 455.0927, and the molecular formula $C_{21}H_{20}O_{10}Na$ was established by HR-ESI-MS (Figure S2). The molecular formula of a minor product **1c** was determined to be $C_{21}H_{20}O_{10}Na$, by HR-ESI-MS at m/z 455.0943 $[M + Na]^+$ (Figure S2). The evidence suggests that all three metabolites were monoglucosylated sulfuretins. Coelutions of these metabolites with their standards assigned **1a**, **1b** and **1c** to be sulfuretin 3′-, 4′- and 6-glucoside, respectively [29].

2.4. Structural Identification of Sulfuretin Diglucosides

Compounds **1d**, **1e**, **1f**, **1g** and **1h** have the same molecular formula $C_{27}H_{30}O_{15}Na$ with $[M + Na]^+$ ion peaks at m/z 617.1479, 617.1484, 617.1492, 617.1481 and 617.1500, respectively, suggesting their diglucosylation of sulfuretin (Figure S3). Compound **1g** is the major product of these sulfuretin diglucosides. To further determine the structure of **1g**, it was collected using RP-HPLC and subjected to nuclear magnetic resonance (NMR) analyses. The ^{1}H-NMR spectrum (Figure S4 and Table 1) showed signals for the following protons: an olefinic proton at δ 6.77 (1H, s, H-10); and two sets of ABX type aromatic protons at δ 7.71 (1H, d, J = 8.5 Hz, H-4), 6.92 (1H, dd, J = 2.0, 8.5 Hz, H-5), 7.21 (1H, d, J = 2.0 Hz, H-7), and 7.49 (1H, d, J = 2.1 Hz, H-2′), 7.21 (1H, d, J = 8.6 Hz, H-5′), 7.41 (1H, dd, J = 2.1, 8.6 Hz, H-6′). The ^{13}C-NMR and spectroscopic data (Figure S4 and Table 1) indicated 27 carbon resonances, including two glucose moiety carbons, one carbonyl carbon, five aromatic oxygenated carbons, and nine aromatic carbons. The above data revealed that compound **1g** has a typical sulfuretin skeleton. In the HMBC (Figure 4 and Figure S5) spectrum of compound **1g**, long-range correlations between H-1″ and C-6 (δ 164.9), H-1‴ and C-4′ (δ 146.7), demonstrated that the glucosyl group was located at C-6 and C-4′, respectively. The β-configuration of sugars were concluded from the anomeric proton signals at δ 5.18 (1H, d, 7.3 Hz, H-1″), and 4.84 (1H, d, 7.3 Hz, H-1‴) in the ^{1}H-NMR spectrum. Based on these observations, the metabolite 1g was elucidated as sulfuretin 4′,6-diglucoside (Figure 4, Figure S1, S4 and S5 and Table 1). The four other diglucosides were not well characterized due to their trace amount. According to the catalytic behavior of OcUGT1 towards luteolin [13], the four diglucosides should include sulfuretin 3′,6-diglucoside and sulfuretin 3′,4′-diglucoside. Previous study indicated that OcUGT1 was able to attack the hydroxyl group of sugar moiety in monoglucosides, thereby forming disaccharide glycosides [30]. It is therefore reasonable to infer the remaining two diglucosides were disaccharide glucosides of sulfuretin. Thus, OcUGT1-assisted glucosylation of sulfuretin resulted in five diglucosides, namely sulfuretin 4′,6-diglucoside (**1g**) (Figure S1), sulfuretin 3′,6-diglucoside, sulfuretin 3′,4′-diglucoside and two disaccharide glucosides (Figure 1). Of the five diglucosides, the two diglucoside sulfuretin 4′,6-diglucoside (**1g**) and sulfuretin 3′,4′-diglucoside, as well as two disaccharide glucosides of sulfuretin, were new compounds.

Table 1. ^{1}H- and ^{13}C-NMR data of the compound **1g**.

Position	^{13}C	^{1}H
2	146.2,C	
3	181.7,C	
4	125.4,CH	7.71, d (8.5)
5	113.7,CH	6.92, dd (8.5, 2.0)
6	164.9,C	
7	99.4,CH	7.21, d (2.0)
8	167.4,C	
9	115.1,C	
10	111.9,CH	6.77, s
1′	126.3,C	
2′	118.2,CH	7.49, d (2.1)
3′	147.2,C	
4′	146.7,C	
5′	116.0,CH	7.21, d (8.6)
6′	124.0,CH	7.41, dd (8.6,2.1)
	Glc	Glc
1″	101.4,CH	5.18, d (7.3)
2″	73.3,CH	
3″	76.4,CH	
4″	69.9,CH	3.0–3.8, m (overlapped)
5″	77.3,CH	
6″	60.8,CH_2	
	Glc	
1‴	99.7,CH	4.84, d (7.3)
2‴	73.1,CH	
3‴	75.8,CH	
4‴	69.6,CH	3.0–3.8, m (overlapped)
5‴	77.1,CH	
6‴	60.7,CH_2	

2.5. *Structural Identification of Sulfuretin Triglucosides*

The HR-ESI-MS of 1i and 1j displayed molecular ion $[M + Na]^+$ peaks at m/z 779.2011 and 779.2031, respectively, both corresponding to the molecular formula of $C_{33}H_{40}O_{20}Na$, which indicated that both compounds were triglucosides of sulfuretin (Figure S6). The structures of the two triglucosides were not well characterized due to their trace amount. According to the catalytic behavior of OcUGT1 [13], one of the triglucosides was sulfuretin 3′,4′,6-triglucoside. The other triglucoside could not been deduced from the HR-ESI-MS data. To the best of our knowledge, the two triglucosides were also new compounds.

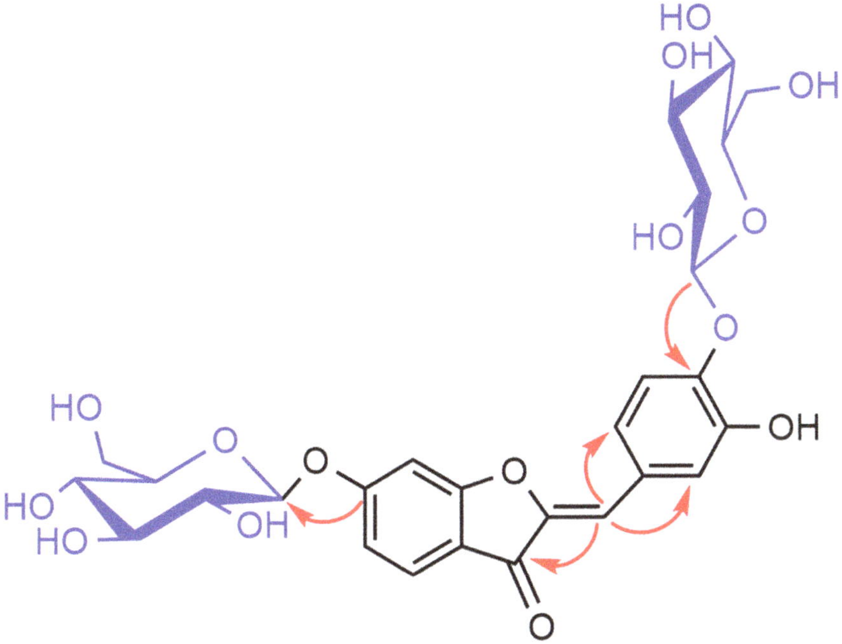

Figure 4. Selected HMBC (arrows) correlations of **1g**.

Overall, OcUGT1-catalyzed glucosylation of sulfuretin led to the generation of ten glucosides including six new compounds. The data revealed that enzyme-mediated glucosylation is an effective way to diversify glucosides. Previously, glycosyltransferases capable of accepting glycosides for further glycosylation have been reported [3]. However, there are few glycosyltransferases that catalyze the formation of monoglycosides, diglucosides and triglycosides of a single substrate simultaneously. In this study, OcUGT1 has been demonstrated to catalyze sulfuretin to form corresponding monoglycosides, disaccharides and triglycosides simultaneously, indicating that OcUGT1 has a very wide substrate specificity. These results, together with previous reports [13,31,32], indicate that OcUGT1 has potential applications as a biocatalyst in glycodiversification of natural products.

3. Materials and Methods

3.1. Chemicals

Sulfuretin (CAS No.:120-05-8) was purchased from BioBioPha (Kunming, Yunnan, China) (Figure S1). UDP-Glc was obtained from Sigma-Aldrich Co. LLC (St. Louis, MO, USA). The other chemicals were either reagents or analytical grade when available.

3.2. Protein Expression and Purification

Heterologous expression and purification of OcUGT1 was performed as described previously [13]. As introduced by Yuan et al., an expression plasmid pET28a-OcUGT1 and a chaperone plasmid pKJE7 (Takara, Dalian, China) were co-transformed into *E. coli* strain BL21 (DE3) for soluble expression. Total protein extracts from isopropyl-β-D-thiogalactoside (IPTG)-induced bacterial cells were separated by SDS-PAGE. The expressed recombinant protein with His-Tag were purified by affinity chromatography. The concentration of the purified protein was determined based on the procedure introduced by Yin et al. [33]. The resultant purified OcUGT1 was applied as the biocatalyst for the glycosylation towards sulfuretin (1) (Figure S1).

3.3. Glycosylation Assay

The reaction mixture and reaction conditions of OcUGT1-catalyzed glycosylation assay was the same as that of our previous reports [13]. In brief, a total of 100 μL phosphate buffer (10 mM, pH 8.0) harboring 10 mg purified OcUGT1, 1 mM sulfuretin and 1 mM UDP-Glc were incubated at 50 °C for 2 h. The glycosylation reaction was monitored by RP-HPLC. The HPLC conditions were the same as previously described by Yuan et al. [13].

3.4. Structural Identification

HR-ESI-MS spectra were recorded on A Triple TOF™ 5600 system (AB SCIEX, CA, USA) with a DuoSpray ionization source operating in the positive ESI mode.

NMR spectroscopic data were obtained as previously described [29,34–36]. Chemical shifts (*d*) and coupling constants (*J*) were provided in ppm and hertz (Hz), respectively.

Supplementary Materials: The following are available online at http://www.mdpi.com/2073-4344/8/10/416/s1, Figure S1: he general position numeration of sulfuretin (1) and sulfuretin 4′,6-diglucoside (**1g**), Figure S2: The mass spectra of **1a** (A), **1b** (B) and **1c** (C) acquired by ESI-HRMS, Figure S3: The mass spectra of **1d** (A), **1e** (B) **1f** (C), **1g** (D) and **1h** (E) acquired by ESI-HRMS, Figure S4: ^{1}H-NMR spectrum (600 MHz, DMSO-*d6*) (A) and ^{13}C-NMR spectrum of **1g** (150 MHz, DMSO-d6) (B), Figure S5: HMBC spectrum of **1g**, Figure S6: The mass spectra of **1i** (A) and **1j** (B) acquired by ESI-HRMS.

Author Contributions: Conceptualization, J.-Q.K.; Investigation, S.Y. and Y.L.X.; Validation, S.Y., Y.Y. and J.-Q.K.; Writing—Original Draft Preparation, J.-Q.K.; Supervision, Writing—review & editing, J.-Q.K.

Funding: This work was supported by the CAMS Innovation Fund for Medical Sciences (CIFMS) (2016-I2M-3-012) and Beijing Natural Science Foundation (7172143).

Conflicts of Interest: The authors declare no conflicts of interest.

Abbreviations

GT	glycosyltransferase
HR-ESI-MS	high-resolution electrospray ionization mass spectrometry
IPTG	isopropyl-β-D-thiogalactoside
NMR	nuclear magnetic resonance
OcUGT1	*Ornithogalum caudatum* UDP-glycosyltransferase
RP-HPLC	reverse phase high performance liquid chromatography
SDS-PAGE	sodium dodecyl sulfate polyacrylamide gel electrophoresis
UDP-Glc	UDP-D-glucose

References

1. Thibodeaux, C.J.; Melancon, C.E., 3rd; Liu, H.W. Natural-product sugar biosynthesis and enzymatic glycodiversification. *Angew. Chem. Int. Ed. Engl.* **2008**, *47*, 9814–9859. [CrossRef] [PubMed]
2. Thibodeaux, C.J.; Melancon, C.E.; Liu, H.W. Unusual sugar biosynthesis and natural product glycodiversification. *Nature* **2007**, *446*, 1008–1016. [CrossRef] [PubMed]
3. Hofer, B. Recent developments in the enzymatic *O*-glycosylation of flavonoids. *Appl. Microbiol. Biotechnol.* **2016**, *100*, 4269–4281. [CrossRef] [PubMed]
4. Xiao, J.; Muzashvili, T.S.; Georgiev, M.I. Advances in the biotechnological glycosylation of valuable flavonoids. *Biotechnol. Adv.* **2014**, *32*, 1145–1156. [CrossRef] [PubMed]
5. Al-Dhabi, N.A.; Arasu, M.V.; Park, C.H.; Park, S.U. An up-to-date review of rutin and its biological and pharmacological activities. *EXCLI J.* **2015**, *14*, 59–63. [PubMed]
6. Sharma, S.; Ali, A.; Ali, J.; Sahni, J.K.; Baboota, S. Rutin: Therapeutic potential and recent advances in drug delivery. *Expert Opin. Investig. Drugs* **2013**, *22*, 1063–1079. [CrossRef] [PubMed]
7. Chua, L.S. A review on plant-based rutin extraction methods and its pharmacological activities. *J. Ethnopharmacol.* **2013**, *150*, 805–817. [CrossRef] [PubMed]

8. Zhao, J.; Luo, D.; Liang, Z.; Lao, L.; Rong, J. Plant natural product puerarin ameliorates depressive behaviors and chronic pain in mice with spared nerve injury (SNI). *Mol. Neurobiol.* **2017**, *54*, 2801–2812. [CrossRef] [PubMed]
9. Yuan, Y.; Fang, M.; Wu, C.Y.; Ling, E.A. Scutellarin as a potential therapeutic agent for microglia-mediated neuroinflammation in cerebral ischemia. *Neuromol. Med.* **2016**, *18*, 264–273. [CrossRef] [PubMed]
10. Sun, L.; Chen, D.; Chen, R.; Xie, K.; Liu, J.; Yang, L.; Dai, J. Exploring the aglycon promiscuity of a new glycosyltransferase from *Pueraria lobata*. *Tetrahedron Lett.* **2016**, *57*, 1518–1521. [CrossRef]
11. Chen, D.; Chen, R.; Wang, R.; Li, J.; Xie, K.; Bian, C.; Sun, L.; Zhang, X.; Liu, J.; Yang, L.; et al. Probing the catalytic promiscuity of a regio- and stereospecific c-glycosyltransferase from *Mangifera indica*. *Angew. Chem. Int. Ed. Engl.* **2015**, *54*, 12678–12682. [CrossRef] [PubMed]
12. Xie, K.; Chen, R.; Li, J.; Wang, R.; Chen, D.; Dou, X.; Dai, J. Exploring the catalytic promiscuity of a new glycosyltransferase from *Carthamus tinctorius*. *Org. Lett.* **2014**, *16*, 4874–4877. [CrossRef] [PubMed]
13. Yuan, S.; Yin, S.; Liu, M.; Kong, J.-Q. Isolation and characterization of a multifunctional flavonoid glycosyltransferase from *Ornithogalum caudatum* with glycosidase activity. *Sci. Rep.* **2018**, *8*, 5886. [CrossRef] [PubMed]
14. Roh, K.; Kim, S.; Kang, H.; Ku, J.M.; Park, K.W.; Lee, S. Sulfuretin has therapeutic activity against acquired lymphedema by reducing adipogenesis. *Pharmacol. Res.* **2017**, *121*, 230–239. [CrossRef] [PubMed]
15. Pariyar, R.; Lamichhane, R.; Jung, H.J.; Kim, S.Y.; Seo, J. Sulfuretin attenuates MPP (+)-induced neurotoxicity through Akt/GSK3beta and ERK signaling pathways. *Int. J. Mol. Sci.* **2017**, *18*, 2753. [CrossRef] [PubMed]
16. Chand, K.; Hiremathad, A.; Singh, M.; Santos, M.A.; Keri, R.S. A review on antioxidant potential of bioactive heterocycle benzofuran: Natural and synthetic derivatives. *Pharmacol. Rep.* **2017**, *69*, 281–295. [CrossRef] [PubMed]
17. Auh, Q.S.; Park, K.R.; Yun, H.M.; Lim, H.C.; Kim, G.H.; Lee, D.S.; Kim, Y.C.; Oh, H.; Kim, E.C. Sulfuretin promotes osteoblastic differentiation in primary cultured osteoblasts and in vivo bone healing. *Oncotarget* **2016**, *7*, 78320–78330. [CrossRef] [PubMed]
18. Song, N.J.; Kwon, S.M.; Kim, S.; Yoon, H.J.; Seo, C.R.; Jang, B.; Chang, S.H.; Ku, J.M.; Lee, J.S.; Park, K.M.; et al. Sulfuretin induces osteoblast differentiation through activation of TGF-beta signaling. *Mol. Cell Biochem.* **2015**, *410*, 55–63. [CrossRef] [PubMed]
19. Kwon, S.-H.; Ma, S.-X.; Lee, S.-Y.; Jang, C.-G. Sulfuretin inhibits 6-hydroxydopamine-induced neuronal cell death via reactive oxygen species-dependent mechanisms in human neuroblastoma SH-SY5Y cells. *Neurochem. Int.* **2014**, *74*, 53–64. [CrossRef] [PubMed]
20. Lee, Y.H.; Shin, M.C.; Yun, Y.D.; Shin, S.Y.; Kim, J.M.; Seo, J.M.; Kim, N.-J.; Ryu, J.H.; Lee, Y.S. Synthesis of aminoalkyl-substituted aurone derivatives as acetylcholinesterase inhibitors. *Bioorgan. Med. Chem.* **2015**, *23*, 231–240. [CrossRef] [PubMed]
21. Rullah, K.; Mohd Aluwi, M.F.F.; Yamin, B.M.; Abdul Bahari, M.N.; Wei, L.S.; Ahmad, S.; Abas, F.; Ismail, N.H.; Jantan, I.; Wai, L.K. Inhibition of prostaglandin E2 production by synthetic minor prenylated chalcones and flavonoids: Synthesis, biological activity, crystal structure, and in silico evaluation. *Bioorg. Med. Chem. Lett.* **2014**, *24*, 3826–3834. [CrossRef] [PubMed]
22. Shin, S.Y.; Shin, M.C.; Shin, J.-S.; Lee, K.-T.; Lee, Y.S. Synthesis of aurones and their inhibitory effects on nitric oxide and PGE2 productions in LPS-induced RAW 264.7 cells. *Bioorg. Med. Chem. Lett.* **2011**, *21*, 4520–4523. [CrossRef] [PubMed]
23. Liu, B.; Zhang, M.; Xie, L.G.; Li, Y.H.; Xu, X. Synthesis, crystal structure and herbicidal activity of aurone derivatives. *Chem. J. Chin. Univ.* **2011**, *32*, 2335–2340.
24. Oberoi, S.; Lalita, L. Isolation and characterization of new plant pigment along with three known compounds from *Butea monosperma* petals. *Arch. Appl. Sci. Res.* **2010**, *2*, 68–71.
25. Ahmed, F.A.; Kim, S.Y.; Kurimoto, S.I. Biflavonoids from flowers of *Butea monosperma* (Lam.) Taub. *Heterocycles.* **2011**, *83*, 2079–2089.
26. Zhu, N.; Li, X.-W.; Liu, G.-Y.; Shi, X.-L.; Gui, M.-Y.; Sun, C.-Q.; Jin, Y.-R. Constituents from aerial parts of *Bidens ceruna* L. and their DPPH radical scavenging activity. *Chem. Res. Chin. Univ.* **2009**, *25*, 328–331.
27. Puri, B.; Seshadri, T.R. Survey of anthoxanthins. Part IX. Isolation and constitution of palasitrin. *J. Chem. Soc.* **1955**, 1589–1592. [CrossRef]

28. Westenburg, H.E.; Lee, K.J.; Lee, S.K.; Fong, H.H.; van Breemen, R.B.; Pezzuto, J.M.; Kinghorn, A.D. Activity-guided isolation of antioxidative constituents of *Cotinus coggygria*. *J. Nat. Prod.* **2000**, *63*, 1696–1698. [CrossRef] [PubMed]
29. Yuan, S.; Liu, M.; Yang, Y.; He, J.-M.; Wang, Y.-N.; Kong, J.-Q. Transcriptome-wide identification of an aurone glycosyltransferase with glycosidase activity from *Ornithogalum saundersiae*. *Genes* **2018**, *9*, 327. [CrossRef] [PubMed]
30. Halbwirth, H.; Wimmer, G.; Wurst, F.; Forkmann, G.; Stich, K. Enzymatic glucosylation of 4-deoxyaurones and 6′-deoxychalcones with enzyme extracts of *Coreopsis grandiflora*, Nutt. I. *Plant Sci.* **1997**, *122*, 125–131. [CrossRef]
31. Yuan, S.; Yin, S.; Liu, M.; He, J.-M.; Kong, J.-Q. OcUGT1-catalyzed glycosylation of testosterone with alternative donor substrates. *Process Biochem.* **2018**, in press. [CrossRef]
32. Yuan, S.; Yang, Y.; Kong, J.-Q. Biosynthesis of 7,8-dihydroxyflavone glycosides via OcUGT1-catalyzed glycosylation and transglycosylation. *J. Asian Nat. Prod. Res.* **2018**. [CrossRef] [PubMed]
33. Yin, S.; Liu, M.; Kong, J.-Q. Functional analyses of OcRhS1 and OcUER1 involved in UDP-L-rhamnose biosynthesis in *Ornithogalum caudatum*. *Plant Physiol. Biochem.* **2016**, *109*, 536–548. [CrossRef] [PubMed]
34. Liu, X.; Kong, J.Q. Steroids hydroxylation catalyzed by the monooxygenase mutant 139-3 from *Bacillus megaterium* BM3. *Acta Pharm. Sin. B* **2017**, *7*, 510–516. [CrossRef] [PubMed]
35. Yin, S.; Kong, J.-Q. Transcriptome-guided gene isolation and functional characterization of UDP-xylose synthase and UDP-d-apiose/UDP-d-xylose synthase families from *Ornithogalum caudatum* Ait. *Plant Cell Rep.* **2016**, *35*, 2403–2421. [CrossRef] [PubMed]
36. Guo, L.; Chen, X.; Li, L.-N.; Tang, W.; Pan, Y.-T.; Kong, J.-Q. Transcriptome-enabled discovery and functional characterization of enzymes related to (2*S*)-pinocembrin biosynthesis from *Ornithogalum caudatum* and their application for metabolic engineering. *Microb. Cell Fact.* **2016**, *15*, 27. [CrossRef] [PubMed]

Article

Laccase Activity as an Essential Factor in the Oligomerization of Rutin

Abel Muñiz-Mouro, Beatriz Gullón, Thelmo A. Lú-Chau, María Teresa Moreira, Juan M. Lema and Gemma Eibes *

Department of Chemical Engineering, Institute of Technology, Universidade de Santiago de Compostela, 15782 Santiago de Compostela, Spain; abel.muniz@usc.es (A.M.-M.); beatriz.gullon@usc.es (B.G.); thelmo.lu@usc.es (T.A.L.-C.); maite.moreira@usc.es (M.T.M.); juan.lema@usc.es (J.M.L.)

* Correspondence: gemma.eibes@usc.es; Tel.: +34-89-8181-6016; Fax: +34-88-1816-702

Received: 9 July 2018; Accepted: 3 August 2018; Published: 6 August 2018

Abstract: The enzyme-mediated polymerization of bioactive phenolic compounds, such as the flavonoid rutin, has gained interest due to the enhanced physico-chemical and biological properties of the products, which increases their potential application as a nutraceutical. In this work, the influence of enzyme activity on rutin oligomerization was evaluated in reactions with low (1000 U/L) and high (10,000 U/L) initial laccase activities. For both reactions, high molecular weight oligomer fractions showed better properties compared to lower weight oligomers. Products of the reaction with low laccase activity exhibited thermal stability and antioxidant potential similar to control reaction, but led to higher inhibitory activity of xanthine oxidase and apparent aqueous solubility. Oligomers obtained in the reaction with high laccase activity showed better apparent aqueous solubility but decreased biological activities and stability. Their low antioxidant activity was correlated with a decreased phenolic content, which could be attributed to the formation of several bonds between rutin molecules.

Keywords: rutin oligomers; laccase activity; aqueous solubility; antioxidant activity; xanthine oxidase inhibition; MALDI-TOF; HPSEC

1. Introduction

Flavonoids are one of the main types of polyphenols commonly identified in plants as secondary metabolites [1]. This type of compound exhibits satisfactory antioxidant capacities [2,3] and pharmacological properties, with potential use in the prevention of various diseases such as diabetes, cancer, cardiovascular, and neurodegenerative diseases [4,5]. Rutin (3′,4′,5,7-tetrahydroxy-flavone-3-rutinoside) is a quercetin *O*-glycoside generally extracted from *Fagopyrum esculentum M.* (Polygonaceae), *Ruta graveolens* L. (Rutaceae), *Sophora japonica* L. (Fabaceae), and *Eucalyptus* spp. (*Myrtaceae*). Concerning food, rutin can be found in vegetables, fruits, and plant-derived beverages. Rutin-rich foods had already been used by traditional Chinese medicine [6], and more recently rutin has been demonstrated to possess beneficial properties for preventing diseases and protecting genome stability [7]. Due to its useful properties, the Dietary Supplement Label Database lists over 1100 currently marketed products that contain rutin [8]. However, both its low solubility in aqueous and non-toxic organic solvents, which imply scarce bioavailability, and poor thermal stability [9–11], restrict its application. Enzymatic polymerization of polyphenolic monomers has been shown to produce higher molecular weight polyphenols with improved solubility, thermostability, and superior antioxidant properties [12–16].

Laccases (EC 1.10.3.2) and peroxidases (EC 1.11.1) are enzymes that oxidize phenolic substrates to form phenoxyl radicals. The pathway followed by these radicals depends on various factors such as the nature of the substrate or the composition of the reaction medium. One of the possibilities relies on an oxidative coupling that can lead to the formation of oligomers/polymers from the initial substrate [17]. The main advantage of laccase compared to peroxidases is that it requires oxygen as a final electron acceptor rather than hydrogen peroxide [18]. Laccases have been used in the food, pharmaceutical, and cosmetics sectors and have proved their usefulness as catalysts for the oligomerization or polymerization of phenolic compounds [14–16].

More specifically, the production of oligomers from flavonoids with improved properties using laccase has been demonstrated. Kurisawa et al. [14] reported the oligomerization of rutin by laccase, obtaining products with higher superoxide scavenging activity and aqueous solubility compared with the rutin monomer. Moreover, the improved capacity of rutin oligomers to inhibit the xanthine oxidase enzyme [19] and their improved antigenotoxic activity [20] compared to the rutin monomer were also demonstrated.

Although the properties of oligorutin are of great interest, reaction conditions should be optimized to make the process more efficient, both economically and environmentally. Methanol is often used as a co-solvent to allow for higher rutin solubility, but because of its toxicity, it would be advisable to avoid its use, which may involve lower costs in subsequent purification of the products, as methanol cannot be present especially in cosmetics and food applications. Ethanol is a valid alternative organic solvent as it increases the solubility of rutin and is considered safe according to the European Food Safety Authority (EFSA).

The aim of this work is to evaluate the effect of laccase activity on the rutin oligomers produced in a reaction medium compatible with food applications, using ethanol as co-solvent. To our knowledge, there are no studies that comprehensively evaluate the influence of laccase activity on the oligomerization reaction. The products obtained were divided into fractions of different molecular masses by ultrafiltration and subsequently characterized in terms of chemical structure (matrix-assisted laser desorption/ionization-time-of-flight mass spectrometry (MALDI-TOF) and high pressure size exclusion chromatography (HPSEC)), phenolic content, antioxidant activity by different assays (Ferric reducing antioxidant power-FRAP, cupric reducing antioxidant capacity-CUPRAC, 2,2′-azino-bis(3-ethylbenzenothiazoline-6-sulfonic acid) diammonium salt scavenging activity-$ABTS^+$ and xanthine oxidase inhibition) and aqueous solubility. The thermal stability of oligorutins produced in both reactions was also assessed by thermogravimetric analysis (TGA).

2. Results and Discussion

2.1. Polymerization Reaction

The use of EtOH as a co-solvent greatly increased rutin solubility. As shown in Figure S1, rutin solubility increased from 97.1 mg/L (0% EtOH) up to 3370.1 mg/L (50% EtOH). Combining this result with previous studies on laccase stability in the presence of EtOH [15], the value of 50% EtOH: acetate buffer (pH 5, 10 mM) was chosen as the appropriate reaction medium for this work.

Color changes were observed during the experiments, being more gradual in reaction A than in reaction B. Reaction A started to take an orange tone during the first few hours of reaction, eventually turning into an amber-colored liquid. Experiment B turned dark brown in less than an hour and continued to darken for the next hour, when visual changes could not be seen for the remaining extent of the reaction. This change in color is in accordance with that found by Sun et al. [21] for the polymerization of catechol and resorcinol by laccase. In contrast, controls maintained a pale-yellow color during the entire experiments. The enzymatic activity and the rutin concentration in the reaction medium over time are showed in Figure 1 (measurements related to their initial value).

Both the greatest loss of enzymatic activity and the highest rutin conversion occurred during the first 5–6 hours of reaction in both experiments. At that time, the losses in laccase activity reached values close to 50% of the initial activity, which is consistent with the decrease due to the presence of ethanol in the reaction medium, as previously reported [15]. Nevertheless, during the first two hours, the decrease in laccase activity in reaction B was lower than that observed in reaction A (14% and 33%, respectively). The enzyme activity continued to decrease until the experiments were completed, with a residual activity of 34% for reaction A and only 13% for reaction B.

The increase in laccase activity from 1000 to 10,000 U/L led to a higher rutin throughput (above 93%). Rutin depletion in reaction A occurred more gradually than in reaction B. After the first five hours, only 42.6% of rutin was consumed in reaction A, whilst reaction B showed a rapid drop in rutin content leaving only 7% of the initial rutin unreacted. Similar yields were obtained by Kurisawa et al. [14] using methanol as co-solvent. In addition, reaction B underwent a slightly higher acidification of the reaction medium. The final pH of reaction B was 5.80, while the pH of reaction A (6.12) was almost identical to that of the controls (6.18).

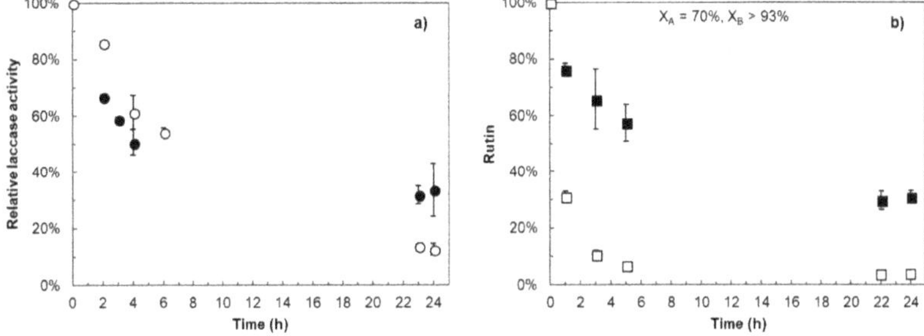

Figure 1. Monitored parameters for the polymerization of 3 g/L rutin in 50% ethanol:acetate buffer (10 mM, pH 5) mixture over 24 h at 25 °C and 1500 rpm for reactions A (1000 U/L of laccase activity) and B (10,000 U/L): (**a**) relative enzymatic activity of reactions A (●) and B (○), (**b**) residual rutin percentage in the reaction medium for reactions A (■) and B (□). X is used to indicate rutin conversion into oligomers.

Both media were divided into three different fractions by ultrafiltration. Although the samples were filtered to reach the same volume for each fraction, once lyophilized, different amounts of solid products were obtained, thus indicating different oligomerization degrees.

2.2. Apparent Solubility of Rutin Oligomers

The apparent solubility of the different fractions of the oligomers obtained in reactions A and B was evaluated. The total flavonoid content (TFC) of sample solutions in both 50% methanol and water supernatants was determined as mg of rutin equivalents per gram of sample (Table S1) and the apparent aqueous solubility was calculated using Equation (1). The results are shown in Table 1.

Table 1. Apparent solubility of oligomer fractions obtained in reactions A and B. The fold is expressed as the improvement of the solubility of the fractions by taking the controls as a reference.

Fraction	Apparent Solubility	Fold
	g/L	Relative to Controls
AF3	6.76 ± 0.04	~58
AF2	5.84 ± 0.40	~50
AF1	2.21 ± 0.43	~19
CA	0.12 ± 0.01	
BF3	10.77 ± 0.21	~71
BF2	9.93 ± 0.28	~65
BF1	*	*
CB	0.15 ± 0.01	

* Recovered lyophilized product was not sufficient to perform the assay. (A: 1000 U/L laccase, B: 10,000 U/L laccase, C: Control, F3 ≥ 10 kDa, F2 ⊂ (10, 1] kDa, F1 < 1 kDa).

The apparent aqueous solubility of CA and CB controls was similar to that obtained for rutin by Krewson et al. [9], thus supporting the validity of the technique used to indirectly assess the solubility of these flavonoids. The results showed that, regardless of the laccase activity used, an increase in the molecular mass of the oligomers led to an improved apparent aqueous solubility, as previously reported for different flavonoid oligomers [16,19,20]. Moreover, higher enzymatic activity during rutin polymerization (reaction B) produced oligomers (F3 fractions) with considerably enhanced apparent solubility, ~70 times greater than its control. Nevertheless, the use of 1000 U/L laccase activity (reaction A) improved the apparent solubility nearly 60 times, which would imply lower production costs.

A molecular modeling study conducted by Anthoni et al. [19] indicated that higher solubility of oligorutins could be attributed to their unfolded molecular structure, which would allow a greater number of intermolecular H-bonds of sugar parts with water molecules.

2.3. *Antioxidant Activities of Rutin Oligomer Fractions and Xanthine Oxidase Inhibitory Potential*

Oligorutins previously obtained by different researchers in methanol: water mixtures have shown good scavenging and antioxidant activity [14]. In addition, their ability to inhibit the enzyme xanthine oxidase, involved in inflammatory-related processes, has also been proven [19]. To evaluate the antioxidant activity of the lyophilized oligorutin fractions produced in reactions A and B, FRAP and CUPRAC methods were used. Moreover, the ABTS assay allowed us to measure their radical scavenging capacity. The results are shown in Table 2.

Table 2. Ferric reducing antioxidant power (FRAP), cupric reducing antioxidant capacity (CUPRAC), $ABTS^+$ scavenging activity and xanthine oxidase inhibitory potential of lyophilized rutin oligomer fractions and their respective controls for the different reaction conditions.

Fraction	FRAP	CUPRAC	$ABTS^+$	Xanthine Oxidase Inhibition
	mg TE [a]/g Sample	mg TE [a]/g Sample	% Inhibition (1 g/L)	IC_{50} (mg/L)
AF3	149.27 ± 22.67	488.30 ± 2.48	98 ± 1	186.36 ± 18.28
AF2	138.02 ± 3.55	415.14 ± 7.43	95 ± 1	198.43 ± 14.90
AF1	81.87 ± 1.93	245.67 ± 12.11	71 ± 2	372.30 ± 30.47
CA	168.79 ± 2.58	526.63 ± 8.81	100 ± 1	259.84 ± 1.45
BF3	59.22 ± 1.29	226.21 ± 22.56	60 ± 3	241.28 ± 24.34
BF2	13.33 ± 0.43	46.08 ± 3.44	11 ± 1	> 400 [b]
BF1	12.11 ± 2.79	27.50 ± 1.10	8 ± 1	> 600 [b]
CB	109.37 ± 2.79	304.24 ± 65.76	85 ± 3	415.12 ± 2.72

A: 1000 U/L laccase, B: 10,000 U/L laccase, C: control, F3 $\geq$ 10 kDa, F2 $\subset$ (10, 1] kDa, F1 < 1 kDa. [a] TE = Trolox equivalents. [b] maximal tested concentration due to the impossibility of measuring higher absorbance.

The differences observed in both control reactions, CA and CB, derive from their different content of inactivated enzyme (since reaction B had 10-fold higher enzymatic activity, inactivated laccase represents a higher mass percentage of the final freeze-dried powder obtained). Experiment B involved 10 times the enzyme activity in A, thus, after lyophilization, the non-rutin content in medium B was higher than in medium A, causing its lower antioxidant activity.

For all fractions, the antioxidant activity was lower than that of their control. AF3 showed the values most similar to those of the controls, reaching about 90% of the CA antioxidant activity for FRAP and CUPRAC methods and almost 100% of its $ABTS^+$ radical scavenging capacity. The use of 10,000 U/L laccase activity caused a remarkable loss in antioxidant activity, with the BF3 fraction reaching values close to 70% of those obtained for CB (except for the FRAP assay, which showed only 54%). Antioxidant activity dropped for smaller Mw fractions, especially for BF2 and BF1 oligomers, where close to 10–15% of CB activities were found, while AF2 and AF1 retained ~80 and ~50%, respectively (except for the ABTS test, which showed 95 and 71% of the CA scavenging activity).

This loss in antioxidant activity did not match the results provided by Kurisawa et al. [14], who observed an improvement in the superoxide scavenging capacity of oligorutin compared to rutin. However, Anthoni et al. [19] did observe this drop in antioxidant activity and, as stated by several authors who studied the effect of phenolic substituents on antioxidant activity [22,23], attributed it to a possible loss of free hydroxyl groups on C4′ and/or C3′. These functional groups, together with the C2–C3 double bond, are the structural features in rutin molecule that contribute the most to its antioxidant activity (the chemical structure of rutin can be seen in Figure S2). Therefore, the phenolic groups in C4′ and C3′ could have taken part in the polymerization reaction, leading to the possible formation of ether and/or carbon–carbon linkages between different rutin molecules.

The enzyme xanthine oxidase is considered to be an important biological source of ROS and catalyzes the oxidation of hypoxanthine and xanthine to uric acid, which plays a crucial role in gout [24]. Xanthine oxidase inhibitors represent an attractive option for the treatment of disorders such as gout, hyperuricemia, ulcers, ischemia, and hypertension, among others [25].

Table 2 shows also the xanthine oxidase inhibitory activity in the oligorutin fractions and their respective controls. The IC_{50} values are the average result of the duplicates, calculated by standard curve regression analysis. The correlation factors (R^2) for each standard curve ranged from 0.964 to 1000. AF3, AF2 and BF3 showed better results than CA and CB, with the IC_{50} of AF3 being the lowest value (186 mg/L). This enhancement was also found for AF2 with a similar IC_{50} but not for AF1, BF2 and BF1, leading to the conclusion that smaller oligomers show the worst xanthine oxidase inhibitory capacity. In fact, for oligomers obtained using higher laccase activity, only BF3 exhibited a lower IC_{50} than its control. This mass-related inhibitory activity is consistent with previous results reported by Anthoni et al. [19], who found that the higher the molecular mass of the oligorutin fractions, the lower the IC_{50} is. However, this is not in agreement with the results obtained by Kurisawa et al. [14], who performed rutin oligomerization in methanol:buffer mixtures using *Myceliophthora* laccase as a biocatalyst.

When comparing the best absolute IC_{50} values, AF3 and AF2 obtained better results than BF3, which means that the excess of laccase activity used in the production of higher molecular mass oligorutin negatively affected the xanthine oxidase inhibitory activity of these fractions.

The structure–activity relationship of flavonoids as inhibitors of xanthine oxidase has been previously studied by several researchers [24,26], who observed that the aromatic hydroxyl groups placed at C3′ and C4′ had little or no effect upon this capacity. In contrast, the OH groups linked to C5 and C7, along with the C2–C3 double bond, were the main causes of their xanthine oxidase inhibitory potential. Thus, the phenolic groups in C5 and C7 should not have been lost by providing a link between the rutin monomers.

In general, the results of the antioxidant activity tests and the xanthine oxidase inhibitory potential assay indicate that oligomerization probably involves phenolic groups in C3′ and/or C4′ carbons to form O–C or C–C bonds.

2.4. Evaluation of the Phenolic Content of Rutin Polymerization Products

Since there are widely studied correlations between the phenolic groups of flavonoids and their effect upon antioxidant activity [23] and the xanthine oxidase inhibition potential [24], the phenolic content of the oligorutin fractions produced was measured using two different protocols, revealing different results (Figure 2) that could be attributed to the chemical basis of the methods. While the Folin–Ciocalteu method measures all the reducing phenolic compounds, the 4-aminoantipyrine (4-AAP) method can only detect phenolic groups that have a free *para* position [27]. Due to the scarce quantity of recovered products for the lower molecular weight fractions, these assays were only performed for the F3 and F2 fractions of both reaction products.

The results provided by the Folin–Ciocalteu assay showed a slight loss of reducing phenolic compounds (RPC) for AF3 of less than 15% compared to CA, indicating that O–C bonding could have occurred. In addition, the reducing phenolics decreased considerably for the BF3 fraction, thus indicating a higher degree of polymerization (DP) through phenolic groups or, alternatively, that several bonds involving those hydroxyl groups were formed between rutin monomers. For F2 fractions, the reduction in RPC was greater, especially for BF2, which retained only about 10% of the phenolic groups present in the control. Therefore, the Folin–Ciocalteu assay proves that the use of higher enzyme activity led to a further decrease in reducing phenolic compounds, and that the smaller fractions of both reaction products suffered a marked drop in comparison with higher Mw fractions.

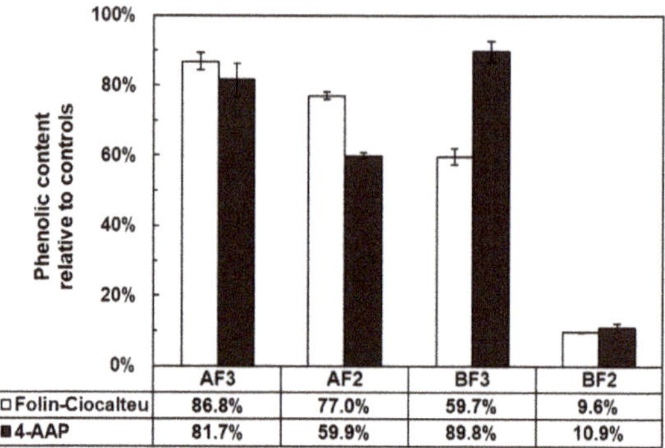

Figure 2. Phenolic content of the different fractions obtained for reaction A and B relative to their respective controls. Values were obtained via Folin–Ciocalteu (□) and 4-AAP (■) assays (A: 1000 U/L laccase, B: 10,000 U/L laccase, F3 $\geq$ 10 kDa, F2 $\subset$ (10, 1] kDa).

The protocol based on the reaction of phenolic compounds and 4-AAP showed different results. Considering the molecular structure of rutin (Figure S2). it can be understood that this method would not measure hydroxyl groups bonded to C4′ and C7, as they do not present a free *para* position.

The values obtained for the fractions show differences in the decrease of phenolic content that may provide insight on the possible bonding occurring between rutin monomers. While AF3 and BF2 did not show major differences for both protocols used to measure phenolic content, AF2 and BF3 did. AF2 showed a higher decrease of non-*para*-substituted phenols in proportion to all the reducing hydroxyl groups bonded to the aromatic part of the rutin molecule, while BF3 showed the opposite effect. Therefore, reaction A shows a general tendency to yield oligomers with a lesser proportion of non-*para*-substituted phenolic groups, while the opposite trend was observed for products of reaction B.

These results lead to the conclusion that the use of higher enzyme activity not only decreased the phenolic content, but also may have affected the phenolic groups involved in the polymerization reaction. However, this information alone does not directly indicate which type of bonding took place between the rutin molecules. A decrease in the phenolic groups detected by the 4-AAP method can mean that the hydroxyl group became part of an O–C bond, or that the carbon atom placed in its *para* position has been involved in a new bond with another rutin molecule (C–O or C–C).

Furthermore, considering the antioxidant and xanthine oxidase inhibitory activities and phenolic content, the hydroxyl group in C3′ appears to be involved in rutin oligomerization when 1000 U/L of laccase is used, while increasing this activity up to 10,000 U/L would imply the greater participation of the phenolic group in C4′ to produce high molecular mass oligomers. This observed trend was indicated in Figure S2, where the atoms marked in blue represent those most likely to be involved in oligomerization under A conditions, whereas those in red would correspond to B conditions.

A similar effect of the enzyme activity has been previously observed on the characteristics of the dehydrogenation polymers (DHP) synthesized from coniferyl alcohol (CA). Mechin et al. [28] demonstrated the influence of enzyme activity/substrate ratio on the structure of DHP's obtained using horseradish peroxidase (HRP). They found a negative correlation between the β-O-4 content of the DHPs and the HRP/CA ratio. Hence, the enzyme activity dose had a significant effect on the structure of the obtained products.

However, if the oligomerization only occurs by C–O and/or C–C bonds, it is not entirely determined by the techniques and methods already applied. Anthoni et al. [19] performed ^{1}H-NMR analysis and confirmed the existence of C–O and C–C bonds between the rutin molecules, not only in the aromatic but also the sugar part. As for the aromatic part, the bonds that might be most likely to form were C2′–C2′, O4′–C6′ and C6′–C6′ [29], the latter two proposed linkages according to our conclusions based on antioxidant and xanthine oxidase inhibitory activities after considering the structure–activity relationships.

2.5. Molecular Weight Characterization of the Different Reaction Products

2.5.1. MALDI-TOF Analysis

The MALDI-TOF analysis of the obtained fractions (Figure S3) showed different high intensity peaks separated by ~608 Da, indicating that two different rutin molecules have lost a hydrogen atom in the formation of a new intermolecular bond.

The highest DP found was 7, only obtained in BF3, thus supporting that the highest laccase activity used in the polymerization reaction implied the longest chain of the oligomers produced, despite the low intensity of the signal. In reaction A, a DP of 6 was found in AF2 and AF3 fractions. This deviation of only one degree of polymerization between AF3 and BF3 would probably not explain the differences in antioxidant and xanthine oxidase inhibitory activities, which might imply that the presence of higher laccase activities may have different effects on the products apart from increasing the length of the oligomer. The results were similar to those obtained by Anthoni et al. [19,30], who found a maximum DP of 6 rutin units using methanol as co-solvent. The intensities read in the AF3 fraction seem to indicate that the R2 and R3 oligomers may be present in the same proportion in this sample. In contrast, BF3 showed a remarkable difference in favor of R3, which coincides with the higher DP displayed by that sample. In addition, all the fractions produced in reaction B show an unidentified compound with a mass close to 1260 Da and intensity similar to that of R2.

The observed masses, intensities and suggested compounds for the peaks found with this technique are listed in Tables S2 and S3, along with the expected theoretical mass of each compound and the deviation of the measured values. Intermediate compounds with lower intensities would correspond to rutin oligomers lacking one or several methyl, OH, glucose, rhamnose, or rutinose substituents. With this in mind, reaction mechanisms similar to those found in esculin oligomerization [15] could be assumed.

The oligomers in reaction B often show the greatest deviations from the theoretical masses of expected polymers. This effect was especially observed for DP greater than 3 and always corresponded to masses lower than expected. The formation of multiple bonds between two rutin molecules in the presence of increased laccase activity would explain this behavior.

2.5.2. HPSEC Analysis

High performance size exclusion chromatography (HPSEC) made it possible to measure the average molecular mass of the oligomers present in the different fractions obtained after ultrafiltration (Figure 3). These results are orientative and can only be used in a comparative study, as the calibration curve was carried out using polystyrene standards (266 to 62,500 Da).

The products of reaction A were well distributed between AF2 and AF3, since high molecular mass compounds were detected in both chromatograms. The highest Mw, registered by BF3, was 8338.51 Da, much higher than that observed in the MALDI-TOF analysis, which corresponds to a maximum degree of polymerization of 14 instead of 7. The same effect but on a smaller scale occurred for all fractions analyzed. The mass underestimation by the MALDI-TOF technique, also reported by Anthoni et al. [30], was attributed to the possibility of a weak ionization for oligomers with a molecular mass greater than 3000 Da.

Compounds of about 4000 and 400 Da were detected in CA and CB, respectively, probably derived from the use of *T. versicolor* laccase. The differences between the chromatograms may be linked to the amount of laccase and rutin present in each control (rutin and laccase HPSEC chromatograms are provided in Figure S4). Moreover, a peak of around 400 Da could also be observed in oligomer fractions and BF2 seems to be enriched in this compound, which shows that the oligomers obtained in reaction B are mainly concentrated in the BF3 fraction, which would explain the lower antioxidant activities and poor xanthine oxidase inhibitory capacity present in BF2 and BF1. The AF3 and AF2 chromatograms did not differ much, despite being AF3 enriched with oligomers of a slightly higher Mw.

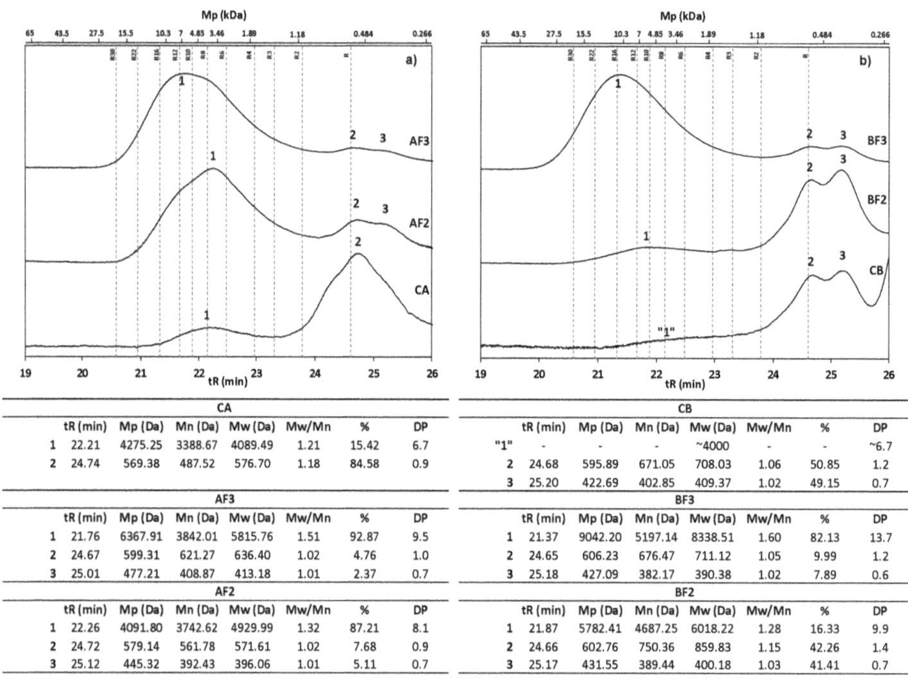

CA							
	tR (min)	Mp (Da)	Mn (Da)	Mw (Da)	Mw/Mn	%	DP
1	22.21	4275.25	3388.67	4089.49	1.21	15.42	6.7
2	24.74	569.38	487.52	576.70	1.18	84.58	0.9
AF3							
	tR (min)	Mp (Da)	Mn (Da)	Mw (Da)	Mw/Mn	%	DP
1	21.76	6367.91	3842.01	5815.76	1.51	92.87	9.5
2	24.67	599.31	621.27	636.40	1.02	4.76	1.0
3	25.01	477.21	408.87	413.18	1.01	2.37	0.7
AF2							
	tR (min)	Mp (Da)	Mn (Da)	Mw (Da)	Mw/Mn	%	DP
1	22.26	4091.80	3742.62	4929.99	1.32	87.21	8.1
2	24.72	579.14	561.78	571.61	1.02	7.68	0.9
3	25.12	445.32	392.43	396.06	1.01	5.11	0.7

CB							
	tR (min)	Mp (Da)	Mn (Da)	Mw (Da)	Mw/Mn	%	DP
"1"	-	-	-	~4000	-	-	~6.7
2	24.68	595.89	671.05	708.03	1.06	50.85	1.2
3	25.20	422.69	402.85	409.37	1.02	49.15	0.7
BF3							
	tR (min)	Mp (Da)	Mn (Da)	Mw (Da)	Mw/Mn	%	DP
1	21.37	9042.20	5197.14	8338.51	1.60	82.13	13.7
2	24.65	606.23	676.47	711.12	1.05	9.99	1.2
3	25.18	427.09	382.17	390.38	1.02	7.89	0.6
BF2							
	tR (min)	Mp (Da)	Mn (Da)	Mw (Da)	Mw/Mn	%	DP
1	21.87	5782.41	4687.25	6018.22	1.28	16.33	9.9
2	24.66	602.76	750.36	859.83	1.15	42.26	1.4
3	25.17	431.55	389.44	400.18	1.03	41.41	0.7

Figure 3. Molecular weight, polydispersity, and degree of polymerization parameter obtained by HPSEC for controls and fractions F3 and F2 of rutin oligomers from reactions A (**a**) and B (**b**) (A: 1000 U/L laccase, B: 10,000 U/L laccase, F3 $\geq$ 10 kDa, F2 $\subset$ (10, 1] kDa). tR: retention time, Mp: peak molecular weight, Mn: number-average, Mw: average molecular weight, Mw/Mn: degree of polydispersity and DP: degree of polymerization. The percentage of each mean molecular mass peak observed in the quantified area of the chromatogram is also shown in the table.

The differences in the degree of polymerization are, according to several authors, related to the antioxidant properties, but it cannot be stated that the properties evaluated in oligorutins derive only from this aspect, and evidence of possible multiple bonding between rutin molecules should be considered. For this reason, and to check whether this hypothesis is true or false, a new set of experiments was designed, with the objective of producing oligomers of similar molecular mass with 1000 and 10,000 U/L laccase activity.

If multiple linkages have been produced for oligomers in reaction B, although the products do not show very different molecular masses, different antioxidant activities are expected. The AF4′ and BF5′ fractions showed several similarities in the HPSEC chromatograms (see Figure S5) and their antioxidant properties were evaluated by FRAP and CUPRAC assays: BF5′ exhibited nearly half of the AF4′ antioxidant activity for both assays: 39.29 $\pm$ 3.02 (FRAP) and 143.44 $\pm$ 2.02 mg TE/g sample

(CUPRAC) for BF5′; 80.52 ± 4.06 (FRAP) and 252.36 ± 3.17 mg TE/g sample (CUPRAC) for AF4′. Despite both products showing similar DP, these comparisons make evident that different laccase activities used for each reaction have led to structurally different oligorutin.

2.6. Thermal Stability of Rutin Oligomers

The thermal stability of high molecular mass oligomer fractions and lyophilized controls for both reactions was assessed by thermogravimetric analysis. The differences in the controls are attributed to their different laccase content, thus decreasing the solid content at 600 °C for CB, since commercial laccase degrades up to around 85% at this temperature. AF3 showed almost the same solid residue at 600 °C as CA, while thermal stability in BF3 was strongly compromised, with only 25.7% of solid residue at the end of the analysis (final solid residue values along with decomposition intervals are presented in Table S4).

The TGA curves (Figure 4) showed a continuous decrease in mass at low temperatures for both AF3 and BF3 at the beginning of the analysis. Nevertheless, while the mass loss in AF3 was comparable to CA and CB (3–4% of the total product), BF3 lost up to 8.6% of its mass. Losses below ~120 °C are often attributed to the moisture content of the samples [31], but, since all samples were manipulated identically, these observed losses could also be associated with the chemical structure of the oligomers, either if the samples are partially decomposing or their hygroscopicity changed, which would relate this major mass loss for BF3 to a higher amount of ambient humidity absorbed after lyophilization.

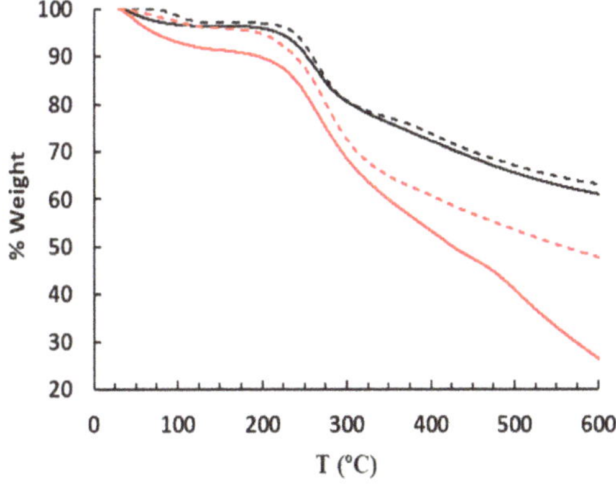

Figure 4. Thermogravimetric analysis (TGA) analysis for high molecular mass rutin oligomers obtained in reactions A and B, compared to control experiments. —AF3, - - CA, —BF3, - - CB (A: 1000 U/L laccase, B: 10,000 U/L laccase, C: control—F3 ≥ 10 kDa).

Unlike AF3, CA and CB, BF3 showed three different decomposition steps (inflection points in the TGA curve) and the solid residue at 600 °C was lower than that of its control, thus showing poor thermal stability, which cannot be attributed to the use of different laccase activities compared to AF3, but to its chemical nature. The properties of oligorutins evaluated in this study were summarized in Table S5.

3. Materials and Methods

3.1. Materials

Laccase from *Trametes versicolor*, rutin hydrate, 2,2′-azino-bis(3-ethylbenzenothiazoline-6-sulfonic acid) diammonium salt (ABTS), potassium persulfate, aluminum chloride hexahydrate, sodium nitrite, iron (III) chloride hexahydrate, 2,4,6-tri(2-pyridyl)-1,3,5-triazine (TPTZ), neocuproine,

6-hydroxy-2,5,7,8-tetramethylchroman-2-carboxylic acid (Trolox), Folin–Ciocalteu's phenol reagent, 4-aminoantipyrine (4-AAP), potassium ferricyanide, xanthine, xanthine oxidase (E.C. 1.17.3.2.), gallic acid, dimethyl sulfoxide (DMSO), absolute ethanol, methanol and polystyrene standards (between 62,500 and 266 Da) were purchased from Sigma (Sigma-Aldrich, St. Louis, MO, USA). Copper (II) chloride dihydrate and phenol were purchased from Merck (Merck KGaA, Darmstadt, Germany). Di-sodium hydrogen phosphate anhydrous was purchased from Panreac (Panreac Química SLU, Barcelona, Spain). 2,6-dimethoxyphenol (2,6-DMP) was purchased from Fluka (Honeywell Specialty Chemicals Seelze GmbH, Seelze, Germany). All reagents were of analytical grade.

3.2. *Laccase Activity*

Laccase activity of the enzyme stock solution was determined using ABTS as substrate [32]. During the oligomerization reaction, the substrate was changed to 2,6-DMP instead of ABTS, to avoid the possible reduction or scavenging of $ABTS^{+}$ by antioxidant compounds present in the reaction medium, such as rutin itself or the oligomers produced, since it is well known that this reverse reaction is possible [33]. This measurement was performed following the protocol defined by Wariishi at al. [34] with slight modifications. Briefly, the oxidation rate of 50 μL of 20 mM 2,6-DMP to 2,2′,6,6′-tetramethoxydibenzo-1,1′-diquinone caused by 50 μL of laccase solution, was monitored at 468 nm (ε_{468} = 49,600 $M^{-1}\cdot cm^{-1}$) in a sodium malonate (250 mM, pH 4.5, 200 μL) and distilled water (700 μL) mixture at room temperature. One unit (U) of activity was defined as the amount of enzyme forming 1 μmol of 2,2′,6,6′-tetramethoxydibenzo-1,1′-diquinone per minute. All spectrophotometric measurements were carried out on a Shimadzu UV-1800 (Shimadzu Europa GmbH, Duisburg, Germany).

3.3. *Synthesis of Rutin Oligomers*

3.3.1. Oligomerization Reaction

Polymerization of rutin was performed in 12 mL polypropylene test tubes with a threaded cap placed inside an MSC-100 thermoshaker incubator. This equipment protected the reaction medium from direct exposure to light. Only 5 mL of each tube was filled with the reaction medium to ensure oxygen excess (thus not limiting the reaction) and correct agitation. The temperature was set to 25 °C and agitation to 1500 rpm. For both experiments, the initial concentration of rutin was 3 g/L and the reaction medium composition was 50% ethanol and 50% acetate buffer (pH 5, 10 mM), with a single initial pulse of the enzyme solution to reach the desired initial laccase activity. In Experiment A, 1000 U/L (ABTS-related units) of laccase was used as a catalyst, whereas 10,000 U/L was used in Experiment B. Control experiments (CA and CB) had exactly the same composition as their reference experiments, but the enzyme was previously thermally inactivated (30 min at 100 °C). After 24 h, all reactions were stopped by adding HCl to acidify the reaction medium at a pH of 1.5.

Rutin concentration was monitored during the oligomerization process by high-performance liquid chromatography (HPLC) in a Jasco XLC HPLC (Jasco Analitica Spain, Madrid, Spain) equipped with a 3110 MD diode array detector (detection at 355 nm) and a Gemini reversed-phase column (150 × 4.6 mm, particle size: 3 μm) (Phenomenex, supplied by Jasco Analitica Spain, Madrid, Spain) at 45 °C. Gradient elution (flow rate of 0.7 $mL\cdot min^{-1}$) started with 10% acetonitrile in water (2% acetic acid), increased to 90% acetonitrile in 8 min, and then decreased back to the initial concentration after 2 min.

3.3.2. Separation and Lyophilization of the Polymers

After completion of the reaction, the samples were divided into three fractions using an Amicon 8010 10-mL ultrafiltration cell (Millipore Corporation, Bedford, MA, USA), with regenerated cellulose membrane discs with a nominal molecular weight limit of 10 and 1 kDa. For each experiment, the fractions F3 (10 kDa retentate), F2 (10 kDa permeate and 1 kDa retentate) and F1 (1 kDa permeate) were obtained and labelled as AF3, AF2, AF1, BF3, BF2 and BF1. Ethanol was removed by evaporation from products and controls (Büchi Rotavapor R-205 BÜCHI Labortechnik AG, Flawil, Switzerland, 50–70 mbar, 40 °C, 60 rpm) and samples were then lyophilized (Labconco FreeZone Benchtop Freeze Dry System (Labconco Corporation, Kansas City, MO, USA), 0.098 Torr, −50 °C).

3.4. Characterization

3.4.1. Apparent Solubility of Rutin Oligomers

Since it was not possible to measure the concentration of rutin oligomers in solution (no standards are available), an indirect protocol for measuring apparent aqueous solubility was developed based on the total flavonoid content (TFC) assay explained by Kim et al. [35] with the modifications proposed by Gullón et al. [36].

With the aim of analyzing the apparent solubility of the target compounds, two different matrices were considered for the preparation of each sample, one in distilled water and one in 50% methanol. For a concentration of 10 g/L of lyophilized fractions and controls, the solution prepared in water showed partial solubilization, which was evidenced by the presence of precipitates and turbidity of the sample, while the solution containing methanol showed complete dissolution. After 1 h of agitation at 1500 rpm and 25 °C in an MSC-100 thermoshaker incubator (LABGENE Scientific SA, Châtel-Saint-Denis Switzerland), each pair of samples was centrifuged at 25 °C and 14,000 rpm using an Eppendorf 5417 R microcentrifuge. The supernatant from each sample was withdrawn and the TFC was measured as described above and expressed as an average of two duplicates. The apparent solubility (g/L) of the products was then calculated as indicated in Equation (1):

$$\text{Apparent solubility} = \frac{TFC_{H_2O}}{TFC_{MeOH:H_2O}} \cdot C \tag{1}$$

where $TFCH_2O$ is the TFC in the supernatant of aqueous solutions, $TFCMeOH{:}H_2O$ is the TFC in methanol:water solutions, and C is the concentration of the tested sample (10 g/L).

3.4.2. Ferric Reducing Antioxidant Power (FRAP), Cupric Reducing Antioxidant Capacity (CUPRAC), $ABTS^+$ Scavenging Activity and Xanthine Oxidase Inhibition Test

FRAP and CUPRAC assays were conducted as indicated in our previous work, where these protocols were applied to measure the antioxidant properties of oligoesculin [15]. The results were expressed as the average value of two replicates. The ABTS assay was carried out to measure the potential of samples (1 g/L solutions) to scavenge $ABTS^+$ radical. This test was performed according to the method proposed by Re et al. [33] but was modified by increasing the incubation time, from 6 min up to 45 min based on previous results obtained by Božiĉ et al. [37], which showed that caffeic acid and gallic acid oligomers increased their scavenging activity over time, reaching values even higher than those obtained for monomer molecules. For FRAP, CUPRAC, and ABTS assays, samples were solubilized in methanol with 15% (*v*/*v*) DMSO.

In order to measure the xanthine oxidase inhibitory activity, the spectrophotometric protocol reported by Chebil et al. [12] was followed with the modifications detailed by Muñiz-Mouro et al. [15].

These protocols were performed using a Shimadzu UV-1800 spectrophotometer (Shimadzu Europa GmbH, Duisburg, Germany).

3.4.3. Assays for the Determination of Phenolic Compounds

Phenolic content was determined using two different protocols: Folin–Ciocalteu assay [38] and 4-aminoantipyrine (4-AAP) assay [39]. The Folin–Ciocalteu protocol was performed with the modifications indicated by Muñiz-Mouro et al. [15], and the values were obtained as the average of duplicates. The percentage of reducing phenolic compounds (RPC) present in each product was calculated in relation to its control.

In the 4-AAP assay, 920 μL of diluted samples or phenol standards (from 0.1 to 4.5 mg/L) were added to 50 μL of sodium–potassium tartrate buffer (pH 9.5), 50 μL of 4-AAP (20 g/L), and 50 μL of potassium ferricyanide (20 g/L). After 5 min of incubation, an absorbance at 510 nm was recorded in a BioTek PowerWave XS2 microplate spectrophotometer (Biotek Germany, Friedrichshall, Germany). The results were expressed in mg of phenol equivalents per g of freeze-dried sample as the average of four replicates. The phenolic content of each product was calculated in relation to its control.

3.4.4. Structural Analysis: MALDI-TOF and HPSEC

Absolute masses were determined by MALDI-TOF based on the protocol described by Anthoni et al. [30], using the equipment described in our previous work [15] and using DHB matrix but choosing the positive ionization and linear operation modes. The lyophilized samples were dissolved (1 g/L) in acetonitrile/water (30:70, *v*/*v*) with 0.1% TFA.

The average molecular weight (Mw) of oligorutin fractions (3 g/L solutions) was analyzed by high-performance size exclusion chromatography (HPSEC) using the method described by Dávila et al. [31] with some modifications: the injection volume was switched to 40 μL, the flow rate to 0.4 mL/min, and the temperature of 50 °C. This technique also made it possible to determine the number average (Mn) and the degree of polydispersity (Mw/Mn). The degree of polymerization (DP) was calculated as the ratio between Mw and the molecular mass of rutin. The chromatograms obtained were processed using the software ChromNAV 2.0 HPLC Software version 1.11.02 (JASCO, Easton, MD, USA).

3.4.5. Production of Oligomers with Similar Molecular Mass

In order to assess whether the properties of the rutin oligomers depend solely on its molecular weight, regardless of the enzyme activity used during oligomerization, reactions A and B were repeated following the same experimental procedure explained in Section 3.3, and the product medium obtained was ultrafiltrated into five different fractions instead of three. In addition, to avoid different protein contents associated to the use of different enzyme activities, Experiment A was supplemented with thermally inactivated laccase until the amount of enzyme was equivalent to that used in Experiment B. The fractions obtained after ultrafiltration were: F5′, retentate of a 30 kDa membrane, F4′, permeate of 30 and retentate of 10 kDa membranes, F3′, permeate of 10 and retentate of 3 kDa membranes, F2′, permeate of 3 and retentate of 1 kDa membranes, and F1′, permeate of 1 kDa membrane. The molecular weights of all lyophilized fractions were determined by HPSEC and similar products of both reactions were selected to assess their antioxidant properties by FRAP and CUPRAC assays.

3.4.6. Thermal Stability Study of Rutin Oligomers

The thermal properties of the oligorutins produced were evaluated by thermogravimetric analysis (TGA). This analysis was performed on a Mettler Toledo TGA/DSC1 thermobalance (Mettler-Toledo, Columbus, OH, USA), with a heating rate at 10 °C·min^{-1} from 25 to 600 °C under a nitrogen flow (20 mL·min^{-1}), placing approximately 2–4 mg of lyophilized samples in an aluminum crucible. For each sample, the decomposition intervals and the maximum decomposition temperature rate (MRDT) were obtained, as well as the final percentage of solids remaining at 600 °C.

4. Conclusions

The effect of the starting enzymatic activity in the laccase-mediated oxidative oligomerization of rutin was studied using the natural enzyme from *T. versicolor* in a food-compatible reaction medium. The rutin oligomers with the best characteristics for its use as a nutraceutical were obtained in the reaction with the lowest laccase activity (1000 U/L). The rutin oligomeric AF3 fraction obtained with this enzyme dosage significantly improved the apparent aqueous solubility and xanthine oxidase inhibitory activity compared to its control reaction, without compromising the antioxidant activity. The thermal stability of rutin oligomers was negatively affected by increasing the enzyme activity involved in the oligomerization reaction. The hypothesis of different molecular structure for rutin oligomers produced under different laccase activities was confirmed by comparing the antioxidant capacity of similar mean molecular mass oligomers produced involving different laccase activities, leading to the conclusion that higher enzyme dosages promoted the formation of multiple intermolecular bonds between rutin units, which negatively affected their antioxidant activity.

To our knowledge, this is the first study that focuses on the effect of laccase activity upon the products obtained in enzymatic oligomerization of the rutin flavonoid as a key parameter to enhance and tailor their physicochemical and biological properties. Further experiments will be required to optimize and scale up the production of this rutin oligomer, to evaluate its bioactive properties more exhaustively, and to provide more insight on the chemical structure of rutin oligomers.

Supplementary Materials: The following are available online at http://www.mdpi.com/2073-4344/8/8/321/s1, Figure S1. Solubility study of rutin in ethanol:acetate buffer mixtures (0.1 M, pH 5), Figure S2. Chemical structure of rutin. Rutin monomer has two para-substituted and two non-para-substituted phenolic groups, one of each type bonded to the A ring and the other to the B ring, Figure S3. MALDI-TOF analysis of rutin oligomer fractions produced in reactions A and B, Figure S4. HPSEC chromatograms for laccase and rutin. Mp: peak molecular weight, tR: retention time, Figure S5. HPSEC chromatograms for rutin oligomer fractions AF5′ and BF4′, Table S1. Total flavonoid content (TFC) in 10 g/L solutions of the different oligomer fractions in 50% MeOH:H2O and in supernatants of oversaturated (10 g/L) aqueous solutions, Table S2. MALDI-TOF results of the different fractions of rutin oligomers produced in reaction A and suggested compounds, detected as the deprotonated compounds, Table S3. MALDI-TOF results of the different fractions of rutin oligomers produced in reaction B and suggested compounds, detected as the deprotonated compounds, Table S4. TGA main parameters for high molecular mass rutin oligomers obtained in reactions A and B, compared to control experiments, Table S5. Summary of tested oligorutin properties.

Author Contributions: G.E., B.G. and J.M.L. conceived and designed the experiments; A.M.-M. performed all the experiments; A.M.-M., G.E., B.G., M.T.M. and T.A.L.-C. analyzed the data; all authors contributed to the writing of the paper.

Funding: This work was financially supported by the Spanish Ministry of Economy and Competitiveness (CTQ2014-58879-JIN). The authors belong to the Galician Competitive Research Group GRC-ED431C 2017/29 and to the strategic group CRETUS (AGRUP2015/02). All these programs are co-funded by FEDER. B. Gullón thanks the Spanish Ministry of Economy and Competitiveness for her postdoctoral fellowship (Grant reference IJCI-2015-25305).

Acknowledgments: The authors would like to acknowledge the use of RIAIDT-USC analytical facilities.

Conflicts of Interest: The authors declare no conflict of interest.

References

1. Romano, B.; Pagano, E.; Montanaro, V.; Fortunato, A.L.; Milic, N.; Borrelli, F. Novel insights into the pharmacology of flavonoids. *Phyther. Res.* **2013**, *27*, 1588–1596. [CrossRef] [PubMed]
2. Rice-Evans, C.A.; Miller, N.J.; Paganga, G. Structure-antioxidant activity relationships of flavonoids and phenolic acids. *Free Radic. Biol. Med.* **1996**, *20*, 933–956. [CrossRef]
3. Borges Bubols, G.; da Rocha Vianna, D.; Medina-Remón, A.; von Poser, G.; Lamuela-Raventos, R.M.; Eifler-Lima, V.L.; Cristina Garcia, S. The Antioxidant Activity of Coumarins and Flavonoids. *Mini-Rev. Med. Chem.* **2013**, *13*, 318–334. [CrossRef]
4. Hosseinzadeh, H.; Nassiri-Asl, M. Review of the protective effects of rutin on the metabolic function as an important dietary flavonoid. *J. Endocrinol. Investig.* **2014**, *37*, 783–788. [CrossRef] [PubMed]
5. Scalbert, A.; Manach, C.; Morand, C.; Rémésy, C.; Jiménez, L. Dietary Polyphenols and the Prevention of Diseases. *Crit. Rev. Food Sci. Nutr.* **2005**, *45*, 287–306. [CrossRef] [PubMed]
6. Rufa, L.; Yunning, Z.; Rui, W.; Jianying, L. A Study on the Extract of Tartary Buckwheat I. Toxicological Safety of the Extract of Tartary Buckwheat. *Small* **2001**, *21*, 602–607.
7. Gullón, B.; Lú-Chau, T.A.; Moreira, M.T.; Lema, J.M.; Eibes, G. Rutin: A review on extraction, identification and purification methods, biological activities and approaches to enhance its bioavailability. *Trends Food Sci. Technol.* **2017**, *67*, 220–235. [CrossRef]
8. NIH Dietary Supplement Label Database. Available online: https://www.dsld.nlm.nih.gov/dsld/index.jsp (accessed on 10 April 2018).
9. Krewson, B.C.F.; Naghskit, J. Some Physical Properties of Rutin. *J. Pharm. Sci.* **1952**, *41*, 582–587. [CrossRef]

10. Rothwell, J.A.; Day, A.J.; Morgan, M.R.A. Experimental Determination of Octanol—Water Partition Coefficients of Quercetin and Related Flavonoids. *J. Agric. Food Chem.* **2005**, *53*, 4355–4360. [CrossRef] [PubMed]
11. Paczkowska, M.; Mizera, M.; Piotrowska, H.; Szymanowska-Powałowska, D.; Lewandowska, K.; Goscianska, J.; Pietrzak, R.; Bednarski, W.; Majka, Z.; Cielecka-Piontek, J. Complex of rutin with β-Cyclodextrin as potential delivery system. *PLoS ONE* **2015**, *10*, e0120858. [CrossRef] [PubMed]
12. Chebil, L.; Rhouma, G.B.; Chekir-Ghedira, L.; Ghoul, M. Enzymatic Polymerization of Rutin and Esculin and Evaluation of the Antioxidant Capacity of Polyrutin and Polyesculin. In *Biotechnology*; Ekinci, D., Ed.; InTech: London, UK, 2015; pp. 117–133, ISBN 978-953-51-2040-7.
13. Chung, J.E.; Kurisawa, M.; Kim, Y.J.; Uyama, H.; Kobayashi, S. Amplification of antioxidant activity of catechin by polycondensation with acetaldehyde. *Biomacromolecules* **2004**, *5*, 113–118. [CrossRef] [PubMed]
14. Kurisawa, M.; Chung, J.E.; Uyama, H.; Kobayashi, S. Enzymatic synthesis and antioxidant properties of poly(rutin). *Biomacromolecules* **2003**, *4*, 1394–1399. [CrossRef] [PubMed]
15. Muñiz-Mouro, A.; Oliveira, I.M.; Gullón, B.; Lú-Chau, T.A.; Moreira, M.T.; Lema, J.M.; Eibes, G. Comprehensive investigation of the enzymatic oligomerization of esculin by laccase in ethanol:water mixtures. *RSC Adv.* **2017**, *7*, 38424–38433. [CrossRef]
16. Ghoul, M.; Chebil, L. Enzymatic polymerization of phenolic compounds by oxidoreductases. In *SpringerBriefs in Molecular Science*; Springer: Dordrecht, The Netherlands, 2012; ISBN 978-94-007-3918-5.
17. Jeon, J.R.; Baldrian, P.; Murugesan, K.; Chang, Y.S. Laccase-catalysed oxidations of naturally occurring phenols: From in vivo biosynthetic pathways to green synthetic applications. *Microb. Biotechnol.* **2012**, *5*, 318–332. [CrossRef] [PubMed]
18. Hollmann, F.; Arends, I.W.C.E. Enzyme Initiated Radical Polymerizations. *Polymers* **2012**, *4*, 759–793. [CrossRef]
19. Anthoni, J.; Lionneton, F.; Wieruszeski, J.M.; Magdalou, J.; Engasser, J.M.; Chebil, L.; Humeau, C.; Ghoul, M. Investigation of enzymatic oligomerization of rutin. *Rasayan J. Chem.* **2008**, *1*, 718–731.
20. Rhouma, G.B.; Chebil, L.; Mustapha, N.; Krifa, M.; Ghedira, K.; Ghoul, M.; Chékir-Ghédira, L. Cytotoxic, genotoxic and antigenotoxic potencies of oligorutins. *Hum. Exp. Toxicol.* **2013**, *32*, 881–889. [CrossRef] [PubMed]
21. Sun, X.; Bai, R.; Zhang, Y.; Wang, Q.; Fan, X.; Yuan, J.; Cui, L.; Wang, P. Laccase-Catalyzed Oxidative Polymerization of Phenolic Compounds. *Appl. Biochem. Biotechnol.* **2013**, *171*, 1673–1680. [CrossRef] [PubMed]
22. Burda, S.; Oleszek, W. Antioxidant and Antiradical Activities of Flavonoids. *J. Agric. Food Chem.* **2001**, *49*, 2774–2779. [CrossRef] [PubMed]
23. Heim, K.E.; Tagliaferro, A.R.; Bobilya, D.J. Flavonoid antioxidants: Chemistry, metabolism and structure-activity relationships. *J. Nutr. Biochem.* **2002**, *13*, 572–584. [CrossRef]
24. Cos, P.; Ying, L.; Calomme, M.; Hu, J.P.; Cimanga, K.; Van Poel, B.; Pieters, L.; Vlietinck, A.J.; Vanden Berghe, D. Structure-activity relationship and classification of flavonoids as inhibitors of xanthine oxidase and superoxide scavengers. *J. Nat. Prod.* **1998**, *61*, 71–76. [CrossRef] [PubMed]
25. Kumar, R.; Darpan; Sharma, S.; Singh, R. Xanthine oxidase inhibitors: A patent survey. *Expert Opin. Ther. Pat.* **2011**, *21*, 1071–1108. [CrossRef] [PubMed]
26. Hayashi, T.; Kazuko, S.; Masaru, K.; Munehisa, A.; Mineo, S.; Naokata, M. Inhibition of Cow's Milk Xanthine Oxidase by Flavonoids. *J. Nat. Prod.* **1988**, *51*, 345–348. [CrossRef] [PubMed]
27. Zhao, C.; Song, J.F.; Zhang, J.C. Determination of total phenols in environmental wastewater by flow-injection analysis with a biamperometric detector. *Anal. Bioanal. Chem.* **2002**, *374*, 498–504. [CrossRef] [PubMed]
28. Méchin, V.; Baumberger, S.; Pollet, B.; Lapierre, C. Peroxidase activity can dictate the in vitro lignin dehydrogenative polymer structure. *Phytochemistry* **2007**, *68*, 571–579. [CrossRef] [PubMed]
29. Chaaban, H.; Ioannou, I.; Chebil, L.; Slimane, M.; Gérardin, C.; Paris, C.; Charbonnel, C.; Chekir, L.; Ghoul, M. Effect of heat processing on thermal stability and antioxidant activity of six flavonoids. *J. Food Process. Preserv.* **2017**, *41*, e13203. [CrossRef]
30. Anthoni, J.; Chebil, L.; Lionneton, F.; Magdalou, J.; Humeau, C.; Ghoul, M. Automated analysis of synthesized oligorutin and oligoesculin by laccase. *Can. J. Chem.* **2011**, *89*, 964–970. [CrossRef]

31. Dávila, I.; Gullón, P.; Andrés, M.A.; Labidi, J. Coproduction of lignin and glucose from vine shoots by eco-friendly strategies: Toward the development of an integrated biorefinery. *Bioresour. Technol.* **2017**, *244*, 328–337. [CrossRef] [PubMed]
32. Zimmermann, Y.-S.; Shahgaldian, P.; Corvini, P.F.X.; Hommes, G. Sorption-assisted surface conjugation: A way to stabilize laccase enzyme. *Appl. Microbiol. Biotechnol.* **2011**, *92*, 169–178. [CrossRef] [PubMed]
33. Re, R.; Pellegrini, N.; Proteggente, A.; Pannala, A.; Yang, M.; Rice-Evans, C. Antioxidant activity applying an improved ABTS radical cation decolorization assay. *Free Radic. Biol. Med.* **1999**, *26*, 1231–1237. [CrossRef]
34. Wariishi, H.; Valli, K.; Gold, M.H. Manganese(II) oxidation by manganese peroxidase from the basidiomycete Phanerochaete chrysosporium: Kinetic mechanism and role of chelators. *J. Biol. Chem.* **1992**, *267*, 23688–23695. [PubMed]
35. Kim, D.-O.; Jeong, S.W.; Lee, C.Y. Antioxidant capacity of phenolic phytochemicals from various cultivars of plums. *Food Chem.* **2003**, *81*, 321–326. [CrossRef]
36. Gullón, B.; Gullón, P.; Lú-Chau, T.A.; Moreira, M.T.; Lema, J.M.; Eibes, G. Optimization of solvent extraction of antioxidants from Eucalyptus globulus leaves by response surface methodology: Characterization and assessment of their bioactive properties. *Ind. Crops Prod.* **2017**, *108*, 649–659. [CrossRef]
37. Božič, M.; Gorgieva, S.; Kokol, V. Laccase-mediated functionalization of chitosan by caffeic and gallic acids for modulating antioxidant and antimicrobial properties. *Carbohydr. Polym.* **2012**, *87*, 2388–2398. [CrossRef]
38. Singleton, V.L.; Rossi, J.A., Jr. Colorimetry of Total Phenolics with Phosphomolybdic-Phosphotungstic Acid Reagents. *Am. J. Enol. Vitic.* **1965**, *16*, 144–158. [CrossRef]
39. EPA, U.S. Phenolics (*Spectrophotometric, Manual 4-AAP with Distillation*). Available online: https://www.epa.gov/hw-sw846/sw-846-test-method-9065-phenolics-spectrophotometric-manual-4-aap-distillation (accessed on 4 August 2018).

Article

Advantageous Preparation of Digested Proteic Extracts from *Spirulina platensis* Biomass

Carlos M. Verdasco-Martín, Lea Echevarrieta and Cristina Otero *

Department of Biocatalysis, Institute of Catalysis and Petroleochemistry, CSIC, C/ Marie Curie 2 L10, Madrid 28049, Spain; c.verdasco@csic.es (C.V.); leaech01@ucm.es (L.E.)
* Correspondence: cotero@icp.csic.es; Tel.: +34-91-5854805/5854800

Received: 26 November 2018; Accepted: 21 January 2019; Published: 2 February 2019

Abstract: *Spirulina* biomass has great nutritional value, but its proteins are not as well adsorbed as animal ones are. New functional food ingredients and metabolites can be obtained from spirulina, using different selective biodegradations of its biomass. Four enzyme-assisted extraction methods were independently studied, and their best operation conditions were determined. Enzymes were employed to increase the yield of easily adsorbed proteic extracts. A biomass pre-treatment using Alcalase® (pH 6.5, 1% v/w, and 30 °C) is described, which increased the extraction yield of hydrophilic biocomponents by 90% w/w compared to the simple solvent extraction. Alcalase® gives rise to 2.5–6.1 times more amino acids than the others and eight differential short peptides (438–1493 Da). These processes were scaled up and the extracts were analyzed. Higher destruction of cell integrity in the case of Alcalase® was also visualized by transmission electron microscopy. The described extractive technology uses cheap, commercial, food grade enzymes and hexane, accepted for food and drug safety. It is a promising process for a competitive biofactory, thanks to an efficient production of extracts with high applied potential in the nutrition, cosmetic, and pharmaceutical industries.

Keywords: microalgae; *Spirulina*; Alcalase®; amino acid; extraction; nutraceutical

---

1. Introduction

New functional products are increasingly demanded by the food, cosmetic, and pharmaceutical industries. Enzyme degrading processes of vegetable biomasses are a 'white biotechnology' for sustainable production of these new functional products [1]. Degradation of the cell membrane before the extraction of biocomponents facilitates the recovery of cytoplasmic products [2]. Since the great part of cyanobacteria biocomponents are inside the cell [3,4], their extraction can be improved via destructive pre-treatment of the cellular and subcellular structures [2,5]. To recover the biomass constituents, enzymatic tools provide energy savings compared to mechanical treatment processes or chemical catalytic hydrolysis at high temperatures. They also compete well in the selectivity of extracted biocomponents. Depending on the type of enzyme, the cellular membrane and cell components may be degraded in different ways. Consequently, the potential of this technology is high, considering the great variety of degradation products that can be obtained through the action of different types of selective catalysts. This fact opens many opportunities for enzyme technology. Examples of advantageous enzymatic processes able to extract high value products from vegetable and animal biomasses are: (i) degradation of cellulose to glucose using cellulases, and (ii) membrane phospholipid hydrolysis by lipases for production of essential fatty acids.

The food chain is based on algae and microalgae products. Microalgae are good protein and metabolite sources. Dietary supplements are obtained from them [6].

Cyanobacteria (also named blue algae) are prokaryotic, photosynthetic, unicellular microorganisms. To this group belong *Arthrospira platensis* and *Arthrospira maxima* (commonly named

Spirulina platensis and *Spirulina maxima*). *Spirulina platensis* grows in Africa, Asia, and South America, with a characteristic helicoidal morphology [7]. *Spirulina* biomass (spirulina) has great potential for its composition and nutritional properties [8]. More than two thousand years ago, it was already being consumed by Aztecs. Spirulina properties are very much appreciated by modern society, being widely consumed for different types of healthy diets. Spirulina is a good source of essential amino acids, vitamins, carbohydrates, macro- and trace minerals, and other nutrients. The composition percent dry weight of spirulina is 64–73% protein, 12–17% carbohydrate, 5–7% lipids, 0.9% P, and 10.3–11.6% N. [9]. Blue pigments (phycocianins) of spirulina contribute to increasing the protein and iron availability [10]. Since 2011, spirulina is considered a safe ingredient in class A diet supplements [11]. Spirulina improves the immune system [12], and it has therapeutic effects against cancer and different virus, microbial, and inflammatory processes [1]. Spirulina provides high amounts of a unique antioxidant amino acid: L-ergothioneine (EGT; 2-mercaptohistidine trimethylbetaine) [13]. Spirulina supplementation provides vegetable proteins to the organism. To increase bioassimilation of its proteic material, extracts of its degraded proteic biocomponents must be obtained [14]. Additionally, the efficient extraction of intracellular components is limited by the cell membrane.

The cell membrane of *Spirulina* is analogous to that of Gram-negative bacteria [15]; they have two lipid membrane layers (cellular and cytoplasmic), separated by the murein layer. Murein is a rigid macromolecular structure formed by complex polymers of peptidoglycans and lipopolysaccharides. Peptidoglycans are covalently linked disaccharides and tetrapeptides. They are placed between cellular and cytoplasmic membranes, and linked to the external membrane layer by lipoproteins. Lipopolysaccharides are formed by a lipid and a complex polysaccharide chains. The cellular membrane has proteins non-covalently linked to lipids, whereas the cytoplasmic membrane is formed by lipoproteins (proteins covalently linked to lipids). The sugar complexes of cell membranes function as energy reserves (e.g., glucogen) [12].

Methods for cell membrane degradation may be physical (ultrasounds, microwaves [16,17], osmotic sock, pulse electric field, heat treatment [18], etc.), but may also be chemical or enzymatic [19,20]. The recovery of biocomponents can be achieved by different methods, such as those based on phase separation with solvents [21,22], supercritical fluids, pervaporation, etc. Safi et al. compared different fragile-cell-walled microalgae using several physical [23] and chemical [24] cell disruption methods. They were attacked according to the following order: *Haematococcus pluvialis* < *Nannochloropsis oculata* < *Chlorella vulgaris* < *Porphyridium cruentum* $\leq$ *Arthrospira platensis*. These authors determined that among the physical methods employed for protein extraction, high-pressure cell disruption was the most efficient one, although it was not enough to recover more than 50% of the proteins from these green microalgae, indicating that more passes are required to completely disrupt their macrostructure, and thus more energy input is necessary [23]. Using several cycles, mechanical treatments released more proteins from all the microalgae compared to chemical treatments. Percentages of protein extracted from *Spirulina* biomass using an alkaline pre-treatment or high-pressure homogenization method were 68% and 75%, respectively [24]. On the other hand, some reports described different enzyme treatments of protein extracts from spirulina, obtaining several bioactive products. In all the cases, the protein extracts were obtained by mechanical methods, prior to being submitted to a given enzyme degradation. More precisely, the iron-chelating peptide Thr-Asp-Pro-Ile(Leu)-Ala-Ala-Cys-Ile(Leu), with a molecular weight of 802 Da, was obtained through the combined action of two proteases [25]. The protein extract was first obtained by homogenization of the cell suspension, centrifugation, and further precipitation of the supernatant with ammonium sulfate. The extracted material was then consecutively submitted to two proteolytic steps by two different proteases. In other study, the antihypertensive peptide Ile-Gln-Pro was prepared via Alcalase® digestion of previously extracted proteins by freeze–thawing and sonication procedures [26]. Similarly, two potential anti-inflammatory peptides (LDAVNR and MMLDF) were obtained by subsequent proteolysis with trypsin, chymotrypsin, and pepsine of the proteins that were previously extracted from spirulina by freeze–thawing and sonication procedures [27]. All these studies were focused on purification and characterization of particular bioactive peptides [25–27]. In these cases, only the fractions of proteins previously extracted by physical or mechanical methods were treated with enzymes, and in none of the cases did the authors take into consideration the advantage that the direct action of the enzymes on the spirulina cells provides: higher recovery of intracellular components (e.g., proteins). Hence, more studies based on direct enzyme extraction protocols are required. In order to take full advantage of the enzyme-assisted extractions of proteins, direct studies of the digestion of *Spirulina* cells by enzymes should be carried out. In that respect, appropriate use of the enzyme technology requires determining for each biocatalyst the influence of different operation parameters on the enzyme degrading activity for each specific biomass.

In this work, the application of different enzymes for obtaining easily adsorbed proteic extracts was studied. Different selective enzyme degradation processes of *Spirulina* biomass were investigated for the extraction of polar spirulina biocomponents. The biomass was enzymatically degraded using four different cheap and easily accessible commercial enzyme preparations. These processes at their corresponding best operation conditions were compared with the extraction done without any prior enzyme treatment of the biomass. In particular, two different enzyme treatments, based on the degradation of membrane proteins, lipoproteins, and peptidoglucan by two proteases (Alcalase® and Flavourzyme®), and two other biomass treatments using endo- and exo-glucanases (Ultraflo® and Vinoflow®) to breakdown the sugar polymer structure, were comparatively studied. The best values of the most important parameters of all the enzymatic pre-treatments for biomass degradation were determined. The corresponding extractive yields in dry weight of aqueous extracts were determined for the four enzyme extraction processes at their respective best conditions, and they were compared with those of the control extract (without enzyme assistance). Changes occurring at the cellular level after the different extraction processes were comparatively analyzed by transmission electron microscopy (TEM). Total amino acid contents of all hydrophilic extracts were compared.

2. Results and Discussion

A commercial dry biomass of the cyanobacteria *Arthrospira platensis* (*Spirulina*) was submitted to different enzyme degradations. *S. platensis* extractions after four different enzyme pre-treatments were independently studied. These processes were carried out in their respective best operation conditions and compared with the control (no enzyme added) extraction. Two distinct proteases and two distinct glucanases were used to favor biocomponent recovery via membrane enzymatic degradation. The four different enzyme extractions were analyzed in terms of both the weight yields of the dry aqueous extracts and their respective amino acids contents.

2.1. Enzyme-Assisted Extraction

The most important parameters of the enzyme-assisted degradation of spirulina were studied. The best operation conditions for each enzyme extraction were determined by following changes in each peak area value of the high performance liquid chromatography with a evaporative light scattering detector (HPLC-ELSD) chromatograms of the aqueous phase. In all cases, the area changes in all the significant peaks were inspected (see below). All significant peaks of the HPLC chromatogram exhibited the same dependence of the studied parameters, that is, the same variation (increase or decrease) of their area values with the studied parameter value, and the same optimal value. Hence, for the sake of clarity, the influences of pH, time, temperature, and enzyme loading are depicted with respect to the total area values of all significant peaks. Nevertheless, a couple of examples of the parameter (pH and time) effect in one peak area are also given.

2.1.1. Influence of pH

Figure 1 represents the change of the total peak areas in the chromatogram (Figure 1A), and the area of a representative individual peak (Figure 1C) obtained after 4h of biomass pre-treatment with Alcalase®. The maximal initial rate (4h) of the biocomponent extraction with Alcalase® was obtained at pH 6.5. The enzyme degradation process was faster at this pH value, giving rise to the maximal total peak area value obtained for a short time pre-treatment (4h, Figure 1A). For the extractions with Flavourzyme®, Ultraflo®, and Vinoflow®, the corresponding best pH values determined were 6.0, 7.0, and 6.5, respectively (not shown).

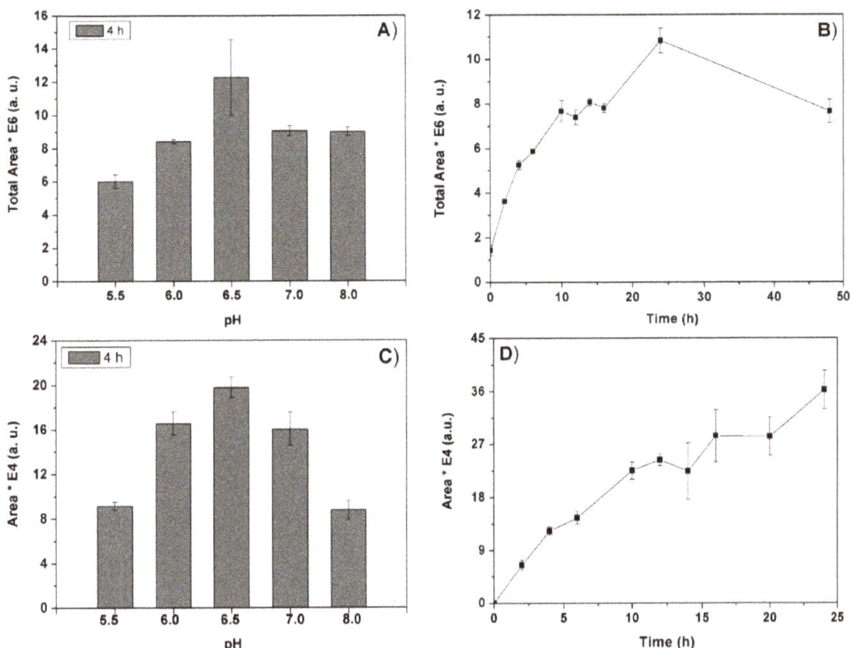

Figure 1. Enzyme-assisted extraction with Alcalase® of the polar spirulina biocomponents. Variations of total peak area in the HPLC-ELSD analyses: (**a**) effect of pH on the total area of peak products extracted at 4 h of enzyme treatment. (**b**) Time course of the extraction at optimal pH 6.5, for the total peak area. (**c**) Variation of the area value of the peak at 67.83 min retention time, for 4 h of enzyme treatment. (**d**) Time course of the extraction at optimal pH 6.5, for the peak at 5.5 min retention time. Other conditions: 1% (*v*/*w*) enzyme loading and 40 °C. A.U.: arbitrary area units.

Lu et al. used a different pH value for the extraction of an anti-hypertensive peptide from spirulina using Alcalase® (pH 8.5 at 50 °C for 10h) after three cycles of freeze–thawing the biomass [26]. Kim et al. also used different operation conditions to obtain the iron-chelating peptide with a multi-enzymatic biomass degradation (Alcalase® pH 8.0 at 50 °C for 1 h + Flavourzyme® pH 7.0 at 50 °C for 8 h) [25]. The obtained products of their peptide degradation were also different from those of this study (work in progress, personal communication).

The dependence of the extraction yield on the pH parameter is a consequence of the effect of pH on the process of biomass degradation by the enzyme. Variations in the total charge of proteins with pH should determine their mutual interaction and interaction with the biocatalyst employed. For the extraction of anti-cancer biocomponents from spirulina, papaine presented an optimal pH of 6.5, while other enzymes exhibited very different optimal pH values (pH 2–8.5) [28]. Optimal values in the basic range of pH were earlier described for the extraction of spirulina antioxidants with different enzymes, although other group of enzymes exhibited low activities under these conditions [29]. Different proteases from those used here were earlier studied to extract spirulina oil, reporting an optimal pH range of 7.5–10 for the best biocatalyst [30].

The time course of the enzyme-assisted extraction at the best pH value was studied for the first 48 h. The results obtained with Alcalase® for all the chromatogram peaks are represented in Figure 1B, and for a representative peak in Figure 1D. Similarly to Alcalase® extraction, the other three types of enzyme extractions required 24 h of enzyme pre-treatment of spirulina to reach maximal recovery of biocomponents (not shown). These results (Figure 1B,D) were obtained via analyses of aliquots taken from the reaction mixtures at the indicated times, while those of Figure 1A,B correspond to analyses of lyophilized extracts. The two studies correspond to samples of different extract concentrations, so the area values of these two figures cannot be compared (see Materials and Methods Section 3.3.1).

2.1.2. Effect of Temperature

The effect of the temperature of the biomass pre-treatment on the extraction of polar biocomponents was carried out at the respective optimal pH values of each enzyme (Alcalase® at pH 6.5; Flavourzyme® at pH 6.0; Ultraflo® at pH 7.0; Vinoflow® at pH 6.5) and an enzyme loading of 1% (*v*/*w*) in the range 30–50 °C after 24 h of biomass treatment (Figure 2). The total area values of peak products of the corresponding aqueous extracts were calculated from the obtained HPLC-ELSD chromatograms. The area value corresponding to 1 g of spirulina was calculated and represented against temperature. Figure 2 depicts the total peaks area values obtained for Alcalase®.

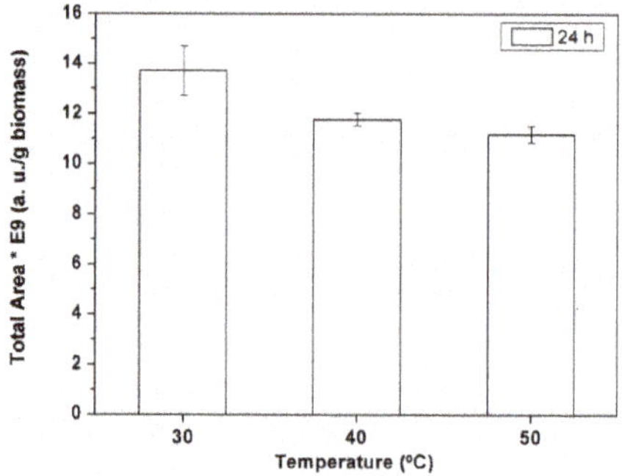

Figure 2. Effect of the temperature on the Alcalase® assisted extraction of the polar spirulina biocomponents. Variation of total peak areas in the HPLC-ELSD analyses relative to 1 g biomass. Conditions: 1% (*v*/*w*) enzyme loading and pH 6.5; A.U.: arbitrary units.

Considering the experimental error, 30 °C was the best temperature determined for the extractions with Alcalase®. Similar results were obtained with the other three enzymes (not shown). From these results, 30 °C was determined to be the best temperature for the two proteases (Alcalase® and Flavourzyme®) and Ultraflo®, the latter one having β-glucanase and xylanase activities as well as several side activities (Cellulase, Hemicellulase, and Pentosanase). The optimal temperature value determined for Vinoflow® was 40 °C (not shown). In all cases, the best temperature to achieve the maximum recovery of biocomponents in the aqueous extract was relatively low. Vitamins, antioxidants, and other thermo-labile components of the extracts were relatively well preserved in this temperature range. At mild temperatures the energy expenses and the product lability were reduced, while the operational stability of the biocatalysts employed was increased. In fact, the decrease observed in the extraction of biocomponents at the higher temperatures (Figure 2), might be explained by a decay of the operational stability of the enzyme.

In the case of proteases, Zhang & Zhang reported higher optimal temperatures for enzyme-assisted extraction of anti-tumor polypeptides from spirulina with trypsin, pepsine, and papain, namely 42 °C, 37 °C, and 55 °C, respectively, than the ones herein determined [28].

2.1.3. Effect of the Enzyme Loading

Catalyst charge affects the speed of biomass biodegradation prior to solvent extraction, determining the extraction yield value obtained at a given time.

The results obtained with different Alcalase® loadings (0–2% *v/w* enzyme) are represented in the Figure 3. These results were compared with those of a control extraction assay, where, instead of the enzyme solution, an equivalent volume of water solution was used (0% enzyme loading in Figure 3).

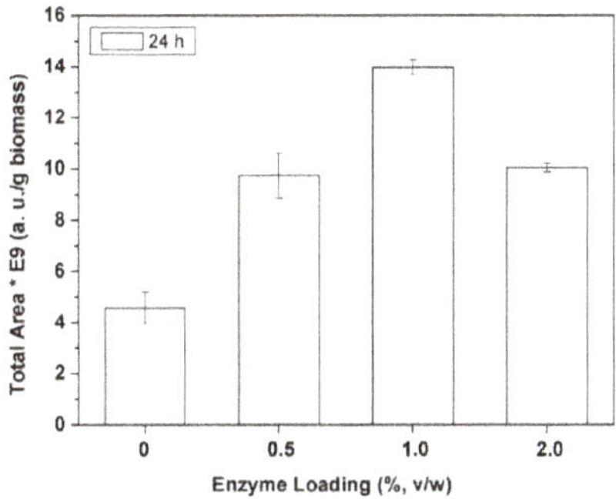

Figure 3. Effect of the enzyme loading on the Alcalase® assisted extraction of the polar spirulina biocomponents. Variation of total peak areas in the HPLC-ELSD analyses relative to 1 g biomass. Conditions: 30 °C and pH 6.5; A.U.: arbitrary units.

The extraction of polar biocomponents increased with the increase of the enzyme loading from 0.5% to 1% (*v/w*) Alcalase®, but it dramatically decreased with an higher enzyme amount (i.e., 2% *v/w*; Figure 3). The same best value was found for Flavourzyme® and Ultraflo®. The best enzyme loading of Vinoflow® was 2% (*v/w*) (not shown). These values correspond to 1.2–2.4% (*w/w*), considering the corresponding density values.

Our results are in good agreement with those of the literature, where the dependence of the extraction yield exhibits a maximum value, with a value of biocatalyst loading above which the process becomes less efficient [31]. The decrease of the product recovery when using high enzyme charges suggests that high enzyme concentrations might favor an excessive biodegradation of the biomass. Different types of biomasses require different amounts of biocatalysts. All these reports and the results of this work indicate that the optimal charge of enzyme requested depends on the type of biomass, type of the biocatalyst, and the biocomponents to be extracted.

The duration of the enzyme pre-treatment is an important parameter, involving important considerations in the process scale up. In this work, 24 h seemed to be necessary to get maximum extraction of intra- and extracellular biocomponents from spirulina. Liang et al. [5] studied an enzyme-assisted extraction of microalgae oil, where optimal yield extraction was found at 12 h of biomass biodegradation. These results, and those of the literature, suggest that the necessary time to degrade the biomass depends on both the enzyme and the biomass studied. It also substantially varies with the method used to degrade the biomass and the product to be extracted. Extractions based on the use of ultrasounds and microwave irradiation for biomass degradation use relatively short times (20 min–4 h) for antioxidant and pigment extractions from microalgae [16].

2.2. Scale Up and Extraction Yields

The four different enzyme-assisted extraction processes were carried out at their corresponding optimal operation conditions at a scale increased by a factor of 2.5. Ten grams of *Spirulina* biomass were used for each experiment. All the extracts were prepared in their respective optimal conditions (pH 6.5, 1% *v/w* and 30 °C for Alcalase®; pH 6.0, 1% *v/w* and 30 °C for Flavourzyme®; pH 7.0, 1% *v/w* and 30 °C for Ultraflo®; pH 6.5, 2% v/w and 40 °C for Vinoflow®). The control extract was obtained in this scale without any enzyme assistance, using milli-Q water instead of the enzyme solution, at 30 °C and 24h. The aqueous extracts obtained with the different enzyme-assisted methods were liophylyzed and kept at –70 °C.

Table 1 summarizes the recovery yields of aqueous extracts obtained. They are expressed in dry weight percent (calculated with respect to the starting *Spirulina* biomass).

Table 1. Extraction yield and total content of amino acids of the aqueous extracts obtained with the enzyme-assisted and control extraction procedures in their respective optimal operation conditions.

Sample	Yield	Total Amino Acid Content				
	(%, *w/w*) [1]	% (*w/w*)	μmol/g Extract	mg/g Biomass	μmol/g Biomass	R [3]
Control [2]	19.20 ± 0.20	34.4 ± 0.5	2986 ± 47	66.0 ± 1.0	573 ± 9	1.0
Alcalase®	36.50 ± 0.10	45.0 ± 7.1	3907 ± 69	164 ± 26.0	1426 ± 25	2.5
Flavourzyme®	31.80 ± 0.10	16.9 ± 0.2	1471 ± 16	53.8 ± 0.5	468 ± 5	0.8
Ultraflo®	19.70 ± 0.10	13.7 ± 0.5	1189 ± 40	27.0 ± 0.91	234 ± 8	0.4
Vinoflow®	26.30 ± 0.10	13.5 ± 0.1	1172 ± 9	35.5 ± 0.3	308 ± 3	0.5

[1] By weight percent with respect to the starting *Spirulina* biomass. [2] No enzyme pre-treatment. [3] Mole ratio of total amino acids obtained per gram of extract with and without enzyme assistance. Mean values of individual amino acid contents of all the extracts resulted statistically not equal in the *t*-test ($p \leq 0.05$).

All the enzyme-assisted extraction processes enabled greater yields of aqueous extracts than the control extraction, being higher the yields obtained with the two proteases (Alcalase® and Flavourzyme®). The highest extraction yield was obtained with Alcalase®. This biocatalyst enables an increase by 1.9 times of the weight of extract obtained, compared to the one obtained in the control extraction process without any enzyme assistance (Table 1).

2.3. Compositional Analysis of Polar Extracts

All dry extracts (obtained in their respective best conditions) were analyzed to determine their amino acid and peptide compositions. The extracts were rich in free amino acids and short peptides. The total values obtained for both hydrolyzed and free amino acids are given in Table 1.

Alcalase® extract had a significantly higher content of amino acids (45% w/w dry extract) than the extract obtained without any enzyme assistance (34% w/w), and also higher than the other three enzymatic extracts (13–17% w/w). Alcalase® extract has higher nutritional interest, and some of the peptides could exhibit bioactivities of therapeutic interest. This value corresponded to most amino acids of the spirulina (50–65% w/w, Table 2).

Table 2. Composition of spirulina from ASN LEADER S.L., provided by the manufacturer.

General Composition					
Proteins	**Lipids**	**Carbohydrates**	**Minerals**	**Fiber**	**Energy**
50–65%	6–7.5%	18–22%	15%	0.2%	390 cal/100 g

Total (hydrolyzed + free) amino acids content values were also calculated per gram of dry spirulina. The ratio between amounts of extracted amino acids for each particular method and the quantity obtained with the control extraction was greater than 1 only in the case of Alcalase®, indicating the superiority of Alcalase® method not only with respect to the control extraction, but also with respect to the other enzymatic methods ($p \leq 0.05$). Remarkably, among the two proteolytic enzymes and the two studied glucanases, only Alcalase® significantly increased the recovery percentage of amino acids with respect to the control assay (853 μmol/g spirulina more). These results suggest that amino acid extraction is not necessarily improved by any type of protease-assisted extraction. However, only the process with Alcalase® yielded a weight of amino acids 2.5 times higher than the control extraction, and 3.1 times higher than the extraction with the other protease (Flavourzyme®). From 1 g biomass, the Alcalase® extraction method obtained 1426 μmol amino acids, a quantity 2.5–6.1 times higher than the total amount obtained with the other methods (control without enzyme and enzyme-assisted methods). The control extraction obtained only 573 μmol amino acids/g biomass.

Liquid chromatography coupled to an electrospray ionization mass spectrometer in positive ionization mode (LC ESI-MS/MS) analyses revealed that the most abundant peptides in the spirulina extracts obtained with four enzymes and without an enzyme pre-treatment (control) were all different from one extract to another. More specifically, the eight peptides present in Alcalase® extract were: MKKIEAIIRPF, LPPL, ALAVGIGSIGPGLGQGQ, TTAASVIAAAL, DFPGDDIPIVS, LELL, WKLLP, and CHLLLSM (438–1493 Da). By contrast, the control extract contained 12 peptides, namely: NGDPFVGHL, VFETGIKVVDL, DFFVDKL, SGPPLDIKL, DVNETVLDNLPKTRTQI, DVNETVLDNLP, DSLISGAAQAVY, GIGNDPLEIQF, GLILLPHLATL, GLILLPHLA, AVLGAGALFHTF, and DVNETVLDNLP (851–1955 Da).

2.4. Biomass Analyses by TEM

Structural changes in the spirulina were investigated by TEM analysis (Figures 4 and 5).

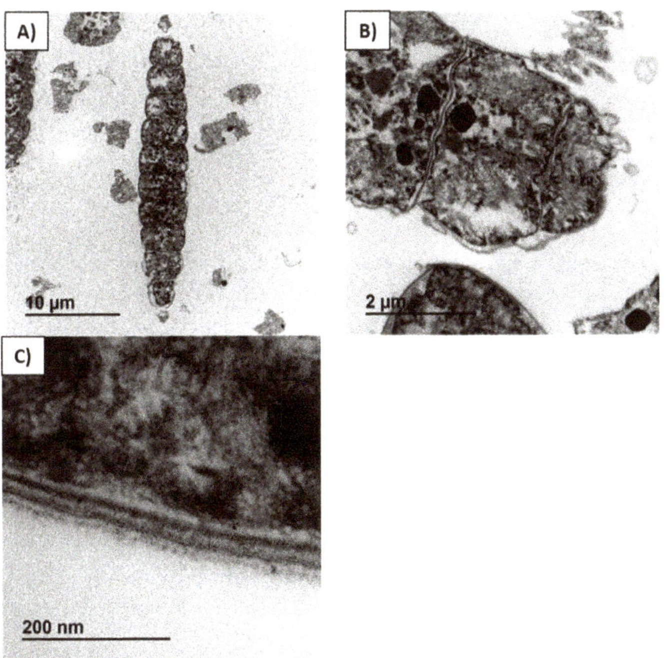

Figure 4. Micrographs of *Arthrospira (Spirulina) platensis* dry biomass, neither treated with any enzyme nor extracted by any solvent; (**A**) a longitudinal cut trichome and several transversally cut trichomes; (**B**) three contiguous cyanobacteria of a trichome separated by their cell walls; (**C**) detail of the cell wall.

Figure 4A is a micrograph of commercial dry biomass not treated with enzymes nor extracted with any solvent, where a longitudinal section through a trichome with twelve cells is visualized. Complete integrity of the cellular material, including the membrane, is clearly visualized in Figure 4B,C. The four layers of cell membrane were identified (Figure 4C). In the starting spirulina, the cytoplasm and thylakoid system were compressed against the internal cell membrane.

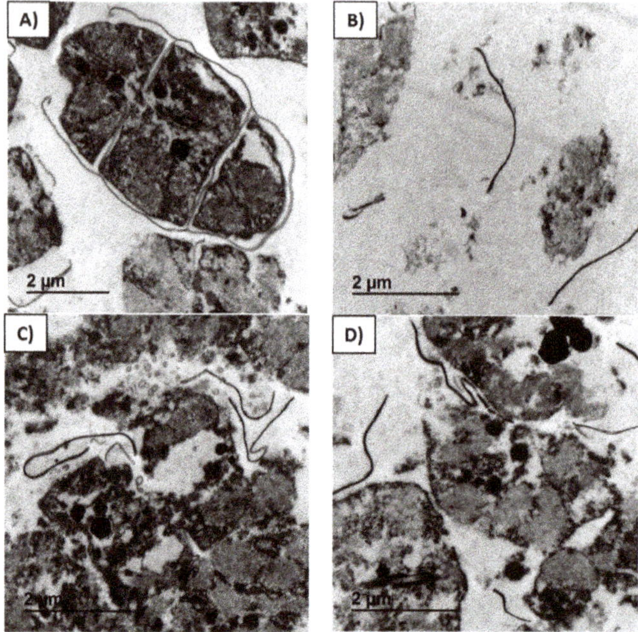

Figure 5. Transmission electron microscope (TEM) micrographs of residual biomasses of *Arthrospira p.*, obtained after the control extraction (**A**) and the extractions with Alcalase® (**B**), Ultraflo® (**C**), and Vinoflow® (**D**) pre-treatments. Extracts were obtained in their optimal conditions (pH 6.5, 1% *v*/*w* Alcalase® and 30 °C; pH 7.0, 1% *v*/*w* Ultraflo® and 30 °C; pH 6.5, 2% *v*/*w* Vinoflow® and 40 °C), and the control experiment was done with milli-Q water instead of the enzyme solution at 30 °C and 24 h.

In Figure 5, the micrographs of residual biomass obtained after the control extraction (Figure 5A), and residual biomasses obtained after extraction with Alcalase® (Figure 5B), Ultraflo® (Figure 5C), and Vinoflow® (Figure 5D) assistance are compared at the same scale. Trichomes were still observed after the control extraction of the biomass, although their thylakoid system was not any more compressed against the internal cell membrane (Figure 5A). Cells after the control extraction exhibited a swelling phenomenon, and a few were detached from the trichome. By contrast, after all the enzyme extractions studied, the number of trichomes observed was minimal and cellular degradation was evident (Figure 5B–D). A dispersion of remaining (non-extracted) intracellular material and membrane rests was obtained after biomass treatment with Alcalase® (Figure 5B). After extraction with Alcalase® assistance, most of the cellular material disappeared. Consequently, the residual biomass is more transparent (Figure 5B) than the other ones (Figure 5A,C,D). This fact is due to the higher recovery of biocomponents with Alcalase® than with the control and other enzyme extractions (Figure 5A,B). These observations agree with previous studies, where cyanobacteria cells modified their shape and size, swelling after a Lysozyme treatment [32]. Less cellular material remained in the residual biomasses extracted with Ultraflo® and Vinoflow® (Figure 5C,D) than in the control experiment (lower observed electronic density). However, in all these cases the amount of cell residues was significantly greater than in the case of the Alcalase® extraction (Figure 3B). These findings are in good correspondence with the values of extraction yields obtained (Table 1), where the Alcalase® extraction process was the most effective one for biocomponent recovery of spirulina.

The micrographs of this study show that Alcalase® pre-treatment of the biomass is the most efficient, resulting in most of trichomes and spheroplasts degraded. Vladimirescu revealed the existence of differences in enzymatic sensitivity of cells [33]. Compared with other bacteria, *Spirulina sp.* has a thicker cell membrane that makes more difficult to detach the cells or spheroplasts from the trichome. Because of that, cell disintegration is not efficiently achieved with lysozyme treatment [34].

The comparative inspection of micrographs of residual biomasses obtained with the different enzyme-assisted extraction processes (Figure 5) reveals that the enzyme pre-treatment of the biomass that allowed the highest biocomponent recovery (Alcalase®) produces the greatest cellular degradation. Compared with the other enzyme-assisted extraction types, the biomass extracted after Alcalase® treatment appeared nearly completely disintegrated and with very low electronic density, as a result of the greater extraction of its biocomponents (Figure 5B–D).

The results here described support the implementation of enzyme technology to replace conventional extraction processes. The Alcalase® extraction herein reported significantly increased the yield and quality of aqueous extracts from *A. (Spirulina) platensis* biomass. The enzyme technology developed uses food grade enzymes and hexane, which are accepted by regulatory agencies for food and drug safety. This enzyme technology looks promising for a more efficient, safe, and environmentally clean industrial production of cyanobacteria extracts with high value in the nutrition, cosmetic, and pharmaceutical industries [14]. The potential of this extraction method will be further clarified once the extracted biocomponents are molecularly and functionally characterized. A complete characterization of the different aqueous extracts obtained with all the enzymes in their respective optimal operation conditions will be reported.

3. Materials and Methods

3.1. Materials

Arthrospira (Spirulina) platensis dry biomass was purchased from ASN Leader S.L. (Murcia, Spain). Composition of the cyanobacteria biomass provided by the manufacturer is given in Table 2. The cyanobacteria biomass was a lyophilized dry powder for nutrition use. Solvents were HPLC grade. N-Hexadecane was used as an internal standard, and sodium sulfate as a desiccant. All of them were from Sigma-Aldrich (Madrid, Spain). Buffer solutions used were CH_3COONa/CH_3COOH (pH 5-6), Na_2HPO_4/NaH_2PO_4 (pH 6.5-8), and $NaCO_3$/ $NaHCO_2$ (pH 8.5-9). Alcalase® 2.4 L FG, Flavourzyme®, Ultraflo® L, and Vinoflow® Max A were generously donated by Novozymes A/S (Kalundborg, Denmark). Alcalase® has an activity of 2.4 AU A/g. Flavourzyme® has at least 1000 LAPU/g (leucine aminopeptidase units/g determined with Leu-pNA), not being the single activity type in this preparation. The main activity of Ultraflo® L is β-glucanase (45 fungal β -glucanase (FBG) per g). In addition, it contains approximately 470 Farbe xylanase units (FXU) per g. Vinoflow® Max A has a declared activity of 46 BGXU/ml.

Alcalase® is a commercial preparation of a serine endo-peptidase (EC. 3.4.21.62) from *Bacillus licheniformis* (mainly subtilisin A). Alcalase® acts as an esterase, catalyzing the stereoselective hydrolysis of esters, and hydrolyzes amino esters including heterocyclic amino esters. Flavourzyme® is a peptidase preparation from *Aspergillus oryzae*, widely and diversely used for protein hydrolysis in industrial and research applications. Eight enzymes have been identified in Flavourzyme®, namely two aminopeptidases, two dipeptidyl peptidases, three endopeptidases, and one α-amylase. Purified Flavourzyme® enzymes were biochemically characterized with regard to pH and temperature profiles and molecular sizes [35]. Ultraflo® L is a multicomponent enzyme preparation that contains 5–10% (*w*/*w*) β-glucanase (endo-1,3(4)-) and 1–5% (*w*/*w*) xylanase (endo-1,4-) as the main active components (EC 3.2.1.6 and EC 3.2.1.8), produced by a strain of *Humicola insolens*. The enzyme preparation is used in the brewing industry to hydrolyze polysaccharide gums and to break down cell wall materials in cereals like beta-glucan and xylans [36]. These two types of enzymatic activities in Ultraflo® are cellulose action and catalysis of the hydrolysis of complex sugars in the amorphous regions of the cellular membrane. Ultraflo® L is also marketed within the European Union as a feed-additive under the name of 'Pentopan/Biofeed Plus.' There is a more abundant enzyme, while a second activity is due to other two enzymes. Vinoflow® Max A is a β-glucanase (exo-1,3-) preparation, used on wine to speed up the aging process [37]. All these enzyme preparations are GRAS type hydrolases.

3.2. *Enzyme-Assisted Extraction of Spirulina Biocomponents*

The enzyme-assisted extraction methods of the cyanobacteria biomass studied using Alcalase® 2.4 L FG, Flavourzyme®, Ultraflo® L, and Vinoflow® Max A differ in the cell degradation step (protein or sugar hydrolysis). All degraded biomasses were next extracted with solvents. These extractions were compared with the same solvent extraction procedure carried out without any enzyme assistance (control experiment).

The influence of the more important parameters of the enzymatic step was studied: pH 5.5–8, temperature 30–50 °C, and enzyme loading 0.5–2 (*v*/*w*, volume of enzyme preparation/weight of enzyme + biomass suspension). Optimal operation conditions of the four enzyme-assisted extraction processes were determined.

A suspension of the biomass (0.2 g/mL aqueous buffer) containing the corresponding loading of enzyme preparation (or an equivalent volume of milli-Q water, in the case of the control experiment), was kept under magnetic agitation for the indicated time at a controlled temperature. Unless indicated, the study was carried out in triplicate at the short and long reaction times (duration of enzyme pre-treatment: 4 and 24 h), as follows:

- Effect of pH: This study was carried out at 40 °C and 1% (*v*/*w*) enzyme solution
- Effect of temperature: This study was carried out at relatively mild temperatures (30–50 °C) at the optimal pH value previously determined.
- Effect of enzyme loading: This study was performed at the optimal values of pH and temperature previously determined in the range of 0.5–2% (*v*/*w*) enzyme solution. The process extraction was followed at 1, 2, 4, 6, 8, and 24 h.
- Solvent extraction step: Each aliquot (0.5 mL) of the enzyme–biomass suspension was dissolved in 1 mL hexane–isopropanol mixture (3:2, *v*/*v*). The resultant solution was centrifuged for 15 min at 10,000 rpm, and allowed to completely separate into the two liquid (aqueous and oil) phases. The residual biomass was then separated from the liquid phase. The liquid solution was then placed in a decantation funnel. After separation of the oil and aqueous phases, the oil phase was collected. Next, the oil phase extraction protocol was repeated two more times by adding 0.25 mL hexane-isopropanol mixture to the decantation funnel. Finally, the two liquid (aqueous and oil) extracts obtained were freeze-dried.
- Scaled up extraction: In order to obtain greater amounts of extracts, all the extraction processes (enzyme-assisted and control extractions) were carried out at a scale factor of 2.5 at their respective optimal conditions for 24 h enzyme pre-treatment under magnetic agitation (500 rpm). Scaled up extraction processes were carried out as follows: pH 6.5, 30 °C and 1% *v*/*w* Alcalase®; pH 6.0, 30 °C and 1% *v*/*w* Flavourzyme®; pH 7.0, 30 °C; and 1% *v*/*w* Ultraflo®; pH 6.5, 40 °C and 2% *v*/*w* Vinoflow®. The extraction without enzyme digestion (control) was carried out using milli-Q water instead of the enzymatic preparation in buffer at 30 °C and 24h. After the indicated time, the enzyme–biomass mixture was centrifuged for 30 min at 14,000 rpm and 10 °C. All corresponding aqueous extracts were lyophilized for 4 days and then weighed. All dry weight values of the aqueous extracts were corrected by subtraction of the corresponding weight of the buffer enzyme solution (lower than 1% *w*/*w*). Residual biomasses obtained were dried under nitrogen. All the experiments were carried out in triplicate.

3.3. *Analysis and Characterization of Aqueous Extracts*

3.3.1. HPLC Analyses

A high-performance liquid chromatography apparatus coupled to an evaporative light scattering detector (HPLC-ELSD) was used for optimization of the enzymatic extraction process. These analyses allowed determination of relative changes of the biocomponent concentrations in the aqueous extracts obtained at different times of enzyme treatment.

Analyses were carried out with a Hitachi D-7000IF apparatus (Germany) with a silica column from Kromasil C18 (5 μm, 250 × 4.6 mm) connected to a Sedex 55 ELSD detector (SEDERE, France). Aqueous solution of the lyophilized samples (20 μL of 50 mg/mL) was injected and analyzed at 30 °C for 83 min with a gradient mobile phase at 1.5 mL/min, phase A being milliQ water (100% *v*/*v*) and phase B being acetonitrile/milliQ water (80:20 *v*/*v*). The composition of the phases (A:B) varied as follows: 96:4 for the first 5 min, increasing to 60:40 in 60 min, followed by a linear increase of Phase B up to 95% in 1 min. Composition was then maintained for the next 7 min (up to min 23), then the mobile phase returned to the first composition (96:4) and remained constant for the rest of the analysis. Analysis of each sample was replicated three times. For analyses of liquid aqueous extracts obtained after 4 h enzyme pre-treatment, 0.5 mL of aqueous phase was mixed with 1 mL distilled water, and 20 μL of the resultant solution was injected into the HPLC.

3.3.2. Amino Acid and Peptide Composition

The extracts obtained in the respective optimal conditions were analyzed to determine their composition in amino acids, using a chromatography procedure developed by Spackman et al. [38]. Solutions of the different aqueous extracts (1–2.6 mg/mL) were prepared in triplicate and placed in the hydrolysis tubes. Norleucine was used as the internal standard in this assay. The analyzer apparatus was calibrated with three tubes of hydrolysis containing known amounts of the standard and norleucine. Standard solutions were submitted to the same hydrolysis treatment as the extract solutions. Finally, all the hydrolysis tubes were vacuum dried in a Speed Vac.

The hydrolysis tubes containing the sample solutions were placed in glass bottles with a valve to make a vacuum, and purged with inert nitrogen gas. To each flask, 200 μL HCl 6N and 50 mg phenol were added. Next, a vacuum was applied to each flask for 20 sec and then they were purged with nitrogen inert gas for 20 sec. This process was repeated three times. Each flask was introduced into an oven at 110 °C for 21 h. After that, the hydrolysis tubes were dried in the Speed Vac. Hydrolyzed samples and standard were dissolved in buffer, and then injected into the analyzer.

Quantitative analysis of amino acids mixtures was carried out in a Biochrom 30 Series Amino Acid Analyzer, with a reproducibility >0.5 CV at 10 nmoles. Biochrom 30 uses the classic methodology for amino acid analysis, based on ion exchange liquid chromatography and a post-column reaction made continuous with ninhydrine, with a sensitivity of ~10 pmol.

All the aqueous extracts were analyzed by liquid chromatography coupled to an electrospray ionization mass spectrometer in positive ionization mode (LC ESI-MS/MS) to identify their respective peptide components. Prior to analysis, the samples were cleaned with C18 tips. LC ESI-MS/MS analyses were carried out in an Ultimate 3000 nanoHPLC (Dionex, Sunnyvale, California) coupled to an ion trap mass spectrometer AmaZon Speed (Bruker Daltonics, Bremen, Germany). The reversed phase analytic column used was an Acclaim C18 PepMap of 75 μm × 15 cm, 3 μm particle size and 100 Å pore size (ThermoScientific, USA). The trap column was a C18 PepMap of 5 μm particle diameter, 100 Å pore size, connected in series with the analytical column. The loading pump flushed a solution of 0.1% trifluoroacetic acid in 98% water/2% acetonitrile (ScharLab, Barcelona, Spain) at 3 μL/ min The nanopump operated at a flow of 300 nL/min in gradient conditions, using 0.1% formic acid (Fluka, Buchs, Switzerland) in water (phase A), and 0.1% formic acid in 80% acetonitrile/20% water (phase B). The scheme of the elution gradient was: isocratic mode with 96% A, 4% B for 5 min, a linear increase to 40% B in 60 min, a linear increase to 95% B in 1 min, isocratic conditions of 95% B for 7 min, and return to initial conditions in 10 min. Five μL of extract solutions (4 μg/μL) were injected, and detected at 214 y 280 nm wavelengths. In a second analysis, 5 μL of extract solutions (10 μg/μL) were injected. The LC system was connected by a CaptiveSpray source (Bruker Daltonics, Bremen, Germany) to the ion trap spectrometer, operating in positive mode with a capillary voltage set of 1400 V. The automatic data acquisition allowed sequential observation of both MS spectra (m/z 350–1500) and the MS CID spectra of the 8 more abundant ions. In the analyses of 10 μg/μL samples, the MS spectra range was 100–1000 m/z. Exclusion dynamics were applied to prevent the isolation of the same m/z for 1 min after its fragmentation.

For peptide identification, MS and MS/MS data of individual fractions of HPLC were processed with DataAnalysis 4.1 (Bruker Daltonics, Bremen, Germany). MS/MS spectra (in the form of generic Mascot files) were analyzed against a data base obtained from NCBInr (National Center for Biotechnology Information) containing 68623 entries of proteins from both *Spirulina* and *Arthrospira*. The database search was carried out with Mascot v.2.6.0 (Matrix Science, London, UK) [39]. Search parameters were set as follows: oxidized methionine as the modification variable without enzyme restriction. Tolerance for peptide mass of 0.3 Da and 0.4 Da in MS and in MS/MS modes, respectively. In most of the cases, a precision of ±0.1–0.2 Da was obtained for both MS and MS/MS spectra.

Additionally, in the case of Alcalase® extract all MS and MS/MS spectra were analyzed using the 'de novo' tool of the Peaks software (Bioinformatics solutions, Inc).This program combines both the unconditioned 'de novo' analysis of MS/MS spectra with the more conventional search against organism-specific (i.e., *Arthrospira–Spirulina)* sequence databases. Only sequence assignments with confidence values equal or superior to 80 have been included, to avoid doubtful sequences. Note that this approach cannot distinguish the following identities: I and L, K and Q, F and M.

3.3.3. Statistical Analyses

The experiments were carried out in triplicate, reporting the results as their corresponding mean values with their standard errors, which were compared at confidence level of 95% ($p \leq 0.05$) using the SPSS program.

3.4. Transmission Electronic Microscopy Analyses, TEM

Fresh and residual (extracted) biomass samples were visualized in a transmission electronic microscope. Morphological changes of the residual biomass after the enzyme-assisted extraction process were visualized and compared by TEM analyses. A Jeol Jem 1010 apparatus (100Kv, Yokyo, Japan), coupled to a digital camera Orius SC200 (Gatan Inc., Pleasanton, CA, USA) and the Digital Micrograph v 3.4 software for image acquisition, was used. Prior to analysis, all the samples were treated as follows: first they were washed three times with 0.1 M sodium phosphate buffer, pH 7.2, then transferred into 2% *w/w* bacteriological agar in buffer and fixed in glutaraldehyde (2.5% *w/w*) for 2 h 40 min, and finally washed with cacodylate buffer 0.1M (pH 7.3). Post-fixation of samples was done on 1–2 mm agar blocks with osmium tetroxide (1% *w/w*) for 1 h 40 min. Samples were then dehydrated in an oven with absolute ethanol and embedded in a durcupan resin, and then were polymerized at 60 °C over 48 h. Samples were cut into ultrafine layers (60 nm) with a Leica ultracut S. Finally the sample slices were dyed with uranyl and lead acetates.

4. Conclusions

Different extraction methods of high value hydrophilic spirulina biocomponents were implemented via four selective enzyme degradations of spirulina biomass. Comparison of the extracts obtained in their optimal operation conditions demonstrated that different products can be obtained through spirulina degradation by different enzyme types. The four enzyme-assisted extraction processes were superior to the corresponding extraction process without enzyme-assistance for prior biomass degradation. Among the two proteases and the endo- and exoglucanases, Alcalase® gave the highest extraction yield of hydrophilic extract, as a result of its effective degradation of membrane proteins, lipoproteins, and peptidoglucan. Both the extract composition and the amount of extracted biocomponents depended on the temperature, enzyme charge and type, pH, and duration of enzymatic pre-treatment of the biomass. Compared to conventional extraction processes, higher extraction yields were obtained in mild conditions; Alcalase® extract was the one with the highest protein content. All the protein extracts obtained could be applied for satiety and muscle building in sports/active nutrition, for geriatric population, convalescent patients, etc.

Author Contributions: This work was designed and supervised by C.O.; C.M.V.-M. and L.E. carried out the experimental work; statistical analyses were done by C.M.V.-M.

Funding: This research was partially supported by the Program of I+D Activities of the Autonomous Community of Madrid, Spain: Tecnología 2013 (Project INSPIRA1-CM Ref. S2013/ABI-2783 "Aplicaciones industriales de la Espirulina" with co-financiation of the European Structural Found (European Social Found -EFS- and European Found for Regional Development- FEDER-), and by MINECO of Spanish Government (grant Ref. CTQ2017-86170-R).

Acknowledgments: The authors wish to thank Rafael Montoro Carrillo (manager of ASN Leader S.L.) for the generous support with a *Spirulina* free sample and information provided. We also thank Ramiro Martínez (Novozymes, Spain) for kindly supplying the enzymes used in this research.

Conflicts of Interest: The authors declare no conflict of interest. The funders had no role in the design of the study; in the collection, analyses, or interpretation of data; in the writing of the manuscript, or in the decision to publish the results.

References

1. Singh, R.; Parihar, P.; Singh, M.; Bajguz, A.; Kumar, J.; Singh, S.; Singh, V.P.; Prasad, S.M. Uncovering potential applications of cyanobacteria and algal metabolites in biology, agriculture and medicine: Current status and future prospects. *Front. Microbiol.* **2017**, *8*. [CrossRef] [PubMed]
2. Fleurence, J. The enzymatic degradation of algal cell walls: A useful approach for improving protein accessibility? *J. Appl. Phycol.* **1999**, *11*, 313–314. [CrossRef]
3. Beveridge, T.J. Structures of Gram-Negative Cell Walls and Their Derived Membrane Vesicles. *J. Bacteriol.* **1999**, *181*, 4725–4733. [PubMed]
4. Zheng, H.; Yin, J.; Gao, Z.; Huang, H.; Ji, X.; Dou, C. Disruption of chlorella vulgaris cells for the release of biodiesel-producing lipids: A comparison of grinding, ultrasonication, bead milling, enzymatic lysis, and microwaves. *Appl. Biochem. Biotechnol.* **2011**, *164*, 1215–1224. [CrossRef] [PubMed]
5. Liang, K.; Zhang, Q.; Cong, W. Enzyme-assisted aqueous extraction of lipid from microalgae. *J. Agric. Food Chem.* **2012**, *60*, 11771–11776. [CrossRef] [PubMed]
6. Vilas, M.V.A.; Hernandez, C.O. *At the Crossroads between Nutrition and Pharmacology*; Bentham Science Publishers: Sharjah, UAE, 2017; Volume 2.
7. Vonshak, A. *Spirulina Platensis Arthrospira: Physiology, Cell-Biology and Biotechnology*; Taylor & Francis: Milton Park, Oxfordshire, UK, 1997.
8. El-Baz, F.K.; El-Senousy, W.M.; El-Sayed, A.B.; Kamel, M.M. In vitro antiviral and antimicrobial activities of Spirulina platensis extract. *J. Appl. Pharm. Sci.* **2013**, *3*, 52–56. [CrossRef]
9. Clement, G. Une nouvelle algue alimentaire:la Spiruline. *Rev. Inst. Pasteur de Lyon* **1971**, *4*, 103–114.
10. Campanella, L.; Crescentini, G.; Avino, P. Chemical composition and nutritional evaluation of some natural and commercial food products based on Spirulina. *Analusis* **1999**, *27*, 533–540. [CrossRef]
11. Hsueh, Y.C.; Wang, B.J.; Yu, Z.R.; Wang, C.C.; Koo, M. Optimization of a continuous preparation method of arthrospira platensis γ-linolenic acid by supercritical carbon dioxide technology using response surface methodology. *Sains Malaysiana* **2015**, *44*, 1739–1744.
12. Singh, S.; Kate, B.N.; Banecjee, U.C. Bioactive compounds from cyanobacteria and microalgae: An overview. *Crit. Rev. Biotechnol.* **2005**, *25*, 73–95. [CrossRef]
13. Pfeiffer, C.; Bauer, T.; Surek, B.; Schömig, E.; Gründemann, D. Cyanobacteria produce high levels of ergothioneine. *Food Chem.* **2011**, *129*, 1766–1769. [CrossRef]
14. Lupatini, A.L.; Colla, L.M.; Canan, C.; Colla, E. Potential application of microalga *Spirulina platensis* as a protein source. *J. Sci. Food Agric.* **2017**, *97*, 724–732. [CrossRef] [PubMed]
15. Palinska, K.A.; Krumbein, W.E. Perforation patterns in the peptidoglycan wall of filamentous cyanobacteria. *J. Phycol.* **2000**, *36*, 139–145. [CrossRef]
16. Bermúdez Menéndez, J.M.; Arenillas, A.; Menéndez Díaz, J.A.; Boffa, L.; Mantegna, S.; Binello, A.; Cravotto, G. Optimization of microalgae oil extraction under ultrasound and microwave irradiation. *J. Chem. Technol. Biotechnol.* **2014**, *89*, 1779–1784. [CrossRef]

17. Hahn, T.; Lang, S.; Ulber, R.; Muffler, K. Novel procedures for the extraction of fucoidan from brown algae. *Process Biochem.* **2012**, *47*, 1691–1698. [CrossRef]
18. Postma, P.R.; Pataro, G.; Capitoli, M.; Barbosa, M.J.; Wijffels, R.H.; Eppink, M.H.M.; Olivieri, G.; Ferrari, G. Selective extraction of intracellular components from the microalga *Chlorella vulgaris* by combined pulsed electric field-temperature treatment. *Bioresour. Technol.* **2016**, *203*, 80–88. [CrossRef] [PubMed]
19. Neves, V.T.D.C.; Sales, E.A.; Perelo, L.W. Influence of lipid extraction methods as pre-treatment of microalgal biomass for biogas production. *Renew. Sustain. Energy Rev.* **2016**, *59*, 160–165. [CrossRef]
20. Rosenthal, A.; Pyle, D.L.; Niranjan, K. Aqueous and enzymatic processes for edible oil extraction. *Enzyme Microb. Technol.* **1996**, *19*, 402–420. [CrossRef]
21. Bligh, E.G.; Dyer, W.J. A rapid method of total lipid extraction and purification. *Can. J. Biochem. Physiol.* **1959**, *37*, 911–917. [CrossRef]
22. Folch, J.; Lees, M.; Sloane Stanley, G.H. A simple method for the isolation and purification of total lipides from animal tissues. *J. Biol. Chem.* **1957**, *226*, 497–509.
23. Safi, C.; Ursu, A.V.; Laroche, C.; Zebib, B.; Merah, O.; Pontalier, P.Y.; Vaca-Garcia, C. Aqueous extraction of proteins from microalgae: Effect of different cell disruption methods. *Algal Res.* **2014**, *3*, 61–65. [CrossRef]
24. Safi, C.; Charton, M.; Ursu, A.V.; Laroche, C.; Zebib, B.; Pontalier, P.Y.; Vaca-Garcia, C. Release of hydro-soluble microalgal proteins using mechanical and chemical treatments. *Algal Res.* **2014**, *3*, 55–60. [CrossRef]
25. Kim, N.H.; Jung, S.H.; Kim, J.; Kim, S.H.; Ahn, H.J.; Song, K.B. Purification of an iron-chelating peptide from spirulina protein hydrolysates. *J. Korean Soc. Appl. Biol. Chem.* **2014**, *57*, 91–95. [CrossRef]
26. Lu, J.; Ren, D.F.; Xue, Y.L.; Sawano, Y.; Miyakawa, T.; Tanokura, M. Isolation of an antihypertensive peptide from alcalase digest of spirulina platensis. *J. Agric. Food Chem.* **2010**, *58*, 7166–7171. [CrossRef] [PubMed]
27. Vo, T.S.; Ryu, B.; Kim, S.K. Purification of novel anti-inflammatory peptides from enzymatic hydrolysate of the edible microalgal *Spirulina maxima*. *J. Funct. Foods* **2013**, *5*, 1336–1346. [CrossRef]
28. Zhang, B.; Zhang, X. Separation and nanoencapsulation of antitumor polypeptide from *Spirulina platensis*. *Biotechnol. Prog.* **2013**, *29*, 1230–1238. [CrossRef] [PubMed]
29. Ismaiel, M.M.S.; El-Ayouty, Y.M.; Piercey-Normore, M. Role of pH on antioxidants production by Spirulina (Arthrospira) platensis. *Braz. J. Microbiol.* **2016**, *47*, 298–304. [CrossRef]
30. In, M.J.; Gwon, S.Y.; Chae, H.J.; Kim, D.C.; Kim, D.H. Production of Spirulina Extract by Enzymatic Hydrolysis. *J. Korean Soc. Appl. Biol. Chem.* **2007**, *50*, 304–307.
31. Mushtaq, M.; Sultana, B.; Anwar, F.; Adnan, A.; Rizvi, S.S.H. Enzyme-assisted supercritical fluid extraction of phenolic antioxidants from pomegranate peel. *J. Supercrit. Fluids* **2015**. [CrossRef]
32. Lindsey, J.K.; Vance, B.D.; Keeter, J.S.; Scholes, V.E. Spheroplast formation and associated ultrastructural changes in a synchronous culture of *Anacystis nidulans* treated with lysozyme. *J. Phycol.* **1971**, *7*, 65–71. [CrossRef]
33. Vladimirescu, A.F. Isolation of permeaplasts and spheroplasts from *Spirulina platensis*. *Romanian Biotechnol. Lett.* **2010**, *15*, 5361–5368.
34. Yi, P.; Zhao, Y.J.; Guo, H.L. Induction of vacuolated spheroplasts and isolation of vacuoles in cyanobacteria. *J. Phycol.* **2005**, *41*, 366–369. [CrossRef]
35. Merz, M.; Eisele, T.; Berends, P.; Appel, D.; Rabe, S.; Blank, I.; Stressler, T.; Fischer, L. Flavourzyme, an Enzyme Preparation with Industrial Relevance: Automated Nine-Step Purification and Partial Characterization of Eight Enzymes. *J. Agric. Food Chem.* **2015**, *63*, 5682–5693. [CrossRef] [PubMed]
36. Smith, J. *Mixed β-Glucanase, Xylanase from Humicola Insolens*; Chemical and Technical Assessment (CTA); FAO: Rome, Italy, 2004; pp. 1–5.
37. Pronk, I.M.E.J.; Leclercq, C. Mixed xylanse, β-glucanase enzyme preparation produced by a strain of humicola insolens. *WHO Food Addit. Ser. JECFA* **2004**, *52*, 1–6.
38. Spackman, D.H.; Stein, W.H.; Moore, S. Automatic Recording Apparatus for Use in the Chromatography of Amino Acids. *Anal. Chem.* **1958**, *30*, 1190–1206. [CrossRef]
39. Perkins, D.N.; Pappin, D.J.C.; Creasy, D.M.; Cottrell, J.S. Probability-based protein identification by searching sequence databases using mass spectrometry data. *Electrophoresis* **1999**, *20*, 3551–3567. [CrossRef]

Article

Preparation of Sterically Demanding 2,2-Disubstituted-2-Hydroxy Acids by Enzymatic Hydrolysis

Andrea Pinto [1,*], Immacolata Serra [1], Diego Romano [1], Martina Letizia Contente [1], Francesco Molinari [1], Fabio Rancati [2], Roberta Mazzucato [2] and Laura Carzaniga [2,*]

1 Department of Food, Environmental and Nutritional Science (DeFENS), University of Milan, 20133 Milan, Italy; Immacolata.serra@unimi.it (I.S.); diego.romano@unimi.it (D.R.); martina.contente@nottingham.ac.uk (M.L.C.); francesco.molinari@unimi.it (F.M.)

2 Chemistry Research and Drug Design Department, Corporate Preclinical R&D, Chiesi Farmaceutici S.p.A., 43122 Parma, Italy; F.Rancati@chiesi.com (F.R.); R.Mazzucato@chiesi.com (R.M.)

* Correspondence: andrea.pinto@unimi.it (A.P.); L.Carzaniga@chiesi.com (L.C.); Tel.: +39-02503-16814 (A.P.); +39-0521-279652 (L.C.)

Received: 18 December 2018; Accepted: 19 January 2019; Published: 24 January 2019

Abstract: Preparation of optically-pure derivatives of 2-hydroxy-2-(3-hydroxyphenyl)-2-phenylacetic acid of general structure 2 was accomplished by enzymatic hydrolysis of the correspondent esters. A screening with commercial hydrolases using the methyl ester of 2-hydroxy-2-(3-hydroxyphenyl)-2-phenylacetic acid (1a) showed that crude pig liver esterase (PLE) was the only preparation with catalytic activity. Low enantioselectivity was observed with substrates 1a–d, whereas PLE-catalysed hydrolysis of 1e proceeded with good enantioselectivity (E = 28), after optimization. Enhancement of the enantioselectivity was obtained by chemical re-esterification of enantiomerically enriched 2e, followed by sequential enzymatic hydrolysis with PLE. The preparation of optically-pure (*S*)-2e was validated on multi-milligram scale.

Keywords: esterase; stereoselective; ester hydrolysis; antimuscarinic agents; pig liver esterase (PLE)

---

1. Introduction

Enzymatic hydrolysis of chiral esters using carboxylesterases is an established method for obtaining kinetic and dynamic resolution [1–5]. A number of stereoselective carboxylesterases is nowadays available, and troublesome application such as the hydrolysis of spatially bulky substrates can be solved by screening and protein engineering [6]. Esters of carboxylic acids with sterically-demanding α-substitutions are not easily hydrolysed by most of the lipases, and protein engineering for making natural enzymes able to accept these substrates is still limited to relatively bulky carboxylic acids [7]. Enzymatic hydrolysis of carboxylic acid esters having an α-quaternary or α-tertiary centre is still a difficult task [8]; in contrast to the broad spectrum of esters with bulky alcohol moieties accepted as substrates [9,10]. Activation by electron-withdrawing (EW) hetero-atoms (e.g., O and N) or by EW-functional groups (e.g., -NO_2, -CN, -CF_3) is often required to observe enzymatic hydrolytic activity [11–13]. α-,α-Disubstituted malonate diesters are among the few α-,α-,α-trisubstituted carboxylic esters accepted by carboxylesterases; in particular, pig liver esterase (PLE) is particularly suited for catalyzing the enantioselective monohydrolysis of differently substituted malonate diesters [14], including ester derivatives, such as dimethyl 3,3-dimethyl-2-methylenecyclohexane-1,1- dicarboxylate, a chiral building block used for the enantioselective total synthesis of ent-kauranoids [15].

In this work, we have studied the enzymatic hydrolysis of esters 1, derivatives of sterically demanding 2,2-diaryl-2-hydroxy acids 2 (Figure 1); these molecules attract great attention for pharmaceutical applications as they can be useful chiral building blocks for the synthesis of compounds exerting muscarinic M3 receptor antagonist activity [16,17]. Antimuscarinic agents have a variety of applications but one of the most well established is their use as inhaled bronchodilators for the treatment of obstructive airway diseases such as asthma and chronic obstructive pulmonary disease (COPD) [18]. The enzymatic hydrolysis of ester 1 has been therefore investigated as a possibly suitable, affordable and sustainable method alternative to classical liquid (LC)/supercritical fluid chromatography (SFC) chiral separation of racemic mixtures or diastereomeric salt crystallization, to obtain the desired active (*S*)-enantiomer 2.

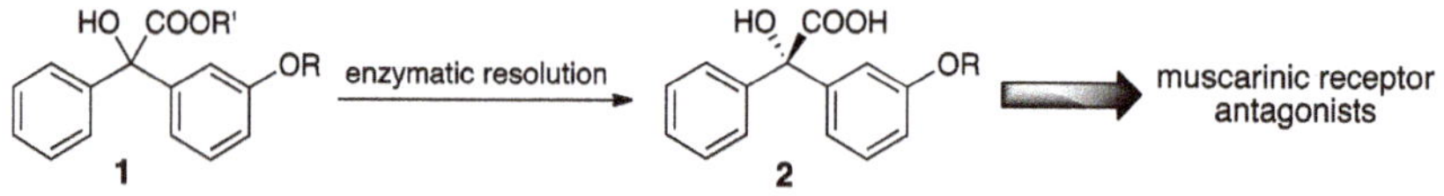

Figure 1. Kinetic resolution of esters of 2,2-diaryl-2-hydroxy acids; optically pure acids are building blocks for the synthesis of muscarinic receptor antagonists.

2. Results

2.1. Screening of Biocatalysts and Substrates

The synthesis of esters 1a-e, used in this work, was realized as described in Scheme 1.

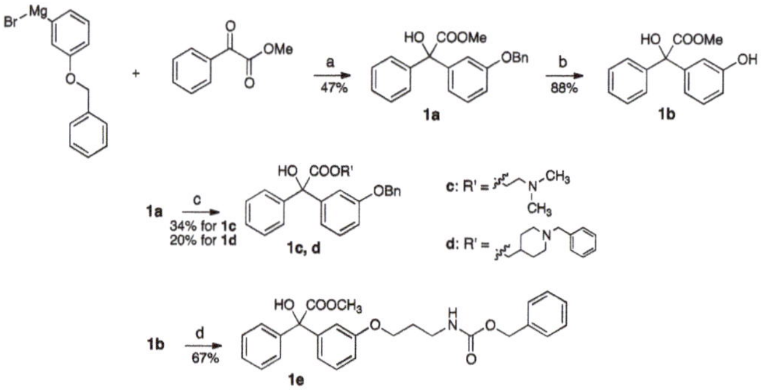

Scheme 1. Synthesis of esters 1**a**-**e**. Reaction conditions: **a**: THF, RT, 16 h; **b**: dihydrotoluene, 10% Pd/C, EtOH, reflux, eight hours; **c**: (i) LiOH, THF/water, RT, 2 h; (ii) CDI, 2-(dimethylamino)ethanol, CH_2Cl_2, 60 °C, 4 h; **d**: N-(benzyloxycarbonyl)-3-amino-1-propyl methanesulfonate, $CsCO_3$, DMF, RT, 16 h.

Hydrolysis of 1a-b was firstly investigated using 20 commercial hydrolases and 15 enzymatic preparations from our laboratory [19–24]; only commercial PLE gave hydrolysis of 1a,b (Scheme 2) with conversions ranging between 50 and 100% after 24 h (Table 1).

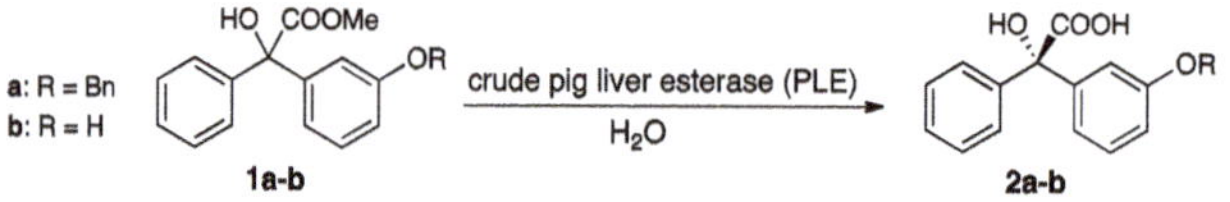

Scheme 2. Enzymatic hydrolysis of esters 1**a**-**b** with pig liver esterase (PLE).

Table 1. Hydrolysis of 1a-b with pig liver esterase (PLE); Conditions: [S] = 2.5 mM, [PLE] = 7.5 mg/mL in 100 mM phosphate buffer at pH = 7.0 and DMSO (5%), 30 °C.

Entry	Substrate	Conv. (%)	ee $_{(R)\text{-ester}}$ (%)	ee $_{(S)\text{-acid}}$ (%)	E	Time (h)
1	1a	52	67	63	8	5
2	1a	>97	<5	<5	-	24
3	1b	50	67	67	10	5
4	1b	> 97	<5	<5	-	24

The reactions occurred with excellent rates, but low enantioselectivity, furnishing the *S*-acid with enantiomeric ratio (E) ranging between 8 and 10. Absolute configurations were assigned by comparison with enantiomerically pure sample previously synthesized [16]. Different (bulkier) alcohol moieties were introduced with the aim of increasing the enantioselectivity, therefore esters 1c,d were synthesized as shown before and used as substrates for the enzymatic hydrolysis with commercial PLE, but enantioselectivity remained quite low (E < 8 in both the cases).

As a strategy for improving enantioselectivity, we synthesized 1e, where a benzyloxy propylcarbamate was introduced as *meta*-substituent for boosting the structural diversity of the two aromatic groups (Scheme 3).

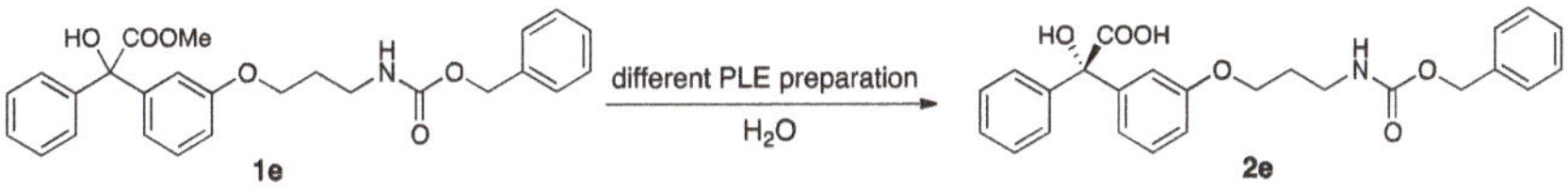

HO COOMe ... 1e —different PLE preparation, H_2O→ HO COOH ... 2e

Scheme 3. Enzymatic hydrolysis of ester 1ewith pig liver esterase (PLE).

In fact, the kinetic resolution of 1e occurred with higher enantioselectivity (E = 21, entry 1, Table 2) than what observed with 1a–d. Commercial PLE preparation is extracted from animal tissues and composed by 6 different isoenzymes, each one potentially leading to different stereoselectivity [14,25]; therefore, we also tested the six isoforms as single recombinant enzymes (commercially available and named ECS-PLE 01–06) for the hydrolysis of 1e (Table 2, entries 2-7).

Table 2. Hydrolysis of 1e with PLE; conditions: [S] = 2.5 mM, [PLE] 7.5 mg/mL in 100 mM phosphate buffer at pH = 7.0 and DMSO (5%), 30 °C.

Entry	Substrate	Conv. (%)	ee $_{(R)\text{-ester}}$ (%)	ee $_{(S)\text{-acid}}$ (%)	E	Time (h)
1	Crude PLEs	30	37	87	21	5
2	ECS-PLE01	<5	-	-	-	24
3	ECS-PLE02	<5	-	-	-	24
4	ECS-PLE03	19	8	30	<5	24
5	ECS-PLE04	8	-	n.d.	-	24
6	ECS-PLE05	<5	-	-	-	24
7	ECS-PLE06	37	51	87	24	24

The highest enantioselectivity was observed with the recombinant isoform ECS-PLE06 (entry 7, Table 3), comparable with the one obtained with crude PLE, which, in turn, showed higher specific activity.

2.2. Optimization

Crude PLE was therefore used for further optimization, carried out using an experimental design (Multisimplex v2.0 (Multisimplex AB, Karlskrona, Sweden), previously used for optimizing the conditions of different biotransformations [26]. The control variables were substrate and enzyme concentration, pH, co-solvent (DMSO) concentration, and temperature. Productivity at 24 h and enantioselectivity were chosen as response parameters. Under optimized conditions ([S] = 3.5 mg/mL (8 mM); [Enz] = 5.0 mg/mL; solvent 0.1 M phosphate buffer/DMSO 8%, pH = 7.0 at 25 °C), where the ratio between substrate and enzyme was reduced, the highest enantioselectivity (E = 28) was obtained, but reaction rate slowed down. Under these conditions, enzymatic hydrolysis gave 2e with an ee of 90% after 24 h, in correspondence of 30% conversion.

As previously reported, the addition of co-solvents, which alter the solubility of the substrate, may affect the enantioselectivity and the reaction rate of reactions catalyzed by crude PLE [14,27]. Consequently, we investigated the activity and the enantioselectivity on the hydrolysis of 1e with crude PLE carried out under optimized conditions in the presence of the solvents listed in Table 3.

Table 3. Hydrolysis of 1e with PLE in the presence of different co-solvents. Conditions: [S] = 8 mM, [PLE] = 5.0 mg/mL in 100 mM phosphate buffer at pH = 7.0 and co-solvents (amounts as indicated in Table), 25 °C. Results after 24 h.

Entry	Co-solvent (% *v*/*v*)	Conv. (%)	ee $_{(R)\text{-ester}}$ (%)	ee $_{(S)\text{-acid}}$ (%)	E
1	none	30	23	90	23
2	EtOH (8)	<5	-	-	-
3	*i*PrOH (8)	<5	-	-	-
4	DMSO (8)	30	39	90	28
5	THF (8)	10	9	79	10
6	acetone (8)	<5	-	-	-
7	Et_2O (30)	<5	-	-	-
8	toluene (30)	<5	-	-	-
9	*n*-heptane (30)	22	25	90	21
10	isooctane (30)	25	30	88	21

Protic water-soluble co-solvents (EtOH and *i*PrOH, entries 2 and 3, Table 3) suppressed enzymatic activity, whereas, DMSO (firstly chosen as co-solvent) was the only polar co-solvent with beneficial effects (entry 4, Table 3). Detected activity and enantioselectivity in the presence of highly hydrophobic solvents (*n*-heptane and isooctane, entries 9 and 10, Table 3) were lower than the ones obtained in water containing 8% DMSO. Reactions performed in the presence of different concentrations of hydrophobic solvents (10, 30, 50% *v*/*v*) did not show any significant differences.

Another way to influence the overall reactivity of organic substrates in aqueous enzymatic reactions involves the use of cyclodextrins (CDX) [28]. CDX can modify the solubility of organic compounds in water, while establishing diastereoisomeric interactions with chiral substrates; for these reasons, different CDX were tested as additive in the enzymatic hydrolysis of 1e (Table 4).

Table 4. Hydrolysis of 1e with PLE in the presence of β-cyclodextrins; conditions: [S] = 8 mM, [PLE] = 5.0 mg/mL, [CDX] 10 mM in 100 mM phosphate buffer and DMSO (8%) at pH = 7.0, 25 °C. Results after 24 h.

Entry	β-Cyclodextrin	Conv. (%)	ee_{ester} (%)	ee_{acid} (%)	E
1	underivatized	45	70	86	28
2	triacetyl	40	59	88	28
3	methyl	33	44	88	24
4	trimethylammonium	25	31	91	28

Cyclodextrins generally improved the reaction rates, with β-CDX showing the highest acceleration (entry 1, Table 4). The screening shown in Table 5 was carried out using a slight excess of CDX over the substrate, so we decided to explore the effect of different stoichiometric ratios between β-CDX and substrate (Table 5), observing that no significant improvements were obtained above 1.25 ratio β-CDX/substrate.

Table 5. Hydrolysis of 1e with PLE in the presence of different amounts of β-cyclodextrin (β-CDX); conditions: [S] = 8 mM [PLE] = 5.0 mg/mL, different amounts of β-CDX in 100 mM phosphate buffer and DMSO (8%) at pH = 7.0, 25 °C. Results after 24 h.

Entry	Ratio [β-CDX]/[S]	Conv. (%)	ee_{ester} (%)	ee_{acid} (%)	E
1	1	41	60	86	24
2	1.25	45	70	86	28
3	1.5	45	70	86	28
4	2	48	78	84	28

2.3. Preparative Biotransformation

A preparative biotransformation was thus performed starting from 150 mg of 1e (Figure 2) using the best reaction conditions (entry 2, Table 5).

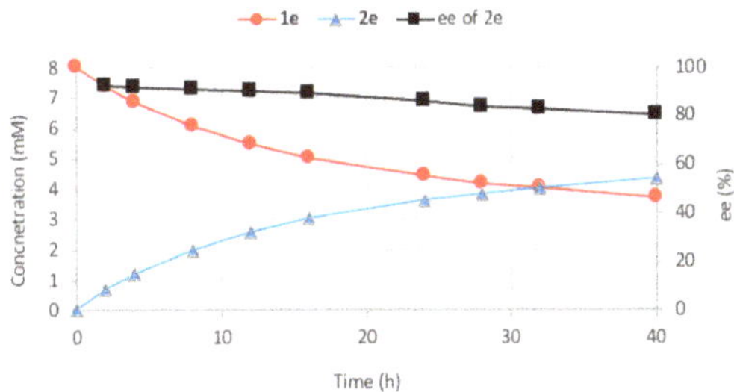

Figure 2. Preparative kinetic resolution of 1e. Conditions. [S] = 8 mM, [PLE] = 5.0 mg/mL, [β-CDX] = 10 mM in phosphate buffer (100 mM) and DMSO (8%) at pH = 7.0, 25 °C.

The reaction was stopped in correspondence of 54% conversion (after 40 h), allowing for the recovery and purification of 2e (67 mg) with an ee of 80%. This batch of optically-enriched 2e was chemically methylated to give optically enriched 1e, which was subsequently hydrolysed with PLE, furnishing 50 mg of optically pure (*S*)-2e. The overall results of this sequential kinetic resolution are given in Scheme 4.

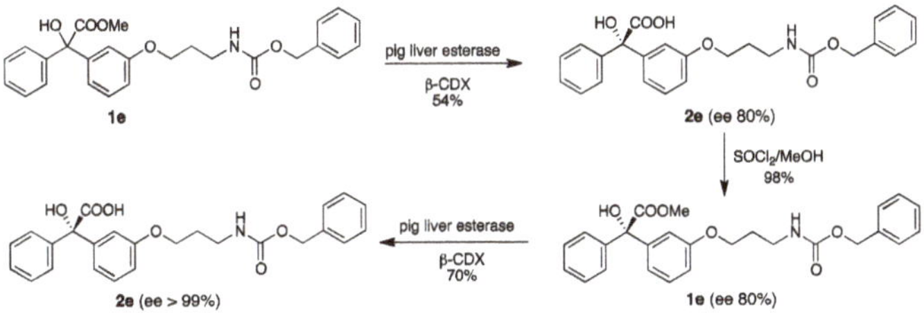

Scheme 4. Preparation of optically pure 2e after sequential enzymatic hydrolysis of 1e with PLE.

3. Discussion

Sterically demanding 2,2-diaryl-2-hydroxy carboxylic acids are valuable chiral building blocks for the synthesis of antimuscarinic agents [9]. Two major problems were encountered in the enzymatic kinetic resolution of these bulky substrates. Firstly, esters having α-quaternary or α-tertiary center show severe steric hindrance that hampers the approach to the active site; in fact, among the different enzymes tested, PLE was the only enzyme active on these substrates. Besides, esters of 2-hydroxy-2-(3-hydroxyphenyl)-2-phenylacetate (the ones considered here as key precursors for antimuscarinic agents preparation) display poor stereo-discrimination due to the presence of two aromatic groups, directly bound to the stereocenter, which differ only for the presence of a meta-substituent on one of the two aromatic rings. Derivative 1e, which bears a benzyloxy propylcarbamate substituent in meta-position, gives sufficient stereo-differentiation for achieving moderate-to-good enantioselectivity (E = 28). Moreover, the biotransformation was optimized by choosing suited co-solvents (DMSO) and additives (β-CDX).

The preparative significance of this method was established by the expedient preparation of optically pure (*S*)-2e on multi-milligram scale, using a sequential kinetic resolution approach.

4. Materials and Methods

All chemicals were from Sigma-Aldrich (Milano, Italy) and/or VWR International (Milano, Italy) and used without further purification unless otherwise stated. Pig liver esterase was purchased from Sigma-Aldrich (Milano, Italy). PLE isoforms were from Enzymicals (Greifswald, Germany). β-Cyclodextrins were provided by Wacker-Chemie GmbH (Munchen, Germany). Anhydrous solvents were purchased from Aldrich and used as received. "Brine" refers to a saturated aqueous solution of NaCl. Unless otherwise specified, solutions of common inorganic salts used in workups are aqueous solutions. Optically pure/enriched compounds, used as HPLC standards, were synthesised as previously described [17].

4.1. Analyticals

HPLC analyses were performed with a Jasco PU-980 pump equipped with a UV–VIS detector Jasco UV-975 (Easton, MD, USA). The NMR of ^{1}H and ^{13}C spectra were recorded in DMSO using Bruker 600 MHz or 400 MHz spectrometer (Karlsruhe, Germany), equipped with a self-shielded z-gradient coil 5 mm $^{1}H/^{n}X$ broad band probehead for reverse detection, deuterium digital lock channel unit, quadrature digital detection unit with transmitter offset frequency shift. Chemical shifts are reported as δ downfield in parts per million (ppm) and referenced to tetramethylsilane (TMS) as the internal standard in the ^{1}H measurements. Coupling constants (J values) are given in hertz (Hz) and multiplicities are reported using the following abbreviation (s = singlet, d = doublet, t = triplet, q = quartet, m = multiplet, br = broad, nd = not determined). The pulse programs were taken from the Varian and Bruker software libraries. The HRMS spectra were recorded on an Agilent instrument (Santa Clara, CA, USA) using the Time-of-Flight Mass Spectrometry (TOF MS) technique. Specific rotation of compounds was measured with a Polarimeter Perkin Elmer (model 241 or 341, Waltham, USA) at sodium D-line (589 nm), 25 °C, 1 dm path length. Reactions were monitored by TLC using 0.25 mm Merck silica gel plates (60 F254, Darmstadt, Germany). For chiral analysis the samples were analysed using a chiral column for the separation of the enantiomers. HPLC analyses were carried out on a Kromasil 5-Amycoat column 4.6 × 250 mm (CPS Analitica, Milan, Italy), 5 μm; mobile phase: n-hexane:isopropanol:TFA 8:2:0.1%, flow rate 1 mL/min, λ = 220 nm. Optically pure/enriched compounds were chemically synthesised as chiral HPLC standards. Column chromatography was performed on Merck silica gel 60 (0.063–0.2 mm).

4.2. Procedure for the Synthesis of methyl 2-(4-(benzyloxy)phenyl)-2-hydroxy-2-phenylacetate (1a)

To a solution of methyl 2-oxo-2-phenylacetate (12.83 mL, 91 mmol) in THF (Volume: 350 mL), (4-(benzyloxy)phenyl)magnesium bromide (100 mL, 100 mmol) was added dropwise at 0 °C over 30 min and stirred overnight at RT. Reaction was partitioned between AcOEt and saturated NaCl, organic phase dried over Na_2SO_4 and evaporated. The oily residue was crystalized in Et_2O to afford methyl 2-(4-(benzyloxy)phenyl)-2-hydroxy-2-phenylacetate (15 g, 43.1 mmol, 47.4 % yield) as white solid. ^{1}H NMR (600 MHz, DMSO-d6; δ ppm 7.24–7.44 (m, 11 H) 6.94–6.98 (m, 2 H) 6.91 (dt, J = 8.08, 1.15 Hz, 1 H) 6.67 (s, 1 H) 5.06 (s, 2 H) 3.71 (s, 3 H); ^{13}C NMR (151 MHz, DMSO-d6) δ ppm 173.97 (s, 1 C) 158.36 (s, 1 C) 145.33 (s, 1 C) 143.63 (s, 1 C) 137.45 (s, 1 C) 129.35 (s, 1 C) 128.88 (s, 1 C) 128.28 (s, 1 C) 128.25 (s, 1 C) 128.15 (s, 1 C) 127.92 (s, 1 C) 127.40 (s, 1 C) 120.06 (s, 1 C) 114.50 (s, 1 C) 113.84 (s, 1 C) 81.05 (s, 1 C) 69.70 (s, 1 C) 52.96 (s, 1 C). HRMS (ESI-TOF): Exact mass calculated for $C_{22}H_{20}O_4$

$[M-H]^-$ = 347.1289, Found: $[M + NH_4]^+$ = 366.1698. ^{1}H NMR and ^{13}C NMR spectra of 1a are reported in Supplementary Materials (Figures S2 and S3, respectively).

4.3. Procedure for the Synthesis of methyl 2-hydroxy-2-(3-hydroxyphenyl)-2-phenylacetate (1b)

Methyl 2-(3-(benzyloxy)phenyl)-2-hydroxy-2-phenylacetate (1a) (200 mg, 0.574 mmol), 1-methylcyclohexa-1,4-diene (129 μl, 1.148 mmol) and Pd/C 10% wet (60 mg, 0.028 mmol) were refluxed in EtOH (2870 μl) in a closed vessel at 80 °C for 8 h, then the mixture filtered on PTFE membrane and evaporated under reduced pressure. The resulting oil was recrystallized in a mixture of cyclohexane/iPr_2O to give methyl 2-hydroxy-2-(3-hydroxyphenyl)-2-phenylacetate (130 mg, 0.503 mmol, 88% yield) as a white powder. ^{1}H NMR (600 MHz, DMSO-d6) δ ppm 9.35 (s, 1 H) 7.32 (d, J = 4.29 Hz, 4 H) 7.25–7.30 (m, 1 H) 7.11 (t, J = 7.87 Hz, 1 H) 6.74–6.77 (m, 1 H) 6.73 (dd, J = 1.67, 0.95 Hz, 1 H) 6.67 (ddd, J = 8.02, 2.41, 0.89 Hz, 1 H) 6.55 (s, 1 H) 3.71 (s, 3 H); ^{13}C NMR ((151 MHz, DMSO-d6) δ ppm 174.14 (s, 1 C) 157.33 (s, 1 C) 145.20 (s, 1 C) 143.82 (s, 1 C) 129.17 (s, 1 C) 128.15 (s, 1 C) 127.81 (s, 1 C) 127.50 (s, 1 C) 118.08 (s, 1 C) 114.89 (s, 1 C) 114.70 (s, 1 C) 81.10 (s, 1 C) 52.87 (s, 1 C); HRMS (ESI-TOF): Exact mass calculated for $C_{15}H_{14}O_4$ $[M-H]^-$ = 257.0819, Found: $[M + Na]^+$ = 281.0783. ^{1}H NMR and ^{13}C NMR spectra of 1b are reported in Supplementary Materials (Figures S4 and S5, respectively).

4.4. Procedure for the Synthesis of 2-(dimethylamino)ethyl 2-(3-(benzyloxy)phenyl)-2-hydroxy-2-phenylacetate (1c)

Methyl 2-(3-(benzyloxy)phenyl)-2-hydroxy-2-phenylacetate (1a) (5 g, 14.35 mmol) and LiOH (1.031 g, 43.1 mmol) were dissolved in THF (Volume: 15 mL)/Water (Volume: 15.00 mL) and stirred for 2 h at RT. Reaction was quenched by the addition on 1M HCl and extracted with AcOEt, the organic phase was washed with aqueous NaCl and dried over Na_2SO_4 before being evaporated under reduced pressure. The desired product 2-(3-(benzyloxy)phenyl)-2-hydroxy-2-phenylacetic acid (2a) (5.1 g, 15.25 mmol, 106 % yield) was obtained as a yellowish foam, not isolated and used as such for the following step. 2-(3-(Benzyloxy)phenyl)-2-hydroxy-2-phenylacetic acid (2a) (2.5 g, 7.48 mmol) and CDI (2.425 g, 14.95 mmol) were reacted for 5 min in DCM (Volume: 8 mL) prior the addition of 2-(dimethylamino)ethanol (2.257 mL, 22.43 mmol). Reaction was stirred for 4 h at 60 °C, then partitioned between AcOEt and sat $NaHCO_3$aq, washed twice with water, dried over Na_2SO_4 and evaporated under reduced pressure. The crude was chromatographed on silica gel by gradient elution from 100 % AcOEt to AcOEt/MeOH (7N NH3) 90/10 in 12 CV to give 2-(dimethylamino)ethyl 2-(3-(benzyloxy)phenyl)-2-hydroxy-2-phenylacetate (1c) (1.02 g, 2.52 mmol, 33.6 % yield) as a yellowish oil. ^{1}H NMR (600 MHz, DMSO-d6) δ ppm 7.23–7.44 (m, 12 H) 6.98–7.01 (m, 1 H) 6.92–6.96 (m, 2 H) 6.58 (s, 1 H) 5.05 (s, 2 H) 4.23 (t, J = 5.57 Hz, 2 H) 2.12 (s, 6 H); ^{13}C NMR (151 MHz, DMSO-d6) δ ppm 173.34 (s, 1 C) 158.33 (s, 1 C) 145.30 (s, 1 C) 143.63 (s, 1 C) 137.47 (s, 1 C) 129.25 (s, 1 C) 128.87 (s, 1 C) 127.88 (s, 1 C) 127.54 (s, 1 C) 127.43 (s, 1 C) 120.21 (s, 1 C) 114.69 (s, 1 C) 113.80 (s, 1 C) 81.06 (s, 1 C) 69.69 (s, 1 C) 63.47 (s, 1 C) 57.42 (s, 1 C) 45.54 (s, 1 C). HRMS (ESI-TOF): Exact mass calculated for $C_{25}H_{27}O_4N$ $[M]^+$ = 405.1940, Found $[M-H]^+$ = 406.2011 ^{1}H NMR and ^{13}C NMR spectra of 1c are reported in Supplementary Materials (Figures S6 and S7, respectively).
^{1}H NMR and ^{13}C NMR spectra of 1c are reported in Supplementary Materials (Figures S6 and S7, respectively).

4.5. Procedure for the Synthesis of (1-benzylpiperidin-4-yl)methyl 2-(3-(benzyloxy)phenyl)-2-hydroxy-2-phenylacetate (1d)

2-(3-(benzyloxy)phenyl)-2-hydroxy-2-phenylacetic acid (2a) (2.5 g, 7.48 mmol) and CDI (2.425 g, 14.95 mmol) were reacted for 5 min in DCM (Volume: 15 mL) prior the addition of (1-benzylpiperidin-4-yl)methanol (2.3 g, 11.20 mmol). Reaction was stirred for 3 h at 60 °C, then DMF was added and the mixture was stirred at 80 °C for 4 h. The mixture was partitioned between AcOEt and an aqueous solution of $NaHCO_3$, washed twice with water, dried over Na_2SO_4 and evaporated under reduced pressure. The crude was chromatographed first on silica gel (gradient elution from 100 % AcOEt to AcOEt/MeOH (7N NH_3) 90/10 in 10 CV), then by flash chromatography on a reverse phase: C18 column 60 g, from 100/0 A/B to 75/25 A/B, A: water/MeCN 95:5 + 0.1% HCOOH B:MeCN/water 95:5 + 0.1% HCOOH, to obtain the desired product (1-benzylpiperidin-4-yl)methyl 2-(3-(benzyloxy)phenyl)-2-hydroxy-2-phenylacetate (1d) (715 mg, 1.371 mmol, 18.3 % yield) as a white oil. ^{1}H NMR (600 MHz, DMSO-d6) δ ppm 7.20–7.42 (m, 16 H) 6.93–6.97 (m, 2 H) 6.90–6.93 (m, 1 H) 6.60 (s, 1 H) 5.05 (s, 2 H) 3.99 (d, J = 6.44 Hz, 2 H) 3.40 (br s, 2 H) 2.73 (br d, J = 10.01 Hz, 2 H) 1.76–1.92 (m, 2 H) 1.49–1.58 (m, 1 H) 1.47 (br d, J = 12.92 Hz, 2 H) 1.12 (br d, J = 12.16 Hz, 2 H); ^{13}C NMR (151 MHz, DMSO-d6) δ ppm 173.40 (s, 1 C) 158.31 (s, 1 C) 145.33 (s, 1 C) 143.63 (s, 1 C) 137.46 (s, 1 C) 129.24 - 129.35 (m, 1 C) 129.16 (br s, 1 C) 128.85 (s, 1 C) 128.50–128.65 (m, 1 C) 128.23 - 128.31 (m, 1 C) 128.18 (s, 1 C) 128.04 (s, 1 C) 127.85–127.94 (m, 1 C) 127.42–127.56 (m, 1 C) 127.17–127.35 (m, 1 C) 120.14 (s, 1 C) 114.63 (s, 1 C) 113.84 (s, 1 C) 81.11 (s, 1 C) 69.79 (s, 1 C) 69.68 (s, 1 C) 62.75 (s, 1 C) 53.04 (s, 1 C) 40.58 (s, 1 C) 35.38 (s, 1 C) 28.62 (s, 1 C). HRMS (ESI-TOF): Exact mass calculated for $C_{34}H_{35}NO_4$ $[M]^+$ = 521.2566, Found $[M+H]^+$ = 522.2642-. To ^{1}H NMR and ^{13}C NMR spectra of 1d are reported in Supplementary Materials (Figures S8 and S9, respectively).

4.6. Procedure for the Synthesis of methyl 2-(3-(3-(((benzyloxy)carbonyl)amino)propoxy)phenyl)-2-hydroxy-2-phenylacetate (1e)

Methyl 2-hydroxy-2-(3-hydroxyphenyl)-2-phenylacetate (2b) (570 mg, 2.207 mmol) and 3-(((benzyloxy)carbonyl)amino)propyl methanesulfonate (761 mg, 2.65 mmol) were dissolved in DMF (6 mL), followed by the addition of $CsCO_3$ (1079 mg, 3.31 mmol). The reaction solution was stirred at room temperature overnight. The reaction mixture was quenched adding water then extracted with EtOAc. The organic phase was dried over Na_2SO_4, filtered and evaporated under reduced pressure. The crude was purified by flash chromatography (25 g silica, from 20 to 50 % EtOAc in heptane) to obtain the desired product methyl 2-(3-(3-(((benzyloxy)carbonyl)amino)propoxy)phenyl)-2-hydroxy-2-phenylacetate (1e) (670 mg, 1.491 mmol, 67.5 % yield) as a colourless oil. ^{1}H NMR (600 MHz, DMSO-d6) δ ppm 7.27–7.39 (m, 11 H) 7.24 (t, J = 8.27 Hz, 1 H) 6.88–6.92 (m, 2 H) 6.82–6.88 (m, 1 H) 6.67 (s, 1 H) 5.02 (s, 2 H) 3.94 (t, J = 6.28 Hz, 2 H) 3.72 (s, 3 H) 3.16 (q, J = 6.54 Hz, 2 H) 1.85 (quin, J = 6.51 Hz, 2 H); ^{13}C NMR (151 MHz, DMSO-d6) δ ppm 173.97 (s, 1 C) 158.36 (s, 1 C) 145.33 (s, 1 C) 143.63 (s, 1 C) 137.45 (s, 1 C) 129.35 (s, 1 C) 128.88 (s, 1 C) 128.28 (s, 1 C) 128.25 (s, 1 C) 128.15 (s, 1 C) 127.92 (s, 1 C) 127.40 (s, 1 C) 120.06 (s, 1 C) 114.50 (s, 1 C) 113.84 (s, 1 C) 81.05 (s, 1 C) 69.70 (s, 1 C) 52.96 (s, 1 C). HRMS (ESI-TOF): Exact mass calculated for $C_{26}H_{27}NO_6$ $[M-H]^-$ = 448.1766, Found: $[M+NH4]^+$ = 467.2177. ^{1}H NMR and ^{13}C NMR spectra of 1a are reported in Supplementary Materials (Figures S10 and S11, respectively).

4.7. Enantiomeric Excess Determination

The enantiomeric excess (ee %) was determined by HPLC with a Kromasil 5-Amycoat column 4.6 × 250 mm, 5 μm, mobile phase: *n*-hexane:isopropanol:TFA 8:2:0.1%, flow rate 1 mL/min, λ = 220 nm. Retention times: (*R*)-1a 10.1 min; (*S*)-1a: 11.5 min; (*R*)-2a 20.7 min; (*S*)-2a: 34.6 min; (*S*)-1b: 11.3 min; (*R*)-1b: 11.5 min; (*S*)-2b: 14.1 min; (*R*)-2b: 16.1 min; (*R*)-1e: 16.3 min; (*S*)-1e: 17.4 min; (*R*)-2e: 23.8 min; (*S*)-2e 29.1 min. Representative HPLC chromatograms are reported in Supplementary Materials (Figure S18).

4.8. General Procedure for Biotransformations

Screening and optimization were carried out by performing reactions in 5 mL screw-capped test tubes with a reaction volume of 2 mL. Preparative biotransformations were carried out at 25 and 150 mL scale. Substrates (2.5–10 mM) were dissolved in DMSO (final concentration 5%) and added to phosphate buffer (100 mM, pH = 7). The reactions were started by the addition of the enzyme. The mixture was then kept at fixed temperature under magnetic stirring. Samples of the biotransformation mixture were withdrawn, diluted with an equal volume of water, acidified with 1 N HCl and extracted with eight volumes of EtOAc. The organic extract was then concentrated and analysed by HPLC.

4.8.1. (S)-2-(3-(Benzyloxy)phenyl)-2-hydroxy-2-phenylacetic acid (2a)

^{1}H NMR (600 MHz, DMSO-d6) δ ppm 12.20–13.89 (m, 1 H) 9.31 (s, 1 H) 7.36–7.39 (m, 2 H) 7.29–7.34 (m, 2 H) 7.24–7.29 (m, 1 H) 7.11 (t, J = 7.89 Hz, 1 H) 6.82 (br s, 1 H) 6.79–6.82 (m, 1 H) 6.66 (ddd, J = 8.05, 2.34, 1.03 Hz, 1 H) 5.48–6.50 (m, 1 H) ^{13}C NMR (151 MHz, DMSO-d6) δ ppm 175.14 (s, 1 C) 157.22 (s, 1 C) 145.55 (s, 1 C) 144.17 (s, 1 C) 129.00 (s, 1 C) 128.01 (s, 1 C) 127.59 (s, 1 C) 118.21 (s, 1 C) 114.86 (s, 1 C) 114.66 (s, 1 C) 80.63 (s, 1 C). HRMS (ESI-TOF): Exact mass calculated for $C_{14}H_{12}O_4$ $[M-H]^-$ = 243.0663; Found: $[M+Na]^+$ = 267.0626.

4.8.2. (S)-2-Hydroxy-2-(3-hydroxyphenyl)-2-phenylacetic acid (2b)

^{1}H NMR (600 MHz, DMSO-d6) δ ppm 13.21 (br s, 1 H) 7.41–7.45 (m, 2 H) 7.36–7.41 (m, 4 H) 7.23–7.35 (m, 5 H) 7.01–7.05 (m, 1 H) 6.93–7.00 (m, 2 H) 6.34 (s, 1 H) 5.05 (s, 2 H); ^{13}C NMR (151 MHz, DMSO-d6) δ ppm 175.01 (s, 1 C) 158.31 (s, 1 C) 145.68 (s, 1 C) 143.96 (s, 1 C) 137.47 (s, 1 C) 129.20 (s, 1 C) 128.88 (s, 1 C) 128.28 (s, 1 C) 128.18 (s, 1 C) 128.13 (s, 1 C) 127.71 (s, 1 C) 127.50 (s, 1 C) 120.23 (s, 1 C) 114.72 (s, 1 C) 113.50 (s, 1 C) 80.61 (s, 1 C) 69.70 (s, 1C). HRMS (ESI-TOF): Exact mass calculated for $C_{21}H_{18}O_4$ $[M-H]^+$ = 333.1132, Found: $[M+NH_4]^-$ = 352.1541.

4.8.3. (S)-2-(3-(3-(((Benzyloxy)carbonyl)amino)propoxy)phenyl)-2-hydroxy-2-phenylacetic acid (2e)

^{1}H NMR (600 MHz, DMSO-d6) δ ppm 13.15 (br s, 1 H) 7.29–7.41 (m, 10 H) 7.25–7.29 (m, 1 H) 7.21–7.24 (m, 1 H) 6.92–6.97 (m, 2 H) 6.77–6.89 (m, 1 H) 6.31 (br s, 1 H) 5.01 (s, 2 H) 3.93 (t, J = 6.22 Hz, 2 H) 3.15 (q, J = 6.63 Hz, 2 H) 1.84 (quin, J = 6.51 Hz, 2 H); ^{13}C NMR (151 MHz, DMSO-d6) δ ppm 174.98 (s, 1 C) 158.50 (s, 1 C) 156.61 (s, 1 C) 145.71 (s, 1 C) 144.13 (s, 1 C) 137.71 (s, 1 C) 129.05–129.19 (m, 1 C) 128.81 (s, 1 C) 128.15–128.36 (m, 1 C) 128.08 (s, 1 C) 127.60–127.69 (m, 1 C) 127.47–127.58 (m, 1 C) 119.97 (s, 1 C) 114.29 (s, 1 C) 112.27–113.50 (m, 1 C) 80.59 (s, 1 C) 65.85 (s, 1 C) 65.12–65.56 (m, 1 C) 37.45–38.24 (m, 1 C) 29.64 (s, 1 C). HRMS (ESI-TOF): Exact mass calculated for $C_{26}H_{27}NO_6$ $[M]^+$ = 435.1682, Found $[M+H]^+$ = 436.1754; $[\alpha]_D = -3.4$ ($CHCl_3$; c=1).

4.9. Procedure for the Synthesis of methyl (S)-2-(3-(3-(((benzyloxy)carbonyl)amino)propoxy)phenyl)-2-hydroxy-2-phenylacetate (optically enriched 1e)

(S)-2-(3-(3-(((Benzyloxy)carbonyl)amino)propoxy)phenyl)-2-hydroxy-2-phenylacetic acid (100 mg, 0,23 mmol) obtained by biotranformation was dissolved in MeOH (1,5 mL) and slowly added $SOCl_2$ (0,5 mL) at 0 °C. The reaction mixture was refluxed at 70 °C for 2 h, after which time it was cooled to RT. MeOH was removed in vacuum and the resulting residue was poured onto ice-H_2O and extracted with EtOAc. The combined organic extracts were washed with 10% $NaHCO_3$, brine, dried over Na_2SO_4, and evaporated to provide the product as a white solid (101 mg, 98% yield). ^{1}H NMR (600 MHz, DMSO-d6) δ ppm 7.20–7.40 (m, 12 H) 6.84–6.86 (m, 1 H) 6.82–6.90 (m, 2 H) 6.65 (s, 1 H) 5.01 (s, 2 H) 3.93 (t, J = 6.22 Hz, 2 H) 3.72 (s, 3 H) 3.15 (q, J = 6.67 Hz, 2 H) 1.84 (quin, J = 6.54 Hz, 2 H); ^{13}C NMR (151 MHz, DMSO-d6) δ ppm 172.50–174.93 (m, 1 C) 158.11–159.19 (m, 1 C) 155.83–157.71 (m, 1 C) 144.39–146.14 (m, 1 C) 143.20–144.08 (m, 1 C) 137.30–137.95 (m, 1 C) 129.20–129.50 (m, 1 C) 128.80 (br d, J = 27.51 Hz, 1 C) 128.24 (s, 1 C) 128.20 (br d, J = 3.30 Hz, 1 C) 127.82–127.98 (m, 1 C) 127.29–127.52 (m, 1 C) 119.48–120.06 (m, 1 C) 113.96–114.24 (m, 1 C) 113.42 (s, 1 C) 81.06 (s, 1 C) 65.65 (s, 1 C) 65.51 (s, 1 C) 52.40 - 53.57 (m, 1 C) 37.02 - 38.35 (m, 1 C) 28.67–29.99 (m, 1 C); HRMS (ESI-TOF): Exact mass calculated for $C_{26}H_{27}NO_6$ $[M-H]^-$ = 448.1766, Found: $[M+H]^+$ = 450.1913. ^{1}H NMR and ^{13}C NMR spectra of optically enriched 1e are reported in Supplementary Materials (Figures S12 and S13, respectively).

Supplementary Materials: The following are available online at http://www.mdpi.com/2073-4344/9/2/113/s1, Table S1: Control variables and initial levels considered for the optimization. Figure S1: Sequential optimization of the PLE-catalysed hydrolysis of 1e. Figure S2: ^{1}H NMR spectrum of 1a. Figure S3: ^{13}C NMR spectrum of 1a. Figure S4: ^{1}H NMR spectrum of 1b. Figure S5: ^{13}C NMR spectrum of 1b. Figure S6: ^{1}H NMR spectrum of 1c. Figure S7: ^{13}C NMR spectrum of 1c. Figure S8: ^{1}H NMR spectrum of 1d. Figure S9: ^{13}C NMR spectrum of 1d. Figure S10: ^{1}H NMR spectrum of 1e. Figure S11: ^{13}C NMR spectrum of 1e. Figure S12: ^{1}H NMR spectrum of (S)-2a. Figure S13. ^{13}C NMR spectrum of (S)-2a. Figure S14: ^{1}H NMR spectrum of (S)-2b. Figure S15. ^{13}C NMR spectrum of (*S*)-2b. Figure S16: ^{1}H NMR spectrum of (*S*)-2e. Figure S17. ^{13}C NMR spectrum of (*S*)-2e. Figure S18: Chiral HPLC of the hydrolysis of 1e to 2e catalysed by PLE.

Author Contributions: Conceptualization, A.P., F.M., D.R. and F.R.; methodology, I.S., M.L.C. and L.C.; software, D.R. and I.S.; validation, A.P., R.M. and L.C.; formal analysis, R.M., M.L.C. and A.P.; investigation, D.R., I.S., R.M. and M.L.C.; resources, F.M. and F.R.; data curation, F.M., A.P., F.R., R.M., L.C. and I.S.; writing—original draft preparation, A.P. and F.M.; writing—review and editing, all the authors; supervision, F.M., D.R., A.P. and F.R.; project administration, A.P., F.M., F.R. and L.C. All authors discussed the results and commented on the manuscript.

Funding: This research received no external funding.

Conflicts of Interest: The authors declare no conflict of interest.

References

1. Bornscheuer, U.T.; Kazlauskas, R.T. *Hydrolases in Organic Synthesis*; Wiley-VCH: Weinheim, Germany, 1999.
2. Gotor-Fernandez, V.; Brieva, R.; Gotor, V. Lipases: Useful biocatalysts for the preparation of pharmaceuticals. *J. Mol. Catal. B.* **2006**, *40*, 111–120. [CrossRef]
3. De Miranda, A.S.; Miranda, L.S.M.; de Souza, R.O.M.A. Lipases: Valuable catalysts for dynamic kinetic resolutions. *Biotechnol. Adv.* **2015**, *33*, 372–393. [CrossRef] [PubMed]
4. Romano, D.; Bonomi, F.; de Mattos, M.C.; de Sousa Fonseca, T.; de Oliveira, M.C.F.; Molinari, F. Esterases as stereoselective biocatalysts. *Biotechnol. Adv.* **2015**, *33*, 547–565. [CrossRef] [PubMed]
5. Carvalho, A.C.; de Sousa Fonseca, T.; de Mattos, M.C.; de Oliveira, M.C.F.; de Lemos, T.L.; Molinari, F.; Romano, D.; Serra, I. Recent advances in lipase-mediated preparation of pharmaceuticals and their intermediates. *Int. J. Mol. Sci.* **2015**, *16*, 29682–29716. [CrossRef] [PubMed]
6. Bornscheuer, U.T. Methods to increase enantioselectivity of lipases and esterases. *Curr. Opin. Biotechnol.* **2002**, *13*, 543–547. [CrossRef]
7. Juhl, P.B.; Doderer, K.; Hollmann, F.; Thum, O.; Pleiss, J. Engineering of *Candida antarctica* lipase B for hydrolysis of bulky carboxylic acid esters. *J. Biotechnol.* **2010**, *150*, 474–480. [CrossRef] [PubMed]
8. Pogorevc, M.; Faber, K. Biocatalytic resolution of sterically hindered alcohols, carboxylic acids and esters containing fully substituted chiral centers by hydrolytic enzymes. *J. Mol. Catal. B.* **2000**, *10*, 357–376. [CrossRef]
9. Kourist, R.; Bornscheuer, U.T. Biocatalytic synthesis of optically active tertiary alcohols. *Appl. Microbiol. Biotechnol.* **2011**, *91*, 505–517. [CrossRef] [PubMed]
10. Rodríguez-Rodríguez, J.A.; Gotor, V.; Brieva, R. Lipase catalyzed resolution of the quaternary stereogenic center in ketone-derived benzo-fused cyclic cyanohydrins. *Tetrahedron: Asymmetry* **2011**, *22*, 1218–1224. [CrossRef]
11. Lalonde, J.J.; Bergbreiter, D.E.; Wong, C.-H. Enzymatic kinetic resolution of α-nitro α-methyl carboxylic acids. *J. Org. Chem.* **1988**, *53*, 2323–2327. [CrossRef]
12. Kometani, T.; Isobe, T.; Goto, M.; Takeuchi, Y.; Haufe, G. Enzymatic resolution of 2-fluoro-2-arylacetic acid derivatives. *J. Mol. Catal. B* **1988**, *5*, 171–174. [CrossRef]
13. Miyazawa, T.; Shimaoka, M.; Yamada, T. Resolution of 2-cyano-2-methylalkanoic acids via porcine pancreatic lipase-catalyzed enantioselective ester hydrolysis: effect of the alcohol moiety of the substrate ester on enantioselectivity. *Biotechnol. Lett.* **1999**, *21*, 309–312. [CrossRef]
14. Domínguez de María, P.; García-Burgos, C.A.; Bargeman, G.; van Gemert, R.W. Pig Liver Esterase (PLE) as biocatalyst in organic synthesis: from nature to cloning and to practical applications. *Synthesis* **2007**, *10*, 1439–1452. [CrossRef]
15. Namiki, Y.; Fujii, T.; Nakada, M. Preparation of chiral building blocks for the enantioselective total synthesis of ent-kauranoids by the pig liver esterase-catalyzed asymmetric hydrolysis of a dialkyl malonate-type prochiral diester. *Tetrahedron: Asymmetry* **2014**, *25*, 718–724. [CrossRef]
16. Rancati, F.; Rizzi, A.; Carzaniga, L.; Linney, I.; Knight, C.; Schmidt, W. Compounds having muscarinic receptor antagonist and beta2 adrenergic receptor agonist activity. Patent WO 2016/128456 Al, August 2016.
17. Rancati, F.; Rizzi, A.; Carzaniga, L.; Linney, I.; Knight, C.; Schmidt, W. Compounds having muscarinic receptor antagonist and beta2 adrenergic receptor agonist activity. Patent WO 2018/011090 A1, January 2018.
18. Montuschi, Z.; Macagno, F.; Valente, S.; Fuso, L. Inhaled muscarinic acetylcholine receptor antagonists for treatment of COPD. *Curr Med Chem.* **2013**, *20*, 1464–1476. [CrossRef] [PubMed]

19. Gandolfi, R.; Marinelli, F.; Lazzarini, A.; Molinari, F. Cell-bound and extracellular carboxylesterases from *Streptomyces*: hydrolytic and synthetic activities. *J. Appl Microbiol* **2000**, *89*, 870–875. [CrossRef] [PubMed]
20. Converti, A.; Del Borghi, A.; Gandolfi, R.; Lodi, A.; Molinari, F.; Palazzi, E. Simplified kinetics and thermodynamics of geraniol acetylation by lyophilized cells of *Aspergillus oryzae*. *Biotechnol. Bioeng.* **2002**, *77*, 232–237. [CrossRef]
21. Molinari, F.; Romano, D.; Gandolfi, R.; Kroppenstedt, R.M.; Marinelli, F. Newly isolated Streptomyces spp. as enantioselective biocatalysts: hydrolysis of 1,2-O-isopropylidene glycerol racemic esters. *J. Appl. Microbiol.* **2005**, *99*, 960–967. [CrossRef]
22. Molinari, F.; Cavenago, K.S.; Romano, A.; Romano, D.; Gandolfi, R. Enantioselective hydrolysis of (RS)-isopropylideneglycerol acetate with *Kluyveromyces marxianus*. *Tetrahedron: Asymmetry* **2004**, *15*, 1945–1947. [CrossRef]
23. Monti, D.; Ferrandi, E.E.; Righi, M.; Romano, D.; Molinari, F. Purification and characterization of the enantioselective esterase from *Kluyveromyces marxianus* CBS 1553. *J. Biotechnol.* **2008**, *133*, 65–72. [CrossRef]
24. De Vitis, V.; Nakhnoukh, C.; Pinto, A.; Contente, M.L.; Barbiroli, A.; Milani, M.; Bolognesi, M.; Molinari, F.; Gourlay, L.; Romano, D. A stereospecific carboxyl esterase from *Bacillus coagulans* hosting nonlipase activity within a lipase-like fold. *FEBS J.* **2018**, *285*, 903–914. [CrossRef] [PubMed]
25. Hummel, A.; Brüsehaber, E.; Böttcher, D.; Trauthwein, H.; Doderer, K.; Bornscheuer, U.T. Isoenzymes of Pig-Liver Esterase reveal striking differences in enantioselectivities. *Angew. Chem. Int. Ed. Engl.* **2007**, *46*, 8492–8494. [CrossRef] [PubMed]
26. Romano, D.; Gandolfi, R.; Guglielmetti, S.; Molinari, F. Enzymatic hydrolysis of capsaicins for the production of vanillylamine using ECB deacylase from *Actinoplanes utahensis*. *Food Chem.* **2011**, *124*, 1096–1098. [CrossRef]
27. Bjorkling, F.; Boutelje, J.; Gatenbeck, S.; Hult, K.; Norin, T. Enzyme catalysed hydrolysis of dialkylated propanedioic acid diesters, chain length dependent reversal of enantioselectivity. *Appl. Microbiol. Biotechnol.* **1985**, *21*, 16–19. [CrossRef]
28. Del Valle, E.M.M. Cyclodextrins and their uses: a review. *Process Biochem.* **2004**, *39*, 1033–1046. [CrossRef]

Article

Synthesis of Ribavirin, Tecadenoson, and Cladribine by Enzymatic Transglycosylation

Marco Rabuffetti [1,†], Teodora Bavaro [2], Riccardo Semproli [2,3], Giulia Cattaneo [2], Michela Massone [1], Carlo F. Morelli [1], Giovanna Speranza [1,*] and Daniela Ubiali [2,*]

1 Department of Chemistry, University of Milan, via Golgi 19, I-20133 Milano, Italy; marco.rabuffetti1@unimi.it (M.R.); michela.massone@unimi.it (M.M.); carlo.morelli@unimi.it (C.F.M.)

2 Department of Drug Sciences, University of Pavia, viale Taramelli 12, I-27100 Pavia, Italy; teodora.bavaro@unipv.it (T.B.); riccardo.semproli01@universitadipavia.it (R.S.); giulia.cattaneo01@universitadipavia.it (G.C.)

3 Consorzio Italbiotec, via Fantoli 15/16, c/o Polo Multimedica, I-20138 Milano, Italy

* Correspondence: giovanna.speranza@unimi.it (G.S.); daniela.ubiali@unipv.it (D.U.); Tel.: +39-02-50314097 (G.S.); +39-0382-987889 (D.U.)

† Present address: Department of Food, Environmental and Nutritional Sciences, University of Milan, via Mangiagalli 25, I-20133 Milano, Italy.

Received: 7 March 2019; Accepted: 8 April 2019; Published: 12 April 2019

Abstract: Despite the impressive progress in nucleoside chemistry to date, the synthesis of nucleoside analogues is still a challenge. Chemoenzymatic synthesis has been proven to overcome most of the constraints of conventional nucleoside chemistry. A purine nucleoside phosphorylase from *Aeromonas hydrophila* (*Ah*PNP) has been used herein to catalyze the synthesis of Ribavirin, Tecadenoson, and Cladribine, by a "one-pot, one-enzyme" transglycosylation, which is the transfer of the carbohydrate moiety from a nucleoside donor to a heterocyclic base. As the sugar donor, 7-methylguanosine iodide and its 2′-deoxy counterpart were synthesized and incubated either with the "purine-like" base or the modified purine of the three selected APIs. Good conversions (49–67%) were achieved in all cases under screening conditions. Following this synthetic scheme, 7-methylguanine arabinoside iodide was also prepared with the purpose to synthesize the antiviral Vidarabine by a novel approach. However, in this case, neither the phosphorolysis of the sugar donor, nor the transglycosylation reaction were observed. This study was enlarged to two other ribonucleosides structurally related to Ribavirin and Tecadenoson, namely, Acadesine, or AICAR, and 2-chloro-N^6-cyclopentyladenosine, or CCPA. Only the formation of CCPA was observed (52%). This study paves the way for the development of a new synthesis of the target APIs at a preparative scale. Furthermore, the screening herein reported contributes to the collection of new data about the specific substrate requirements of *Ah*PNP.

Keywords: Ribavirin; Tecadenoson; Cladribine; purine nucleoside phosphorylase; transglycosylation reaction; 7-methylguanosine iodide; 7-methyl-2′-deoxyguanosine iodide; 7-methylguanine arabinoside iodide

1. Introduction

Nucleoside analogues are well-established drugs in clinical practice; they are mainly used as anticancer and antiviral agents. However, the search for new therapeutically active nucleosides is still a vibrant research area, as witnessed by the approval of the pro-drug Sofosbuvir, marketed as Sovaldi® in 2013, used in the treatment of hepatitis C as an alternative to peginterferon-combined therapies (e.g., in association with Ribavirin and Daclatasvir, Ledipasvir or Simeprevir) [1].

Drug discovery stands alongside the set-up of new synthetic strategies aimed at circumventing the typical constraints of nucleoside chemistry (e.g., multi-step processes, protection/deprotection reactions, lack of selectivity, etc.) [2]. Chemoenzymatic synthesis has been proven to overcome most of these drawbacks. The main advantages of enzymatic methods include high catalytic efficiency, mild reaction conditions (and thus environmentally friendly and safer syntheses), high stereo- and regioselectivity, and fewer numbers of synthetic steps. However, a truly efficient synthesis of nucleoside analogues is often the result of a combination of chemical methods and biochemical transformations [3].

Purine nucleoside phosphorylases (PNPs, EC 2.4.2.1) catalyze the reversible cleavage of the glycosidic bond of purine nucleosides in the presence of inorganic orthophosphate as a co-substrate, to generate the conjugated nucleobase and α-D-pentofuranose-1-phosphate. If a second purine base is in the reaction medium, the formation of a new nucleoside can result by a regio- and stereoselective transglycosylation reaction [3].

Accumulated data about a PNP from *Aeromonas hydrophila* (*Ah*PNP) [4] have clearly shown that this enzyme can be successfully used in the synthesis of a wide range of nucleoside analogues, which are either routinely used as drugs (e.g., arabinosyladenine) [5,6] or can have promising pharmacological activities, such as some 6-substituted purine ribonucleosides [7,8]. This PNP has been shown to have a quite relaxed substrate tolerance toward the purine base, to recognize ribo- and 2′-deoxyribonucleosides as the sugar donor, and, although to a lesser extent, to accept D-arabinose-1-phosphate produced by the phosphorolysis of arabinosyluracil in a bi-enzymatic transglycosylation reaction [4–9].

Ribavirin (Virazole®) is considered the "gold-standard" in the treatment of hepatitis C in association with pegylated interferon-alpha (IFN-α) [10–13]. Tecadenoson and its congeners are selective A1 receptor agonists, which have been investigated for their use against arrhythmia and atrial fibrillation [14–16]. Cladribine (Litak®) has been approved for the treatment of symptomatic tricoleukaemia (hairy-cell leukemia). It can act both as a chemotherapy drug and an immunosuppressive agent. Clinical studies have also suggested its potential usefulness in the treatment of multiple sclerosis [17,18]. Chemical structures of Ribavirin (**1**), Tecadenoson (**2**), and Cladribine (**3**) are reported in Figure 1.

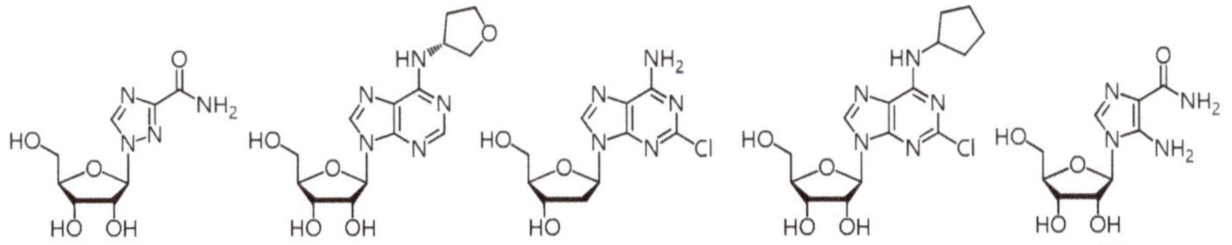

Figure 1. Ribavirin (**1**), Tecadenoson (**2**), Cladribine (**3**), 2-chloro-N^6-cyclopentyladenosine or CCPA (**17**), and Acadesine (**18**).

The synthesis of Ribavirin (**1**) has been achieved by both chemical and enzymatic approaches. The established glycosylation route involves the reaction of peracetylated β-D-ribofuranose with methyl 1,2,4-triazole-3-carboxylate, followed by aminolysis. Despite the good yields (54–83%) as well as the high regio- and stereoselectivity, high temperatures (135–170 °C) and high vacuum (15–55 mmHg) are required for the formation of the glycosydic bond [19,20]. Interestingly, when the reaction was carried out under slightly milder conditions (MW irradiation, 130 °C, 5 min and direct MPLC purification), the overall yield of Ribavirin was only 35% [21].

The synthesis of Ribavirin (**1**) by enzymatic transglycosylation was performed both by using whole cells (e.g., *Escherichia coli*, *A. hydrophila*, *Enterobacter aerogenes*, *Enterobacter gergoviae*) and isolated PNPs [22–33]. Either natural nucleosides or the suitable sugar phosphates were used as the ribose donor, resulting in variable yields (from 19% to 84%).

To date, the only synthetic strategy to obtain Tecadenoson (**2**) consists of nucleophilic substitution on either 2′,3′,5′-tri-*O*-acetyl-6-chloroinosine or 6-chloroinosine with (*R*)-3-aminotetrahydrofurane or its salts followed, when necessary, by deprotection with ammonia (reported yield: 68%) [34–36].

The direct glycosylation of a proper purine base (typically 2,6-dichloropurine) with a protected 1-chloro or 1-acetate ribose is the key step in numerous chemical syntheses of Cladribine (**3**) which can be obtained in variable yields ranging from 24% to 61% [37–39]. More elaborate multistep chemical strategies were also reported: they are based on either the 2′-deoxygenation of preformed protected adenosines [40,41], or the substitution with ammonia sources of purine nucleoside intermediates activated in 6-position [42–46].

Also, for Cladribine, enzymatic glycosylation (i.e., enzyme-catalyzed formation of the glycosydic bond) has been suggested as an alternative to the chemical route. Some examples of transglycosylation based on the use of PNPs (from *E. coli* and *Geobacillus thermoglucosidasius*) in mono- or bi-enzymatic processes have been reported [47,48].

While this research was ongoing, an *E. coli* PNP-catalyzed transglycosylation for the preparation of 2′-deoxynucleosides, including Cladribine, was developed by Mikhailov and co-workers who exploited a 7-methyl purine nucleoside iodide as the sugar donor [49].

The aim of this work was to enzymatically prepare Ribavirin, Tecadenoson, Cladribine, and some congeners (2-chloro-N^6-cyclopentyladenosine or CCPA, **17**, and Acadesine or AICAR, **18**, see Figure 1), by exploiting the well-established relaxed substrate specificity of *Ah*PNP [4–8] through a "one-pot, one-enzyme" transglycosylation based on the use of 7-methylguanosine (**7**) or 7-methyl-2′-deoxyguanosine iodide (**8**) as the sugar donor. It is worth noting that this route represents the first enzymatic synthesis of Tecadenoson.

In this context, we explored the use of 7-methylguanine arabinoside iodide (**9**) (Scheme 1) as the sugar donor for the synthesis of arabinosyl purine analogues such as the antiviral drug Vidarabine, as an alternative scheme to both the conventional chemical synthesis and the bi-enzymatic transglycosylation reaction [6].

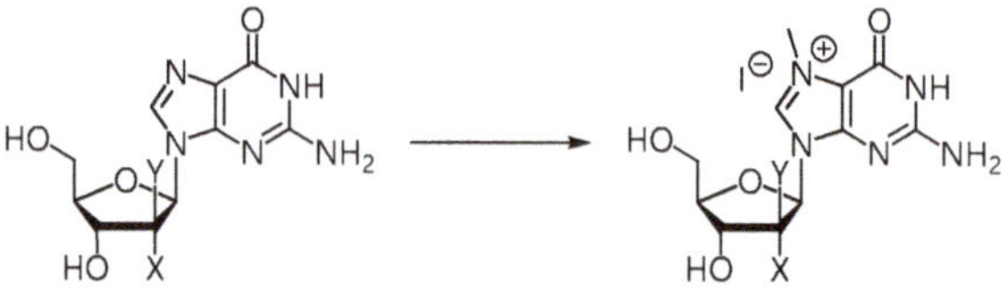

4 (X = OH, Y = H)
5 (X = Y = H)
6 (X = H, Y = OH)

7 (X = OH, Y = H)
8 (X = Y = H)
9 (X = H, Y = OH)

Scheme 1. Reagents and conditions (yield): CH_3I, DMF/DMSO (**7**: 87%; **9**: 89%) or CH_3I, DMSO, 20 °C (**8**: 80%).

2. Results and Discussion

2.1. Synthesis of the Sugar Donors

The three sugar donors, i.e., 7-methylguanosine iodide (**7**), 7-methyl-2′-deoxyguanosine iodide (**8**), and 7-methylguanine arabinoside iodide (**9**), were prepared by methylation of the corresponding nucleosides (Scheme 1).

The choice of CH_3I as the methylating agent and of the proper solvent (DMSO or a DMF/DMSO mixture) led to the selective formation of the iodide salts at N-7 (Scheme 1). Light exposure during the reaction at room temperature was avoided to prevent the decomposition of the final products, which had to be stored at −20 °C until use. The stability of 7-methylated nucleosides in DMSO or DMSO/DMF mixtures was found to depend on the nature of the sugar moiety, and posed a serious issue in the synthesis of 7-methyl-2′-deoxyguanosine iodide (**8**), as highlighted by ^{13}C NMR spectra registered in DMSO-d_6 (see Supplementary Materials, Figure S1). The methylation reaction was thus performed by modifying a previously reported protocol [50] under controlled temperature (20 °C), short reaction time, and an excess of CH_3I in order to avoid any decomposition of **8** in DMSO. Higher temperatures (25–30 °C) and longer reaction times led to product decomposition.

No synthesis of 7-methylguanine arabinoside iodide (**9**) has been reported in the literature to date; thus, the same strategy (CH_3I in a DMSO/DMF mixture) was successfully applied to the methylation of arabinosylguanine (**6**), thus affording **9** in 83% yield. In contrast to **8**, 7-methylguanine arabinoside iodide (**9**) was as stable as the corresponding ribo-derivative (**7**). No decomposition products were detected both in DMSO/DMF and under bioconversion conditions (see below).

2.2. Synthesis of the Base Acceptors

The synthesis of 2-chloro-6-aminopurine (**11**), the base acceptor to prepare Cladribine (**3**), was achieved by treatment of 2,6-dichloropurine (**10**) with NH_3/MeOH under MW irradiation. Following the same approach, the base acceptors of Tecadenoson (**2**) and its congener, i.e., CCPA, were synthesized starting from 2,6-dichloropurine (**10**) or 6-chloropurine (**13**) and (*R*)-3-aminotetrahydrofuran hydrochloride/LiOH/EtOH or cyclopentylamine/*n*-BuOH, respectively (Scheme 2). Products were purified either by precipitation or by flash column chromatography in 35–69% yield. 1,2,4-Triazole-3-carboxamide (**15**) and 5-amino-1*H*-imidazole-4-carboxamide (**16**) (see Scheme 3) were commercially available.

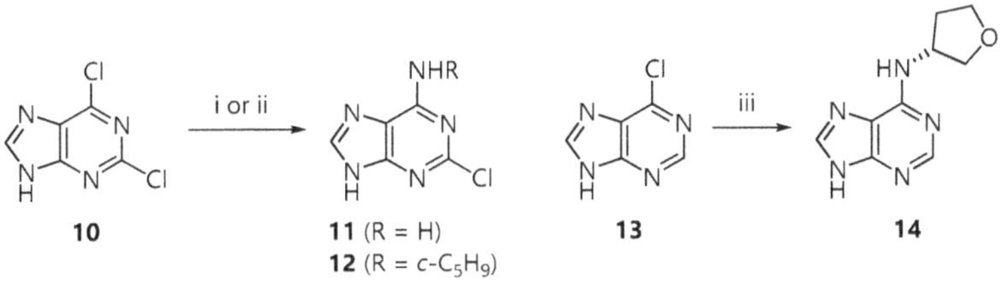

10 → (i or ii) → **11** (R = H), **12** (R = *c*-C_5H_9); **13** → (iii) → **14**

Scheme 2. Synthesis of adenine acceptors (i: NH_3, MeOH, 110 °C, MW, 66%; ii: *c*-$C_5H_9NH_2$, *n*-BuOH, reflux, 69%; iii: (*R*)-3-aminotetrahydrofuran hydrochloride, LiOH, EtOH, reflux, 35%).

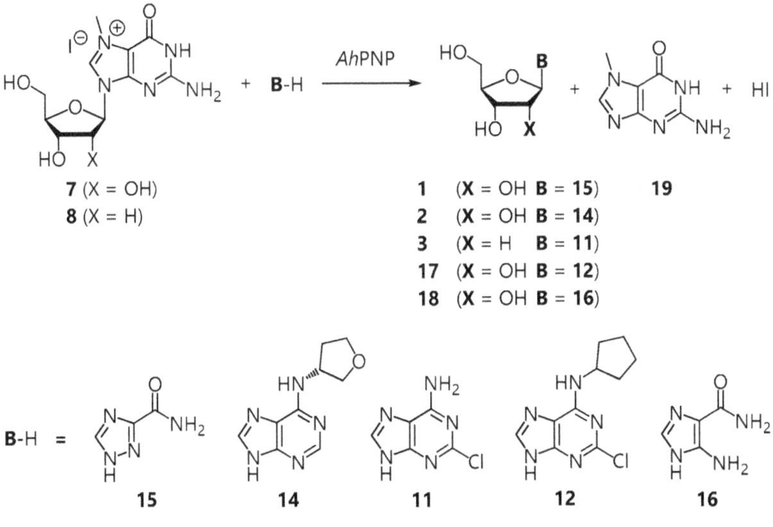

7 (X = OH), **8** (X = H) + **B**-H → (*Ah*PNP) → **1** (**X** = OH **B** = **15**), **2** (**X** = OH **B** = **14**), **3** (**X** = H **B** = **11**), **17** (**X** = OH **B** = **12**), **18** (**X** = OH **B** = **16**) + **19** + HI

B-H = **15**, **14**, **11**, **12**, **16**

Scheme 3. Synthesis of nucleoside analogues **1–3** and **17–18** by enzymatic transglycosylation.

2.3. "One-Pot, One-Enzyme" Transglycosylations

"One-pot one-enzyme" transglycosylations were carried out starting from either 7-methylguanosine iodide (**7**) or 7-methyl-2′-deoxyguanosine iodide (**8**) as the sugar donors for the synthesis of the APIs **1–2** and **3**, respectively (see Scheme 3). As previously reported for nucleoside **7** [4,7], also the phosphorolysis reaction of **8** was almost complete and irreversible. In fact, the conjugated nucleobase of these nucleosides (7-methylguanine, **19**, see Scheme 3) was not recognized by *Ah*PNP as a substrate, thus assisting the shift of the reaction equilibrium toward the product formation.

Transglycosylation reactions occur under very mild conditions, generally in phosphate buffer at room temperature. The typical drawback of enzymatic reactions is the need to conjugate the poor solubility of substrates in aqueous media with the stability of the biocatalysts in organic solvent. Starting from a reaction set-up established in our labs, glycerol was used as the co-solvent in order to improve the substrate solubility and to preserve the enzyme activity (glycerol is routinely used as a protein preservative). Only in the case of Cladribine (**3**), DMSO was added as a second co-solvent besides glycerol (1 mL, 5% of the total volume) to overcome the very poor solubility of 2-chloro-6-aminopurine (**11**). As previously reported [8], the use of DMSO is quite well tolerated by *Ah*PNP.

As for the solubility issue, 7-methyl purine nucleosides are highly water soluble and their use as the sugar donors is, indeed, a true advantage. On the other hand, the conjugated base of these nucleosides (**7–9**), i.e., 7-methylguanine (**19**), is poorly water soluble and easily separates out the reaction, thus giving a further contribution to drive the reaction equilibrium. No less important, a further strength-point of 7-methyl purine nucleosides as the sugar donor relies on their straightforward and high-yielding preparation (see Section 2.1).

All reactions were carried out in 50 mM phosphate buffer (pH 7.5) containing 20% of glycerol (*v*/*v*) at room temperature (Scheme 3). Bioconversions were performed at an analytical scale (1 mM substrate concentration) by using a 1:1 donor/acceptor ratio. Reactions were monitored both by measuring the depletion of the nucleobase acceptor as well as the formation of the new nucleoside (see Materials and Methods). Conversions (end-point: 24 h) are reported in Table 1. All the HPLC peaks were assigned on the basis of the pure reference compounds, either purchased or synthesized (see Materials and Methods).

Table 1. Synthesis of the target nucleosides **1–3** and **17–18** by enzymatic transglycosylation. [1]

X [2]	B-H [2]	Product	Conversion
OH	**15**	**1 (Ribavirin)**	67%
OH	**14**	**2 (Tecadenoson)**	49%
H	**11**	**3 (Cladribine)**	56% [3]
OH	**12**	**17 (CCPA)**	52% [3]
OH	**16**	**18 (Acadesine)**	n.d.

[1] Experimental conditions: 50 mM KH_2PO_4, pH 7.5, and glycerol (20%), (substrate) = 1 mM, 1:1 donor/acceptor ratio, r.t., *Ah*PNP (21.5 mg mL^{-1}; 39 IU mg^{-1}) = 1.15 IU or 80 IU; time monitoring (HPLC): 1, 3, 6, and 24 h, endpoint = 24 h. [2] X and B-H as in Scheme 3. [3] For the synthesis of **3** and **17**, DMSO (5% or 10% *v*/*v*, respectively) was used as the second co-solvent besides glycerol. n.d. = not detected.

As recalled in the Introduction, the synthesis of nucleoside-based APIs by an enzymatic transglycosylation reaction has been investigated by many authors (for a comprehensive review, see References [3,51]; for a recent example see Reference [52]). However, the enzymatic synthesis of Tecadenoson has been reported herein for the first time (see Supplementary Materials, Figure S7). Taking into account this result and the evidence that *Ah*PNP can accept a wide array of 6-substituted purines as substrates, we also successfully synthesized CCPA (**17**), the congener of Tecadenoson (Scheme 3). In this case, as the base acceptor (**12**) was not soluble in the buffer–glycerol mixture even in the presence of 5% DMSO (as applied in the synthesis of Cladribine), the biotransformation was carried out in a sort of "fed batch" mode. A 10 mM stock solution of **12** in DMSO was progressively added to the reaction upon monitoring the rate of phosphorolysis and the transglycosylation reaction (see Supplementary Materials, Figure S5). The final percentage of DMSO was 10% *v*/*v*. It is worth reporting that a large excess of *Ah*PNP was used in this case, as the formation of the target nucleoside was hardly detectable when using 1.15 U of enzyme. This result prompted us to further extend this approach to the synthesis of other ribo-derivatives such as Acadesine (or AICAR, **18**), a congener of Ribavirin. This reaction was performed under standard conditions (glycerol–buffer) by using the same excess of *Ah*PNP as for CCPA. At this stage of the project, in fact, the goal was to assess whether the biocatalyst could synthesize Acadesine. Surprisingly, whereas in the case of Ribavirin a conversion of 67% was registered after 24 h, the formation of Acadesine was not observed (see Supplementary Materials, Figures S4 and S6).

As it is well known, the chemical route to purine 2′-deoxyribonucleosides is even more challenging than that to its ribo-counterparts [2]. Therefore, the availability of 7-methyl-2′-deoxyguanosine iodide would represent a valuable tool for an alternative synthetic approach, as proven by the bioconversion of Cladribine (see Table 1). However, this sugar donor was found to be less stable than 7-methylguanosine iodide in aqueous medium, in agreement with a very recent report by Mikhailov et al. [49]. We have found that this molecule is unstable also in DMSO at room temperature, as indicated from the appearance of extra signals in the ^{13}C NMR spectrum recorded at two-hour intervals in DMSO, clearly showing that the decomposition of the nucleoside had occurred (see Supplementary Materials, Figure S1). Dimethyl sulfoxide was used in the synthesis of Cladribine (**3**) as the second co-solvent. The evidence of the poor stability of 7-methyl-2′-deoxyguanosine iodide both in buffer solutions and in DMSO makes this molecule a substrate which is difficult to handle in preparative applications.

As a natural continuation of this study, our efforts were then focused on the synthesis of arabinosyl purine nucleosides by using the newly prepared 7-methylguanine arabinoside iodide (**9**) as the sugar donor. Vidarabine (arabinosyladenine) was selected as the target API as its enzymatic synthesis, although through a bi-enzymatic approach [6,9], it was successfully achieved even at a preparative scale with a good yield and purity (3.5 g/L, 53% yield, 98.7% purity). Surprisingly, when using 7-methylguanine arabinoside iodide (**9**), neither phosphorolysis, nor transglycosylation, thereof, were observed. This result suggests the need for a deeper understanding of the structural requisites for the enzyme-substrate molecular recognition; this is a necessary step to rationalize all the data collected over the years about the substrate specificity of *Ah*PNP as well as to further widen the exploitation of this enzyme in the bio-catalyzed synthesis of modified nucleosides, also at a preparative scale. In this regard, taking into account that the synthesis of Ribavirin (**1**), Tecadenoson (**2**), and Cladribine (**3**) were performed under screening conditions by using a 1:1 donor/acceptor ratio, conversions values were remarkable (≥50%) and foresee considerable room for improvement.

3. Materials and Methods

3.1. General

3.1.1. Chemicals

Solvents and reagents were purchased from Sigma–Aldrich (Milano, Italy), Fluorochem (Hadfield, Derbyshire, UK), Fluka (Milwaukee, WI, USA), Merck (Darmstadt, Germany), and were used without any further purification, unless stated otherwise. Dichloromethane (CH_2Cl_2), chloroform ($CHCl_3$), acetone, methanol (MeOH), and ethanol (EtOH) were distilled before use. All other solvents were of HPLC grade.

Purine nucleoside phosphorylase from *Aeromonas hydrophila* (*Ah*PNP) was provided by Gnosis S.p.A. (Desio, MB, Italy). Specific activity toward inosine was 39 IU·mg^{-1} (stock solution 21.5 mg·mL^{-1}) [4]. One IU corresponded with an amount of enzyme that converts one mmol of inosine into hypoxanthine per min.

3.1.2. Methods

Analytical TLC was performed on silica-gel F254 precoated aluminum sheets (0.2 mm layers, Merck, Darmstadt, Germany). Elution solvent: CH_2Cl_2–MeOH, 9:1. Detection: UV lamp (λ 254 nm) and 4.5% *w*/*v* $CeSO_4/(NH_4)_6Mo_7O_{24}\cdot 4H_2O$ solution or 5% *w*/*v* ninhydrin solution in EtOH followed by heating at 150 °C.

Flash column chromatography was performed using silica gel 60, 40–63 μm (Merck, Darmstadt, Germany). Reaction performed by microwave (MW) irradiation (300 W) were run in a Biotage Initiator + apparatus (Biotage, Uppsala, Denmark).

^1H and ^{13}C spectra were recorded at 400.13 and 100.61 Hz, respectively, on a Bruker AVANCE 400 spectrometer equipped with a TOPSPIN software package (Bruker, Karlsruhe, Germany) at 300 K, unless stated otherwise. ^1H and ^{13}C chemical shifts (δ) are given in parts per million and were referenced to the solvent signals (δ_H 3.31–δ_C 49.00, δ_H 2.50–δ_C 39.52 ppm from tetramethylsilane (TMS) for CD_3OD and DMSO-d_6, respectively). The ^{13}C NMR signal multiplicities were based on APT (attached proton test) spectra. The ^{13}C NMR signals were assigned with the aid of ^1H-^{13}C correlation experiments (heteronuclear multiple quantum correlation spectroscopy, HMQC, and heteronuclear multiple bond correlation spectroscopy, HMBC).

Electrospray ionization mass spectra (ESI-MS) were recorded on a ThermoFinnigan LCQ Advantage spectrometer (Hemel Hempstead, Hertfordshire, UK).

The pH measurements were performed by using a 718 Stat Titrino pHmeter from Metrohm (Herisau, Switzerland).

Enzymatic reactions were monitored by HPLC using a Merck Hitachi L-7000 La-Chrom liquid chromatographer equipped with a UV-Vis detector, an autosampler (injection volume: 20 μL), and a column oven (instrument 1), or a Chromaster 600 bar system, Merck Hitachi VWR equipped with a UV-Vis detector, an autosampler (injection volume: 20 μL), and a column oven (instrument 2).

Chromatographic conditions: column, Phenomenex Gemini C_{18} (5 μm, 250 × 4.6 mm, Supelco) or SepaChrom C_{18}-Extreme (5 μm, 250 × 4.6 mm); flow rate: 1.0 mL·min^{-1}; λ: 260 nm (225 nm for Ribavirin synthesis); temperature: 35 °C; eluent: 50 mM K_2HPO_4 buffer pH 4.5 (A) and MeOH (B); method: from 3% to 65% B (20 min), 65% B (5 min), from 65% to 3% B (0.1 min), 3% B (15 min). Under these conditions the following retention times (t_R) were registered: 7-methylguanosine iodide (**7**) (6.64 min); 7-methyl-2′-deoxyguanosine iodide (**8**) (7.20 min); Ribavirin (**1**) (4.53 min); 1,2,4-triazole-3-carboxamide (**15**) (3.25 min); Tecadenoson (**2**) (19.32 min); 6-(3-aminotetrahydrofuranyl)purine (**14**) (17.46 min); Cladribine (**3**) (18.62 min); 2-chloro-6-aminopurine (**11**) (15.16 min); CCPA (**17**) (26.39 min); 2-chloro-N^6-cyclopentyladenine (**12**) (27.60 min); Acadesine or AICAR (**18**) (6.47 min); 5-amino-1*H*-imidazole-4-carboxamide (**16**) (3.72 min); 7-methylguanine (**19**) (7.97–8.06 min). Retention times of CCPA (**17**) and 7-methylguanine (**19**) were assigned by exclusion, upon analyzing the profile of each chromatogram. The samples from the enzymatic reactions were analyzed after filtering off the enzyme through centrifugal filter devices (10 kDa MWCO, VWR International, Milano, Italy).

3.2. Chemical Synthesis of Sugar Donors

7-Methylguanosine iodide (**7**). The title compound was synthesized in 87% yield as previously reported [4]. 7-Methyl-2′-deoxyguanosine iodide (**8**). The title compound was prepared following a published procedure with some modifications [50]. Under inert atmosphere, a solution of **5** (160 mg, 0.60 mmol, 1.00 equivalent), and CH_3I (0.30 mL, 4.82 mmol, 8.03 equivalent) in dry DMSO (1.20 mL) was stirred at 20 °C for 4 h 30′. The mixture was continuously protected from light exposure, and as the solution slowly turned brown-red, the substrate disappearance was monitored by TLC (EtOH–H_2O, 4:1; R_f = 0.80). Cold $CHCl_3$ (15 mL) was added to precipitate a pale-yellow powder and the suspension was decanted at 0 °C for 2 h. The precipitate was filtered, washed with cold $CHCl_3$, and dried to get **8** as an off-white powder, which was stored at −20 °C (197 mg, 0.48 mmol, 80%). R_f: 0.21 (EtOH–H_2O, 4:1). HPLC t_R: 7.20 min. ^1H NMR (DMSO-d_6, 400 MHz): δ (ppm) 11.67 (s, 1H, NH^1), 9.28 (s, 1H, H^8), 7.18 (br s, 2H, N^2H_2), 6.20 (t, *J* = 6.0 Hz, 1H, $H^{1'}$), 5.40 (br s, 1H, $OH^{3'}$), 4.98 (br s, 1H, $OH^{5'}$), 4.37 (dd, *J* = 9.1, 4.6 Hz, 1H, $H^{3'}$), 4.00 (s, 3H, N^7CH_3), 3.93 (q, *J* = 3.9 Hz, 1H, $H^{4'}$), 3.62 (dd, *J* = 12.1, 4.2 Hz, 1H, $H^{5'a}$, partially overlapped with $H^{5'b}$), 3.57 (dd, *J* = 12.1, 4.2 Hz, 1H, $H^{5'b}$, partially overlapped with

$H^{5'a}$), 2.55–2.47 (m, 1H, $H^{2'a}$, partially covered by DMSO), 2.40 (ddd, *J* = 13.4, 6.2, 4.8 Hz, 1H, $H^{2'b}$). ^{13}C NMR (DMSO-d_6, 100 MHz): δ (ppm) 156.0 (C^2), 153.9 (C^6), 149.3 (C^4), 136.7 (C^8), 108.1 (C^5), 89.1 ($C^{4'}$), 85.8 ($C^{1'}$), 70.2 ($C^{3'}$), 61.3 ($C^{5'}$), 40.5 ($C^{2'}$, partially covered by DMSO), 36.1 (N^7CH_3). MS (ESI^+): *m/z* calcd. for $[C_{11}H_{16}N_5O_4]^+$: 282.12; found: 166.3 [M-2′-deoxyribosyl]$^+$, 282.1 $[M]^+$, 305.3 $[M + Na]^+$, 563.7 $[2M]^+$. MS (ESI^-): *m/z* calcd. for $[I]^-$: 126.91; found: 127.2 $[M]^-$.

7-Methylguanine arabinoside iodide (**9**). Under inert atmosphere, a solution of **6** (100 mg, 0.35 mmol. 1.00 equivalent), synthesized as previously reported [53], and CH_3I (0.14 mL, 2.25 mmol, 6.43 mmol) in a dry DMF/DMSO mixture (3:1 *v*/*v*, 1.00 mL) was stirred at room temperature for 6 h. The mixture was continuously protected from light exposure, and as the solution slowly turned yellow, the substrate disappearance was monitored by TLC (EtOH–H_2O, 7:3; R_f = 0.83). The mixture was diluted with H_2O (50 mL) and freeze-dried until complete removal of DMSO. The resulting pale-yellow crude was suspended in dry acetone (2.5 mL), filtered, washed with few cold dry acetone, and dried to get **9** as an off-white powder, which was stored at −20 °C (133 mg, 0.31 mmol, 89%). R_f: 0.53 (EtOH–H_2O, 7:3). ^1H NMR (DMSO-d_6, 400 MHz):δ (ppm) 11.68 (br s, 1H, NH^1), 9.22 (s, 1H, H^8), 7.21 (br s, 2H, N^2H_2), 6.17 (d, *J* = 4.3 Hz, 1H, $H^{1'}$), 5.82 (d, *J* = 5.3 Hz, 1H, $OH^{2'}$), 5.64 (d, *J* = 4.3 Hz, 1H, $OH^{3'}$), 5.04 (t, *J* = 5.4 Hz, 1H, $OH^{5'}$), 4.17 (dd, *J* = 8.6, 4.4 Hz, 1H, $H^{2'}$), 4.11 (dd, *J* = 7.5, 3.7 Hz, 1H, $H^{3'}$), 4.06 (s, 3H, N^7CH_3), 3.90 (dd, *J* = 8.9, 5.1 Hz, 1H, $H^{4'}$), 3.70 (dd, *J* = 11.5, 5.0 Hz, 1H, $H^{5'a}$, partially overlapped with $H^{5'b}$), 3.64 (dd, *J* = 11.6, 5.4 Hz, 1H, $H^{5'b}$, partially overlapped with $H^{5'a}$). ^{13}C NMR (DMSO-d_6, 100 MHz): δ (ppm) 156.2 (C^2), 153.8 (C^6), 149.6 (C^4), 137.8 (C^8), 107.6 (C^5), 86.5 ($C^{4'}$), 86.0 ($C^{1'}$), 75.5 ($C^{2'}$), 75.4 ($C^{3'}$), 61.3 ($C^{5'}$), 36.1 (N^7CH_3). MS (ESI^+): *m/z* calcd. for $[C_{11}H_{16}N_5O_5]^+$: 298.11; found: 166.2 [M-arabinosyl]$^+$, 297.1 $[M - H]^+$, 298.0 $[M]^+$. MS (ESI^-): *m/z* calcd. for $[I]^-$: 126.91; found: 127.1 $[M]^-$.

3.3. Chemical Synthesis of Base Acceptors and Tecadenoson

2-Chloro-6-aminopurine (**11**). The title compound was prepared following a published procedure with some modifications [54]. Under inert atmosphere, a solution of **10** (113 mg, 0.60 mmol, 1.00 equivalent) in a 30% aq. NH_3–MeOH mixture (2:3 *v*/*v*, 11.3 mL) was stirred under MW irradiation at 110 °C for 14 h. The solution was evaporated and the resulting light-blue solid was suspended in H_2O. The precipitate was filtered and dried to get **11** as a white powder (67 mg, 0.40 mmol, 66%).^1H NMR (DMSO-d_6, 400 MHz): δ (ppm) 8.13 (s, 1H, H^2), 7.62 (s, 2H, NH_2). ^{13}C NMR (DMSO-d_6, 100 MHz): δ (ppm) 156.1, 153.8, 140.8. MS (ESI^+): *m/z* calcd. for $[C_5H_4ClN_5]^+$: 169.02; found: 507.31 $[3M]^+$. MS (ESI^-): *m/z* calcd. for $[C_5H_4ClN_5]^-$: 169.02; found: 168.13 $[M - H]^-$.

2-Chloro-N^6-cyclopentyladenine (**12**). The title compound was prepared following a published procedure with some modifications [55]. Under inert atmosphere, a suspension of **10** (142 mg, 0.75 mmol, 1.00 equivalent), cyclopentylamine (0.22 mL, 2.25 mmol, 3.00 equiv) in dry *n*-BuOH (1.50 mL) was refluxed for 4 h. The solvent was evaporated and the resulting crude was purified by flash chromatography (CH_2Cl_2–MeOH, 9.7:0.3) to get **12** as a white powder (191 mg, 0.52 mmol, 69%). R_f: 0.33 (CH_2Cl_2–MeOH, 9.7:0.3). ^1H NMR (CD_3OD, 400 MHz): δ (ppm) 8.04 (s, 1H, H^8), 4.56 (br s, 1H, H^a), 2.33–2.05 (m, 2H, H^b), 1.83 (qd, *J* = 10.5, 8.8, 5.0 Hz, 2H, H^b), 1.78–1.68 (m, 2H, H^c), 1.63 (ddt, *J* = 14.0, 8.1, 5.2 Hz, 2H, H^c). ^{13}C NMR (CD_3OD, 100 MHz): δ (ppm) 153.5, 139.6, 112.2, 52.0, 32.4, 23.9. MS (ESI^+): *m/z* calcd. for $[C_{10}H_{12}ClN_5]^+$: 237.08; found: 259.99 $[M + Na]^+$.

6-(3-Aminotetrahydrofuranyl)purine (**14**). Under inert atmosphere, a suspension of **13** (154 mg, 1.00 mmol, 1.00 equivalent), (*R*)-3-aminotetrahydrofuran hydrochloride (595 mg, 6.00 mmol, 6.00 equivalent) and LiOH (250 mg, 6.00 mmol, 6.00 equivalent) in EtOH (3.5 mL) was refluxed for 6 h. The solution was evaporated and the residue was purified by flash chromatography (CH_2Cl_2–MeOH, 9.3:0.7) to get **14** as a white powder (72 mg, 0.35 mmol, 35%). ^1H NMR (CD_3OD, 400 MHz): δ (ppm) 8.28 (s, 1H, H^2), 8.10 (s, 1H, H^8), 4.92–4.79 (m, 1H, H^a), 4.16–3.99 (m, 2H, H^b, H^c), 3.91 (td, *J* = 8.4, 5.5 Hz, 1H, H^c), 3.81 (dd, *J* = 9.2, 3.4 Hz, 1H, H^b), 2.41 (ddt, *J* = 13.0, 8.2, 7.2 Hz, 1H, H^d), 2.04 (dddd, *J* = 13.0, 7.6, 5.4, 3.7 Hz, 1H, H^d). ^{13}C NMR (CD_3OD, 100 MHz): δ (ppm) 152.3, 73.0, 66.7, 51.4, 32.4. MS (ESI^+): *m/z* calcd. for $[C_9H_{11}N_5O]^+$: 205.10; found: 206.05 $[M + H]^+$, 228.06 $[2M + Na]^+$.

Tecadenoson (**2**). The title compound was prepared following a published procedure with some modifications [35]. Under inert atmosphere, a suspension of 6-chloroinosine (143 mg, 0.50 mmol, 1.00 equivalent), (*R*)-3-aminotetrahydrofuran hydrochloride (148 mg, 1.50 mmol, 3.00 equivalent) and triethylamine (0.21 mL, 1.50 mmol, 3.00 equivalent) in EtOH (4.0 mL) was refluxed for 4 h. The solution was evaporated and the residue was purified by flash chromatography (CH_2Cl_2–MeOH, 9:1) to get **2** as a white powder (93 mg, 0.28 mmol, 55%). ^1H NMR (DMSO-d_6, 400 MHz): δ (ppm) 8.39 (s, 1H, H^2), 8.24 (s, 1H, H^8), 8.14–7.93 (m, 1H, N^6H), 5.90 (d, J = 6.1 Hz, 1H, $H^{1'}$), 5.45 (s, 1H, $OH^{2'}$), 5.37 (t, J = 5.6 Hz, 1H, $OH^{5'}$), 5.19 (s, 1H, $OH^{3'}$), 4.72 (s, 1H, NH^6), 4.61 (t, J = 5.5 Hz, 1H, $H^{2'}$), 4.15 (s, 1H, $H^{3'}$), 4.10 (q, J = 5.2 Hz, 3H, H^a, H^b, H^c), 4.01–3.83 (m, 1H, $H^{3'}$), 3.84–3.43 (m, 3H, $H^{4'}$, H^b, H^c), 2.20 (dq, J = 14.6, 7.6 Hz, 1H, H^d), 2.04 (br, 1H, H^d). ^{13}C NMR (DMSO-d_6, 100 MHz): δ (ppm) 153.2, 151.9, 87.5, 85.9, 72.4, 71.0, 66.5. MS (ESI$^+$): m/z calcd for $[C_{14}H_{19}N_5O_5]^+$: 337.14; found: 360.04 $[M + Na]^+$.

3.4. Enzymatic Synthesis of Nucleoside Analogues: General Procedure of Transglycosylation Reactions

Purine nucleoside phosphorylase from *A. hydrophila* (1.15 or 80 IU) was added to a solution of 50 mM KH_2PO_4 buffer pH 7.5 and 20% (*v*/*v*) glycerol containing **7** or **8** (1 mM) and the modified nucleobase **B**-H (1 mM, see Table 1). In the case of Cladribine (**3**) and CCPA (**17**), DMSO was added as the second co-solvent (5% or 10% *v*/*v*, respectively). The final reaction volume was 20 mL or 5 mL. The mixture was gently stirred (rolling shaker) at room temperature. Aliquots (200 μL) of the reaction mixture were withdrawn at fixed times (1, 3, 6, 24 h), and filtered by centrifugation (MWCO 10 kDa, 5 min, 12,000 rpm). The supernatant was diluted 1:4 with the mobile phase and analyzed by HPLC (injection volume: 20 μL). Conversions were estimated by Equations (1) and (2):

$$\text{Base consumption } (\%) = \frac{\text{base area}}{\text{base area} + \text{nucleoside area}} \times 100 \quad (1)$$

$$\text{Nucleoside formation } (\%) = \frac{\text{nucleoside area}}{\text{base area} + \text{nucleoside area}} \times 100 \quad (2)$$

Supplementary Materials: The following are available online at http://www.mdpi.com/2073-4344/9/4/355/s1, Figure S1: ^{13}C-NMR monitoring of 7-methyl-2′-deoxyguanosine iodide (**8**) stability, Figure S2: ^1H-NMR and ^{13}C-NMR spectra of 7-methylguanine arabinoside iodide (**9**), Figure S3: ^1H-NMR spectrum of Tecadenoson (**2**), Figure S4: HPLC monitoring of the enzymatic synthesis of Ribavirin (**1**), Figure S5: HPLC monitoring of the enzymatic synthesis of CCPA (**17**), Figure S6: HPLC monitoring of the enzymatic synthesis of Acadesine (**18**), Figure S7: HPLC monitoring of the enzymatic synthesis of Tecadenoson (**2**).

Author Contributions: Conceptualization, G.S. and D.U.; Methodology, M.R., T.B., R.S., G.C., G.S., D.U.; Investigation, M.R., T.B., R.S., G.C., M.M., C.F.M.; Resources, G.S. and D.U.; Data Curation, M.R., T.B., R.S., G.S., D.U.; Writing-Original Draft Preparation, G.S. and D.U. with the assistance of M.R.; Writing-Review and Editing, G.S. and D.U. with the assistance of M.R., T.B., and R.S.; Supervision, G.S. and D.U.

Funding: This research received no external funding.

Acknowledgments: This manuscript is dedicated to A. M. Albertini.

Conflicts of Interest: The authors declare no conflict of interest.

References

1. McConachie, S.M.; Wilhelm, S.M.; Kale-Pradhan, P.B. New direct-acting antivirals in hepatitis C therapy: A review of sofosbuvir, ledipasvir, daclatasvir, simeprevir, paritaprevir, ombitasvir and dasabuvir. *Expert Rev. Clin. Pharmacol.* **2016**, *9*, 287–302. [CrossRef]
2. Vorbrueggen, H.; Ruh-Pohlenz, C. Synthesis of nucleosides. *Org. React.* **2000**, *55*. [CrossRef]
3. Mikhailopulo, I.; Miroshnikov, A.I. Biologically important nucleosides: Modern trends in biotechnology and application. *Mendeleev Commun.* **2011**, *21*, 57–68. [CrossRef]
4. Ubiali, D.; Serra, C.D.; Serra, I.; Morelli, C.F.; Terreni, M.; Albertini, A.M.; Manitto, P.; Speranza, G. Production, characterization and synthetic application of a purine nucleoside phosphorylase from *Aeromonas hydrophila*. *Adv. Synth. Catal.* **2012**, *354*, 96–104. [CrossRef]

5. Serra, I.; Ubiali, D.; Piškur, J.; Christoffersen, S.; Lewkowicz, E.S.; Iribarren, A.M.; Albertini, A.M.; Terreni, M. Developing a collection of immobilized nucleoside phosphorylases for the preparation of nucleoside analogues: Enzymatic synthesis of arabinosyladenine and 2′,3′-dideoxyinosine. *ChemPlusChem* **2013**, *78*, 157–165. [CrossRef]
6. Serra, I.; Daly, S.; Alcantara, A.R.; Bianchi, D.; Terreni, M.; Ubiali, D. Redesigning the synthesis of Vidarabine via a multienzymatic reaction catalyzed by immobilized nucleoside phosphorylases. *RSC Adv.* **2015**, *5*, 23569–23577. [CrossRef]
7. Ubiali, D.; Morelli, C.F.; Rabuffetti, M.; Cattaneo, G.; Serra, I.; Bavaro, T.; Albertini, A.; Speranza, G. Substrate specificity of a purine nucleoside phosphorylase from *Aeromonas hydrophila* toward 6-substituted purines and its use as a biocatalyst in the synthesis of the corresponding ribonucleosides. *Curr. Org. Chem.* **2015**, *19*, 2220–2225. [CrossRef]
8. Calleri, E.; Cattaneo, G.; Rabuffetti, M.; Serra, I.; Bavaro, T.; Massolini, G.; Speranza, G.; Ubiali, D. Flow-synthesis of nucleosides catalyzed by an immobilized purine nucleoside phosphorylase from *Aeromonas hydrophila*: Integrated systems of reaction control and product purification. *Adv. Synth. Catal.* **2015**, *357*, 2520–2528. [CrossRef]
9. Cattaneo, G.; Rabuffetti, M.; Speranza, G.; Kupfer, T.; Peters, B.; Massolini, G.; Ubiali, D.; Calleri, E. Synthesis of adenine nucleosides by transglycosylation using two sequential nucleoside phosphorylase-based bioreactors with on-line reaction monitoring by using HPLC. *ChemCatChem* **2017**, *9*, 4614–4620. [CrossRef]
10. Leyssen, P.; De Clercq, E.; Neyts, J. Molecular strategies to inhibit the replication of RNA viruses. *Antivir. Res.* **2008**, *78*, 9–25. [CrossRef]
11. Broder, C. Henipavirus outbreaks to antivirals: The current status of potential therapeutics. *Curr. Opin. Virol.* **2012**, *2*, 176–187. [CrossRef]
12. Lau, J.Y.N.; Tam, R.C.; Liang, T.J.; Hong, Z. Mechanism of action of ribavirin in the combination treatment of chronic HCV infection. *Hepatology* **2002**, *35*, 1002–1009. [CrossRef]
13. Chung, R.T.; Gale, M., Jr.; Polyak, S.J.; Lemon, S.M.; Liang, T.J.; Hoofnagle, J.H. Mechanisms of action of interferon and ribavirin in chronic hepatitis C: Summary of a workshop. *Hepatology* **2008**, *47*, 306–320. [CrossRef] [PubMed]
14. Joosen, M.J.; Bueters, T.J.; van Helden, H.P. Cardiovascular effects of the adenosine A_1 receptor agonist N^6-cyclopentyladenosine (CPA) decisive for its therapeutic efficacy in sarin poisoning. *Arch. Toxicol.* **2004**, *78*, 34–39. [CrossRef] [PubMed]
15. Peterman, C.; Sanoski, C.A. Tecadenoson: A novel, selective A_1 adenosine receptor agonist. *Cardiol. Rev.* **2005**, *13*, 315–321. [CrossRef] [PubMed]
16. Balakumar, P.; Singh, H.; Reddy, K.; Anand-Srivastava, K.; Madhu, B. Adenosine-A_1 receptors activation restores the suppressed cardioprotective effects of ischemic preconditioning in hyperhomocysteinemic rat hearts. *J. Cardiovasc. Pharm.* **2009**, *54*, 204–212. [CrossRef] [PubMed]
17. Juliusson, G.; Samuelsson, H. Hairy cell leukemia: Epidemiology, pharmacokinetics of cladribine, and long-term follow-up of subcutaneous therapy. *Leuk. Lymphoma* **2011**, *52*, 46–49. [CrossRef]
18. Giovannoni, G.; Comi, G.; Cook, S.; Rammohan, K.; Rieckmann, P.; Sørensen, P.S.; Vermersch, P.; Chang, P.; Hamlett, A.; Musch, B.; et al. A placebo-controlled trial of oral cladribine for relapsing multiple sclerosis. *N. Eng. J. Med.* **2010**, *362*, 416–426. [CrossRef]
19. Liu, W.Y.; Li, H.Y.; Zhao, B.X.; Shin, D.S.; Lian, S.; Miao, J.Y. Synthesis of novel ribavirin hydrazone derivatives and anti-proliferative activity against A549 lung cancer cells. *Carbohydr. Res.* **2009**, *344*, 1270–1275. [CrossRef]
20. Li, Y.S.; Zhang, J.J.; Mei, L.Q.; Tan, C.X. An improved procedure for the preparation of Ribavirin. *Org. Prep. Proced. Int.* **2012**, *44*, 387–391. [CrossRef]
21. Bookser, B.C.; Raffaele, N.B. High-throughput five minute microwave accelerated glycosylation approach to the synthesis of nucleoside libraries. *J. Org. Chem.* **2007**, *72*, 173–179. [CrossRef]
22. Shirae, H.; Yokozeki, K.; Kubota, K. Enzymatic production of Ribavirin. *Agric. Biol. Chem.* **1988**, *52*, 295–296. [CrossRef]
23. Shirae, H.; Yokozeki, K.; Uchiyama, M.; Kubota, K. Enzymatic production of Ribavirin from purine nucleosides by *Brevibacterium acetylicum* ATCC 954. *Agric. Biol. Chem.* **1988**, *52*, 1777–1783. [CrossRef]
24. Hennen, W.J.; Wong, C.H. A new method for the enzymic synthesis of nucleosides using purine nucleoside phosphorylase. *J. Org. Chem.* **1989**, *54*, 4692–4695. [CrossRef]

25. Shirae, H.; Yokozeki, K.; Kubota, K. Adenosine phosphorolyzing enzymes from microorganisms and ribavirin production by the application of the enzyme. *Agric. Biol. Chem.* **1991**, *55*, 605–607. [CrossRef]
26. Barai, V.N.; Zinchenko, A.I.; Eroshevskaya, L.A.; Kalinichenko, E.N.; Kulak, T.I.; Mikhailopulo, I.A. A universal biocatalyst for the preparation of base- and sugar-modified nucleosides via an enzymatic transglycosylation. *Helv. Chim. Acta* **2002**, *85*, 1901–1908. [CrossRef]
27. Trelles, J.A.; Fernandez, M.; Lewkowicz, E.S.; Iribarren, A.M.; Sinisterra, J.V. Purine nucleoside synthesis from uridine using immobilized *Enterobacter gergoviae* CECT 875 whole cells. *Tetrahedron Lett.* **2003**, *44*, 2605–2609. [CrossRef]
28. Konstantinova, I.D.; Leont'eva, N.A.; Galegov, G.A.; Ryzhova, O.I.; Chuvikovskii, D.V.; Antonov, K.V.; Esipov, R.S.; Taran, S.A.; Verevkina, K.N.; Feofanov, S.A.; et al. Ribavirin: Biotechnological synthesis and effect on the reproduction of *Vaccinia* virus. *J. Bioorg. Chem.* **2004**, *30*, 553–560. [CrossRef]
29. Nobile, M.; Terreni, M.; Lewkowicz, E.; Iribarren, A.M. *Aeromonas hydrophila* strains as biocatalysts for transglycosylation. *Biocatal. Biotransfor.* **2011**, *28*, 395–402. [CrossRef]
30. Ding, Q.B.; Ou, L.; Wei, D.Z.; Wei, X.K.; Xu, Y.M.; Zhang, C.Y. Enzymatic synthesis of nucleosides by nucleoside phosphorylase co-expressed in *Escherichia coli*. *J. Zhejiang Univ. Sci. B* **2010**, *11*, 880–888. [CrossRef]
31. De Benedetti, E.C.; Rivero, C.W.; Trelles, J.A. Development of a nanostabilized biocatalyst using an extremophilic microorganism for ribavirin biosynthesis. *J. Mol. Catal. B Enzym.* **2015**, *121*, 90–95. [CrossRef]
32. Rivero, C.W.; De Benedetti, E.C.; Lozano, M.E.; Trelles, J.A. Bioproduction of ribavirin by green microbial biotransformation. *Process Biochem.* **2015**, *50*, 935–940. [CrossRef]
33. Shirae, H.; Yokozeki, K.; Kubota, K. Enzymic production of Ribavirin from pyrimidine nucleosides by *Enterobacter aerogenes* AJ 11125. *Agric. Biol. Chem.* **1988**, *52*, 1233–1237. [CrossRef]
34. Palle, V.P.; Varkhedkar, V.; Ibrahim, P.; Ahmed, H.; Li, Z.; Gao, Z.; Ozeck, M.; Wu, Y.; Zeng, D.; Wu, L.; et al. Affinity and intrinsic efficacy (IE) of 5′-carbamoyl adenosine analogues for the A1 adenosine receptor—Efforts towards the discovery of a chronic ventricular rate control agent for the treatment of atrial fibrillation (AF). *Bioorg. Med. Chem. Lett.* **2004**, *14*, 535–539. [CrossRef] [PubMed]
35. Ashton, T.D.; Aumann, K.M.; Baker, S.P.; Schiesser, C.H.; Scammells, P.J. Structure-activity relationships of adenosines with heterocyclic N^6-substituents. *Bioorg. Med. Chem. Lett.* **2007**, *17*, 6779–6784. [CrossRef]
36. Petrelli, R.; Scortichini, M.; Belardo, C.; Boccella, S.; Luongo, L.; Capone, F.; Kachler, S.; Vita, P.; Del Bello, F.; Maione, S.; et al. Structure-based design, synthesis, and in vivo antinociceptive effects of selective A1 adenosine receptor agonists. *J. Med. Chem.* **2018**, *61*, 305–318. [CrossRef]
37. Robins, M.J.; Robins, R.K. Purine nucleosides. XI. The synthesis of 2′-deoxy-9-α- and -β-D-ribofuranosylpurines and the correlation of their anomeric structure with proton magnetic resonance spectra. *J. Am. Chem. Soc.* **1965**, *87*, 4934–4940. [CrossRef]
38. Christensen, L.F.; Broom, A.D.; Robins, M.J.; Bloch, A. Synthesis and biological activity of selected 2,6-disubstituted-(2-deoxy-α-and-β-D-erythro-pentofuranosyl) purines. *J. Med. Chem.* **1972**, *15*, 735–739. [CrossRef]
39. Kazimierczuk, Z.; Cottam, H.B.; Revankar, G.R.; Robins, R.K. Synthesis of 2′-deoxytubercidin, 2′-deoxyadenosine, and related 2′-deoxynucleosides via a novel direct stereospecific sodium salt glycosylation procedure. *J. Am. Chem. Soc.* **1984**, *106*, 6379–6382. [CrossRef]
40. Xia, R.; Chen, L.S. Efficient synthesis of Cladribine via the metal-free dioxygenation. *Nucleosides Nucleotides Nucleic Acids* **2015**, *34*, 729–735. [CrossRef]
41. Xu, S.; Yao, P.; Chen, G.; Wang, H. A new synthesis of 2-chloro-2′-deoxyadenosine (Cladribine), CdA. *Nucleosides Nucleotides Nucleic Acids* **2011**, *30*, 353–359. [CrossRef] [PubMed]
42. Matyasovsky, J.; Perlikova, P.; Malnuit, V.; Pohl, R.; Hocek, M. 2-Substituted dATP derivatives as building blocks for polymerase-catalyzed synthesis of DNA modified in the minor groove. *Angew. Chem. Int. Ed.* **2016**, *55*, 15856–15859. [CrossRef] [PubMed]
43. Matsuda, A.; Shinozaki, M.; Suzuki, M.; Watanabe, K.; Miyasaka, T. A convenient method for the selective acylation of guanine nucleosides. *Synthesis* **1986**, *1986*, 385–386. [CrossRef]
44. McGuinness, B.F.; Nakanishi, K.; Lipman, R.; Tomasz, M. Synthesis of guanine derivatives substituted in the O^6 position by Mitomycin C. *Tetrahedron Lett.* **1988**, *29*, 4673–4676. [CrossRef]
45. Zhong, M.; Nowak, I.; Robins, M.J. Regiospecific and highly stereoselective coupling of 6-(substituted-imidazol-1-yl) purines with 2-deoxy-3,5-di-*O*-(*p*-toluoyl)-α-D-erythro-pentofuranosyl chloride. Sodium-salt glycosylation in binary solvent mixtures: Improved synthesis of Cladribine. *J. Org. Chem.* **2006**, *71*, 7773–7779. [CrossRef]

46. Satishkumar, S.; Vuram, P.K.; Relangi, S.S.; Gurram, V.; Zhou, H.; Kreitman, R.J.; MartínezMontemayor, M.M.; Yang, L.; Kaliyaperumal, M.; Sharma, S.; et al. Cladribine analogues via O^6-(benzotriazolyl) derivatives of guanine nucleosides. *Molecules* **2015**, *20*, 18437–18463. [CrossRef] [PubMed]
47. Hironori, K.; Araki, T. Efficient chemo-enzymatic syntheses of pharmaceutically useful unnatural 2′-deoxynucleosides. *Nucleosides Nucleotides Nucleic Acids* **2005**, *24*, 1127–1130. [CrossRef]
48. Zhou, X.; Szeker, K.; Jiao, L.Y.; Oestreich, M.; Mikhailopulo, I.A.; Neubauer, P. Synthesis of 2,6-dihalogenated purine nucleosides by thermostable nucleoside phosphorylases. *Adv. Synth. Catal.* **2015**, *357*, 1237–1244. [CrossRef]
49. Drenichev, M.S.; Alexeev, C.S.; Kurochkin, N.N.; Mikhailov, S.N. Use of nucleoside phosphorylases for the preparation of purine and pyrimidine 2′-deoxynucleosides. *Adv. Synth. Catal.* **2018**, *360*, 305–312. [CrossRef]
50. Voegel, J.J.; Altorfer, M.M.; Benner, S.A. The donor-acceptor-acceptor purine analog: Transformation of 5-aza-7-deaza-1*H*-isoguanine (=4-aminoimidazo-[1,2-α]-1,3,5-triazin-2(1*H*)-one) to 2′-deoxy-5-aza-7-deaza-isoguanosine using purine nucleoside phosphorylase. *Helv. Chim. Acta* **1993**, *76*, 2061–2069. [CrossRef]
51. Kamel, S.; Yehia, H.; Neubauer, P.; Wagner, A. Enzymatic synthesis of nucleoside analogues by nucleoside phosphorylases. In *Enzymatic and Chemical Synthesis of Nucleic Acid Derivatives*, 1st ed.; Fernández-Lucas, J., Camarasa Rius, M.J., Eds.; Wiley-VCH Verlag GmbH & Co. KGaA: Weinheim, Germany, 2018; Chapter 1; pp. 1–20.
52. Acosta, J.; del Arco, J.; Martinez-Pascual, S.; Clemente-Suárez, V.X.; Fernández-Lucas, J. One-pot multi-enzymatic production of purine derivatives with application in pharmaceutical and food industry. *Catalysts* **2018**, *8*, 9. [CrossRef]
53. Gruen, M.; Becker, C.; Beste, A.; Siethoff, C.; Scheidig, A.J.; Goody, R.S. Synthesis of 2′-iodo- and 2′-bromo-ATP and GTP analogues as potential phasing tools for X-ray crystallography. *Nucleosides Nucleotides Nucleic Acids* **1999**, *18*, 137–151. [CrossRef]
54. Borrmann, T.; Abdelrahman, A.; Volpini, R.; Lambertucci, C.; Alksnis, E.; Gorzalka, S.; Knospe, M.; Schiedel, A.C.; Cristalli, G.; Mueller, C.E. Structure-activity relationships of adenine and deazaadenine derivatives as ligands for adenine receptors, a new purinergic receptor family. *J. Med. Chem.* **2009**, *52*, 5974–5989. [CrossRef] [PubMed]
55. Thompson, R.D.; Secunda, S.; Daly, J.W.; Olsson, R.A. N^6-9-Disubstituted adenines: Potent, selective antagonists at the A_1 adenosine receptor. *J. Med. Chem.* **1991**, *34*, 2877–2882. [CrossRef] [PubMed]

Article

Development of Biotransamination Reactions towards the 3,4-Dihydro-2*H*-1,5-benzoxathiepin-3-amine Enantiomers

Daniel González-Martínez [1], Nerea Fernández-Sáez [2], Carlos Cativiela [3], Joaquín M. Campos [2,4,*] and Vicente Gotor-Fernández [1,*]

[1] Organic and Inorganic Chemistry Department, University of Oviedo, Avenida Julián Clavería 8, 33006 Oviedo, Spain; daniel.dgm9@gmail.com

[2] Departamento de Química Farmacéutica y Orgánica, Facultad de Farmacia, c/Campus de Cartuja s/n, 18071 Granada, Spain; nefdez_p73@hotmail.com

[3] Departamento de Química Orgánica, Instituto de Síntesis Química y Catálisis Homogénea (ISQCH), CSIC–Universidad de Zaragoza, 50009 Zaragoza, Spain; cativiela@unizar.es

[4] Instituto de Investigación Biosanitaria ibs.GRANADA, Complejo Hospitalario Universitario de Granada/Universidad de Granada, 18071 Granada, Spain

* Correspondence: jmcampos@ugr.es (J.M.C.); vicgotfer@uniovi.es (V.G.-F.); Tel.: +34-958-243-850 (J.M.C.); +34-985-103-454 (V.G.-F.)

Received: 7 September 2018; Accepted: 16 October 2018; Published: 19 October 2018

Abstract: The stereoselective synthesis of chiral amines is an appealing task nowadays. In this context, biocatalysis plays a crucial role due to the straightforward conversion of prochiral and racemic ketones into enantiopure amines by means of a series of enzyme classes such as amine dehydrogenases, imine reductases, reductive aminases and amine transaminases. In particular, the stereoselective synthesis of 1,5-benzoxathiepin-3-amines have attracted particular attention since they possess remarkable biological profiles; however, their access through biocatalytic methods is unexplored. Amine transaminases are applied herein in the biotransamination of 3,4-dihydro-2*H*-1,5-benzoxathiepin-3-one, finding suitable enzymes for accessing both target amine enantiomers in high conversion and enantiomeric excess values. Biotransamination experiments have been analysed, trying to optimise the reaction conditions in terms of enzyme loading, temperature and reaction times.

Keywords: amine transaminases; asymmetric synthesis; benzoxathiepins; biocatalysis; biotransamination; stereoselective synthesis

1. Introduction

We have reported several (*RS*)-benzo-fused seven-membered rings with oxygen and sulfur atoms in 1,5 relative positions with interesting anti-proliferative activities against the MCF-7 cancer cell line. The most active compounds are **1** and **2** [1] (Figure 1). Other compounds, such as **3** [2] and **4** [3], exhibited more potent anti-ischemic effects than reference compounds, whilst **5** can be the prototype for the design of more potent anti-proliferative agents [4] (Figure 1). The (3*R*)-3,4-dihydro-2*H*-1,5-benzoxathiepin-3-amine core appears in red in compounds **3–5** (Figure 1). Such a (3*R*)-amino-1,5-benzoxathiepin scaffold has been obtained from L-cystine ((2*R*)-2-amino-3-[[(2*R*)-2-amino-2-carboxyethyl]disulfanyl]propanoic acid) [4,5]. The incorporation of α-amino acids into heterocyclic structures is an effective strategy for generating numerous peptidomimetics and combinatorial library scaffolds.

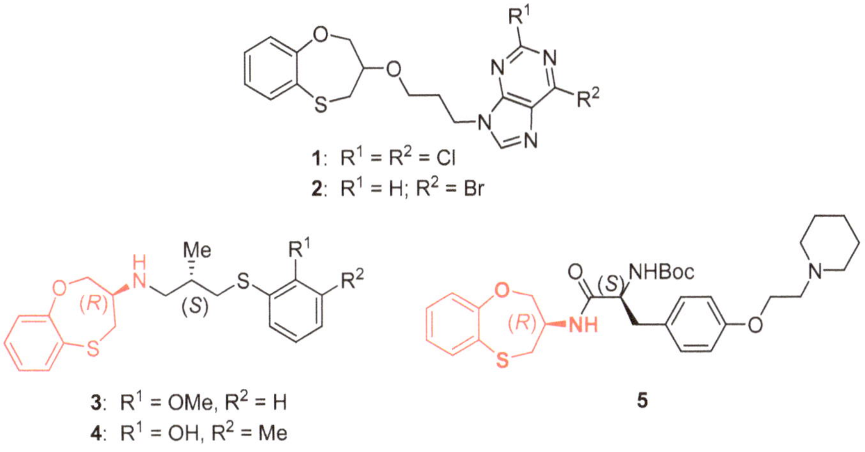

Figure 1. Benzo-fused seven-membered rings with oxygen and sulphur atoms in 1,5 relative positions (**1–5**) with interesting biological properties [1–5]. The (3*R*)-3,4-dihydro-2*H*-1,5-benzoxathiepin-3-amine core appears in red in compounds **3–5**.

Due to the fact that the primary amine is a key functional group in all areas of chemistry, methods to generate molecules containing primary amine groups are of intense interest and impact on many research fields. The use of enzymes in organic synthesis has gained maturity in the last few decades, since the advances in enzyme immobilisation, modification and rational design allow for the application of improved biocatalysts for the development of a wide variety of stereoselective transformations [6–10]. In this context, the synthesis of chiral amines is particularly challenging, with the conversion of prochiral ketones into optically active amines receiving great attention in recent years [11–14] by using mainly imine reductases [15–17] and amine transaminases (ATAs) [18–23]. Taking into account the potential of ATAs in the single biotransamination of cyclic ketones [24–32], even as part of multienzymatic sequences [33–38], but especially since they have served as valuable biocatalysts in the production of pharmacologically active products [39–43], we have focused herein our efforts in the pursuit of an efficient biotransamination protocol for 3,4-dihydro-2*H*-1,5-benzoxathiepin-3-one (**6**).

2. Results and Discussion

The synthesis of the benzo-fused seven-membered ketone **6** is depicted in Scheme 1. 2-Mercaptophenol was alkylated with two equivalents of ethyl bromoacetate in refluxing acetone in the presence of dry potassium carbonate to give diester **7** (83%). Examination of the Dieckmann reaction of **3** showed that the reaction occurred smoothly when sodium ethoxide/ethanol was used as a base in dry tetrahydrofuran (THF) to give ethyl 3-oxo-3,4-dihydro-2*H*-1,5-benzothiepin-4-carboxylate **8** as the sole cyclised product in 90% yield. Decarboxylation of the β-ketoester **4** in boiling acetic acid containing aqueous sulfuric acid gave the 3,4-dihydro-2*H*-1,5-benzothiepin-3-one (**6**, 60%). Regioselectivity of the Dieckmann cyclisation was deduced based on the ^{1}H-NMR ($CDCl_3$) spectral data of the resulting product **8**, which exhibited two doublets (integrating each one for 1H) at δ 4.88 and 4.59 ppm (J = 17.5 Hz) assignable to the geminal methylene protons adjacent to the oxygen atom in the seven-membered ring. Compounds **7** and **8** have not been described previously, whilst ketone **6** was reported formerly by Sugihara et al. [44].

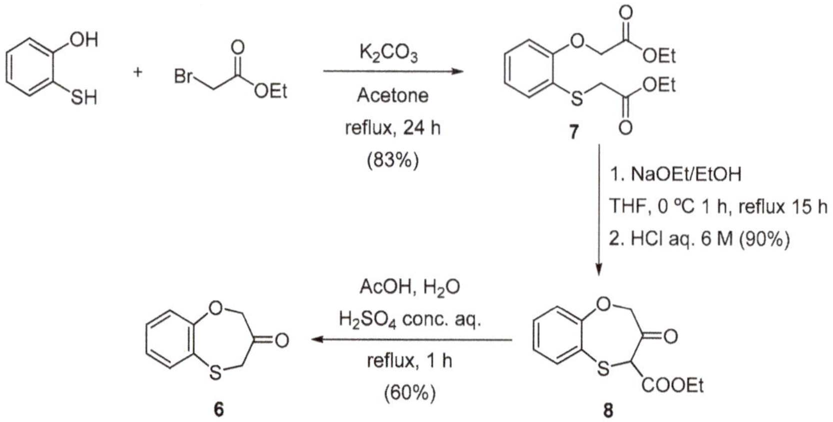

Scheme 1. Chemical synthesis of 3,4-dihydro-2*H*-1,5-benzoxathiepin-3-one **6**.

Due to the amine transaminases catalytic mechanism, which involves two pairs of ketones and amines in equilibrium, the reductive amination of the substrate must be thermodynamically favoured in order to obtain high yields of the desired product [45,46]. In order to displace the equilibrium towards amine formation, the removal of the generated co-products by coupling different multienzyme networks is often required [20], but also worth noting is the use of sacrificial substrates, which normally range from the use of a large excess of a commercially available amine donor, typically isopropylamine [47], to "smart cosubstrates", mainly diamines, in a stoichiometric amount that are able to drive equilibrium by spontaneous cyclisation or aromatisation reactions [31,48–50]. Promisingly, we have found a favourable ΔG of -31.0 kJ/mol (calculated at M06-2X/6-311++G(3df,2p) level; see Section 3.8) for the transamination of **6** to **9** when using isopropylamine and acetone is formed as a by-product, probably due to ring strain instability. Figure 2 shows the charge density of the optimised geometry of the ketone **6**, where steric and electronic differences between the two substituents of the carbonyl group can be observed. This prompted us to study the biocatalytic process in depth.

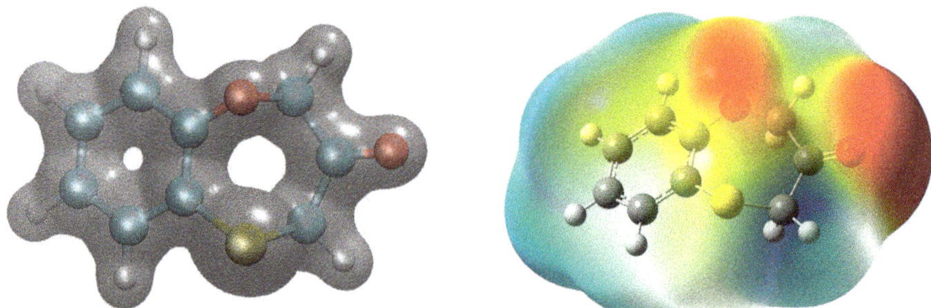

Figure 2. Optimised geometry of 3,4-dihydro-2*H*-1,5-benzoxathiepin-3-one (**6**): electronic isodensity contour (**left**); colour-mapped with the electrostatic potential (**right**), where red and blue zones are related to the electrophilic and nucleophilic zones of the molecule, respectively.

The biotransamination of 3,4-dihydro-2*H*-1,5-benzoxathiepin-3-one (**6**, 20 mM) was then studied in standard conditions previously employed in our research group [46,51]. These settings include the use of a large excess of isopropylamine as amine donor (1 M), pyridoxal 5′-phosphate (PLP, 1 mM) as cofactor, a 100 mM phosphate buffer pH 7.5 with acetonitrile (5% v/v) as cosolvent to favour the ketone solubility, at 30 °C and 250 rpm for 20 h (Scheme 2). Three different types of enzymes were employed:

(a) lyophilised *Escherichia coli* cells containing overexpressed ATAs; (b) commercially available ATAs from Codexis Inc.; (c) commercially available ATAs from Enzymicals AG.

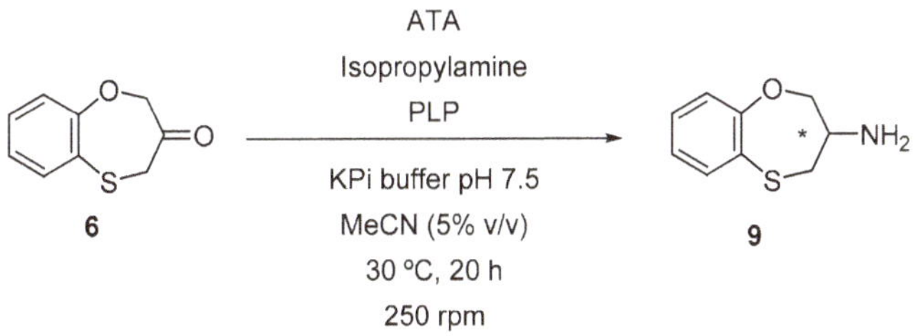

Scheme 2. Biotransamination of 3,4-dihydro-2*H*-1,5-benzoxathiepin-3-one (**6**) into amine **9**, using ATAs.

Initially, for the biotransamination experiments made in house ATAs were used, all of them overexpressed in *Escherichia coli*. Some of them, such as the ones from *Chromobacterium violaceum* [52] or *Arthrobacter* species [53], displayed very low activity (<5%), while others such as *Arthrobacter citreus* [54] or the *Arthrobacter* species evolved variant named ArRmut11 [55] provided almost quantitative conversion but moderate (74% *ee*) or negligible stereoselectivity, respectively. Trying to improve both activity and selectivity values, commercially available ATAs were employed from two different commercial sources (Codexis Inc. and Enzymicals AG).

To start with, 30 Codexis enzymes were employed (Table 1), and we found that 19 of them led to the complete disappearance of the starting ketone. Remarkably, four enzymes from this kit provided the desired amine **9** in optical purities over 80% *ee*, the ATA-200 conducting to the (*S*)-**9** (entry 8), while the TA-P1-B04, TA-P1-F03 and TA-P1-G05 gave access to its amine antipode (entries 23, 24 and 26).

Table 1. Biotransamination of ketone **6** using Codexis ATAs [a].

Entry	Enzyme	Conversion (%) [b]	*ee* (%) [c]
1	ATA-7	<1	n.d.
2	ATA-13	30	n.d.
3	ATA-24	93	<1
4	ATA-25	96	<1
5	ATA-33	>99	<1
6	ATA-113	13	n.d.
7	ATA-117	2	n.d.
8	ATA-200	>99	85 (*S*)
9	ATA-217	6	n.d.
10	ATA-234	4	n.d.
11	ATA-237	>99	41 (*S*)
12	ATA-238	4	n.d.
13	ATA-251	>99	72 (*S*)
14	ATA-254	>99	56 (*S*)
15	ATA-256	>99	63 (*S*)
16	ATA-260	>99	79 (*S*)
17	ATA-301	>99	7 (*S*)
18	ATA-303	>99	<1
19	ATA-412	>99	55 (*S*)
20	ATA-415	>99	<1
21	TA-P1-A01	>99	62 (*R*)
22	TA-P1-A06	>99	50 (*R*)
23	TA-P1-B04	>99	82 (*R*)
24	TA-P1-F03	>99	90 (*R*)
25	TA-P1-F12	>99	28 (*R*)

Table 1. *Cont.*

Entry	Enzyme	Conversion (%) [b]	*ee* (%) [c]
26	TA-P1-G05	>99	93 (*R*)
27	TA-P1-G06	>99	67 (*R*)
28	TA-P2-A01	4	n.d.
29	TA-P2-A07	60	16 (*S*)
30	TA-P2-B01	99	19 (*R*)

[a] For reaction details, see Section 3.6. [b] Conversion values measured by GC analyses of the reaction crudes. [c] Enantiomeric excess (*ee*) of amine **9** determined by HPLC analyses after derivatisation of the reaction crude. These *ee* values were calculated for those reactions with conversions over 30% (n.d.: not determined).

Using the best found enzyme, TA-P1-G05 (entry 26, >99% conversion and 93% *ee*), the transamination of **6** was followed over time using two enzyme loadings (90% and 45% *w*/*w* enzyme vs. ketone); we observed a very fast conversion in the first 2 h and then a slower rate until complete depletion of the substrate occurred, after 6 h or 24 h, respectively (Figure 3).

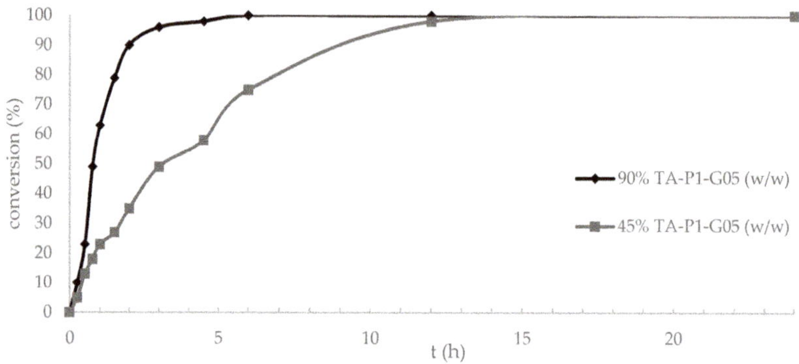

Figure 3. Study of the enzymatic transamination of ketone **6** with TA-P1-G05 over time employing: (◆) 90% of enzyme loading (*w*/*w*) or (■) 45% of enzyme loading (*w*/*w* vs. **6**).

Eight enzymes from Enzymicals AG were employed (Table 2), finding in three cases an amine with over 90% *ee* (entries 3, 7 and 8). Interestingly, the ATA08 from *Silicibacter pomeroyi* allowed the quantitative conversion of the ketone into the amine (*R*)-**9** (entry 7).

Table 2. Biotransamination of ketone **6** using Enzymicals AG ATAs [a].

Entry	Enzyme	Conversion (%) [b]	*ee* (%) [c]
1	ATA01 *Aspergillus fumigatus*	9	n.d.
2	ATA02 *Gibberella zeae*	<1	n.d.
3	ATA03 *Neosartorya fischeri*	29	90 (*S*)
4	ATA04 *Aspergillius oryza*	2	n.d.
5	ATA05 *Aspergillius terreus*	8	n.d.
6	ATA06 *Penicillium chrysogenum*	<1	n.d.
7	ATA07 *Mycobacterium vanbaalenii*	20	95 (*S*)
8	ATA08 *Silicibacter pomeroyi*	>99	91 (*R*)

[a] For reaction details, see Section 3.6. [b] Conversion values measured by GC analyses of the reaction crudes. [c] Enantiomeric excess of amine **9** determined by HPLC analyses after derivatisation of the reaction crude. These values were calculated for those reactions with conversions over 20% (n.d.: not determined).

In order to improve the conversion values towards the amine (*S*)-**9** the ATA03 *Neosartorya fischeri* (entry 3) and ATA07 *Mycobacterium vanbaalenii* (entry 7) were selected for optimization studies. So, new experiments were developed that includes the decrease of the substrate concentration, the use of longer reaction times, higher temperatures and enzyme loadings, and the performance of the biotransaminations without an organic cosolvent (Table 3). Interestingly, the best results were found when no cosolvent was employed, suggesting a deactivation of both enzymes in the presence of even low amounts of the organic solvent (MeCN, 5% *v*/*v*). In particular, the reduction of the substrate concentration from 20 to 10 mM of ketone **6** allowed higher conversions, although this limited its practical application. In addition, prolonged reaction times led to better conversions, while the use of higher temperatures led to a significant deactivation of the enzyme.

Table 3. Optimisation of the biotransamination of ketone **6** using selected enzymes [a].

ATA
Isopropylamine
PLP
KPi buffer pH 7.5
(Cosolvent)
30-45 °C, 20-65 h
250 rpm

6 (10-20 mM) → (*S*)-**9**

Entry	Enzyme	[6] (mM)	Cosolvent [b]	T (°C)	t (h)	c (%) [c]
1	ATA03 *Neosartorya fischeri*	20	MeCN (5%)	30	20	29
2	ATA03 *Neosartorya fischeri*	10	none	30	48	86
3	ATA07 *Mycobacterium vanbaalenii*	20	MeCN (5%)	30	20	20
4	ATA07 *Mycobacterium vanbaalenii*	10	none	30	20	73
5	ATA07 *Mycobacterium vanbaalenii*	20	none	45	48	43
6 [d]	ATA07 *Mycobacterium vanbaalenii* [d]	20	none	30	65	91

[a] For reaction details, see Section 3.6. [b] Concentration values in *v*/*v* % indicated in brackets. [c] Conversion values measured by GC analyses of the reaction crudes. [d] Double the amount of enzyme was used (4 mg, 180% *w*/*w*).

Focusing on the scaling up of the biotransformations, we decided to move to higher substrate concentrations (50 mM of ketone) in order to produce a significant amount of the optically active amine (*R*)-**9**, which is a precursor of organic molecules with interesting biological profiles [2–5]. In this case, 225 mg of **6** were used, selecting TA-P1-G05 (entry 26, Table 1) as the ideal candidate since in standard conditions the amine (*R*)-**9** was formed in complete conversion and good selectivity (93% *ee*). The enzyme loading was reduced from an initial 90% *w*/*w* enzyme vs. substrate ratio to 33% to improve the economy of the process, and after 22 h quantitative conversion was also achieved, maintaining the selectivity and isolating the desired amine in 98% yield after a simple liquid-liquid extraction protocol (Scheme 3). Measurement of the optical rotation for the pure amine and its corresponding hydrochloride salt allowed us to unequivocally assign the absolute configuration by comparison with previously reported data [4,5].

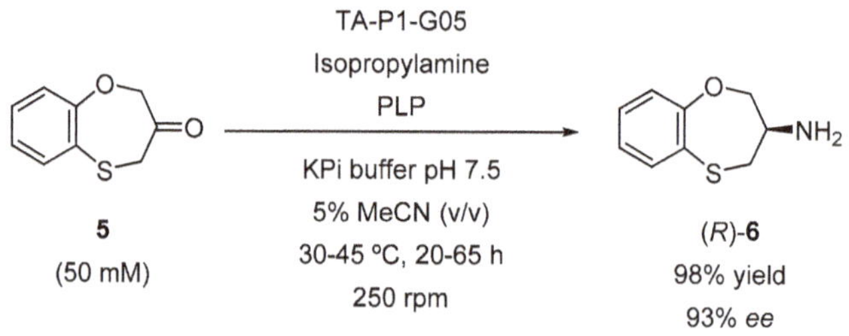

TA-P1-G05
Isopropylamine
PLP
KPi buffer pH 7.5
5% MeCN (v/v)
30-45 °C, 20-65 h
250 rpm

5 (50 mM) → (*R*)-**6** 98% yield 93% *ee*

Scheme 3. Scale-up of the biotransamination towards the (3*R*)-3,4-dihydro-2*H*-1,5-benzoxathiepin-3-amine (*R*-**9**).

3. Materials and Methods

3.1. General Materials and Methods

2-Mercaptophenol, ethyl bromoacetate and sodium ethoxide were purchased from Sigma-Aldrich, now Merck (Madrid, Spain). PLP as enzyme cofactor, other chemical reagents and solvents were obtained with the highest quality available from Sigma-Aldrich-Fluka (Steinheim, Germany). Amine transaminases were obtained from Codexis Inc. (Redwood City, CA, USA) and Enzymicals AG (Greifswald, Germany). Transaminases from *Chromobacterium violaceum* (2.1 U/mg), *Arthrobacter citreus* (0.9 U/mg), *Arthrobacter* species (0.6 U/mg) and the evolved ArRmut11 were overexpressed in *E. coli* and used as lyophilised cell lysates, as previously reported [26,56].

Melting point of compound **6** was measured in an open capillary in an Electrothermal digital melting point IA9200 apparatus (Cole-Parmer, Stone, UK) and is uncorrected. Elemental analyses were performed on a Thermo Scientific Flash 2000 analyzer (Thermo Flash & Carlo Erba Analyzers, Pennsauken, NJ, USA) and the measured values were indicated with the symbols of the elements or functions within ±0.4% of the theoretical values. NMR spectra were recorded on a Bruker AV300 MHz spectrometer (Bruker Co., Faellanden, Switzerland). All chemical shifts (δ) are given in parts per million (ppm) and referenced to the residual solvent signal as internal standard. High-resolution mass spectroscopy (HRMS) was performed on a VG AutoSpec Q high-resolution mass spectrometer (Fision Instrument, Milford, MA, USA). Measurement of the optical rotation values was carried out at 590 nm on an Autopol IV Automatic polarimeter (Rudolph Research Analytical, Hackettstown, NJ, USA).

Gas chromatography (GC) analyses were performed for the determination of conversion values using a Hewlett-Packard HP-6890 chromatograph (Hewlett Packard, Palo Alto, CA, USA). A non-chiral HP-1 column (Agilent Technologies, Inc., Wilmington, DE, USA) was used with the following temperature programme: 90 °C (2 min) then 10 °C/minutes and finally 180 °C (0 min). The reaction crudes were analysed, obtaining the following retention times: 9.3 min for ketone **6** and 10.5 min for amine **9**.

High-performance liquid chromatography (HPLC) analyses were performed for enantiomeric excess value measurements using an Agilent 1260 Infinity chromatograph with UV detector at 210 nm (Agilent Technologies, Inc., Wilmington, DE, USA). A Chiralpak IA (25 cm × 4.6 mm) was used as chiral column at 30 °C (Chiral Technologies, Mainz, Germany), employing a mixture of n-hexane/2-propanol (90:10) as eluent with a 0.8 mL/min flow. The reaction crudes were derivatised as acetamides, obtaining the following retention times: 11.2 min for the (*R*)-**10** and 12.6 min for the (*S*)-**10** enantiomer (Figure 4).

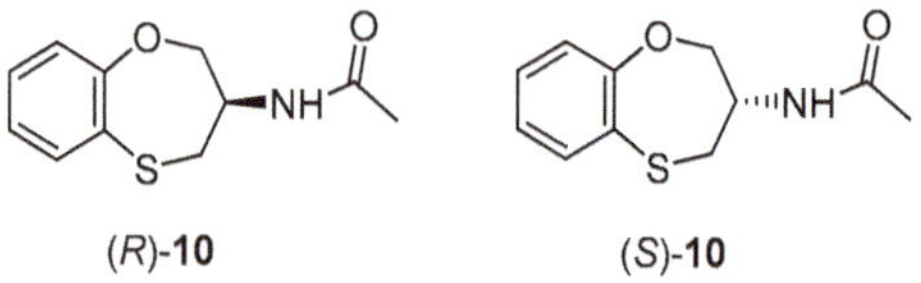

(*R*)-**10** (*S*)-**10**

Figure 4. Structures of (*R*)- and (*S*)-**10**.

Thin-layer chromatography (TLC) analyses were conducted with Merck Silica Gel 60 F254 precoated plates (Merck KGaA, Darmstadt, Germany). They were visualised with UV and potassium permanganate stain. Column chromatography purifications were performed using Merck Silica Gel 60 (230–400 mesh, Merck KGaA, Darmstadt, Germany).

3.2. *General Procedure for the Synthesis of Ethyl 2-Ethoxycarbonylmethylthiophenoxyacetate* (**7**)

A mixture of 2-mercaptophenol (1 g, 7.925 mmol), ethyl bromoacetate (1.93 mL, 17,4 mmol) and dry K_2CO_3 (3.3 g, 23.8 mmol) in anhydrous acetone (20 mL) was added under argon atmosphere, and then stirred under reflux. After 24 h the solvent was evaporated under reduced pressure and the residue purified by column chromatography (EtOAc/n-hexane, 1:8), obtaining **7** as a colourless syrup. Yield 83%. ^{1}H NMR (300.13 MHz, $CDCl_3$) δ 7.43 (dd, J_{HH} = 7.7, 1.7 Hz, 1H), 7.21 (td, J_{HH} = 7.9, 1.7 Hz, 1H), 6.95 (td, J_{HH} = 7.5, 1.2 Hz, 1H), 6.76 (dd, J_{HH} = 8.2, 1.2 Hz, 1H), 4.70 (s, 2H), 4.26 (q, J_{HH} = 7.1 Hz, 2H), 4.11 (q, J_{HH} = 7.1 Hz, 2H), 3.72 (s, 2H), 1.28 (t, J_{HH} = 7.1 Hz, 3H), 1.18 (t, J_{HH} = 7.1 Hz, 3H) ppm. HRMS (ESI-TOF) (*m/z*) calcd. for $C_{14}H_{19}O_5S$ $(M + H)^+$ 299.0953, found 299.0955. Anal. Calcd for $C_{14}H_{18}O_5S$: C, 56.36; H, 6.08; S, 10.75. Found: C, 56.45; H, 5.89; S, 10.55.

3.3. *General Procedure for the Synthesis of Ethyl 3-Oxo-3,4-dihydro-2H-1,5-benzoxathiepin-4-carboxylate* (**8**)

To a mixture of diester **7** (1.37 g, 4.59 mmol) in THF (40 mL) at 0 °C, a solution of NaOEt (21% wt, 1.78 g, 5.51 mmol) in EtOH (2.06 mL) was added. The mixture was stirred at 0 °C for 1 h and then refluxed for 15 h. Solvent was evaporated under reduced pressure and the residue cooled to 0 °C, quenched first with water, and later with an aqueous HCl 6 M solution up to pH 6. The mixture was extracted with EtOAc (2 × 40 mL) and the organic fractions combined, dried over anhydrous Na_2SO_4, filtered and the solvent was removed under reduced pressure. Compound **8** was purified by column chromatography (EtOAc/n-hexane, 1:7) as a colourless oil. Yield, 90%. ^{1}H NMR (300.13 MHz, $CDCl_3$) δ 7.22–7.12 (m, 1H), 7.08–6.85 (m, 3H), 4.88 (d, J_{HH} = 17.5, 1H), 4.75 (s, 1H), 4.59 (d, J_{HH} = 17.5, 1H), 4.29–4.15 (m, 2H), 1.21 (td, J_{HH} = 7.1, 1.4 Hz, 3H) ppm. HRMS (ESI-TOF) (m/z) calcd. for $C_{12}H_{11}O_4S$ $(M - H)^+$ 251.0378, found 251.0376. Anal. Calcd for $C_{12}H_{12}O_4S$: C, 57.13; H, 4.79; S, 12.71. Found: C, 56.99; H, 4.98; S, 12.72.

3.4. *General Procedure for the Synthesis 3,4-Dihydro-2H-1,5-benzoxathiepin-3-one* (**6**)

A mixture of keto ester **8** (2.5 g, 11.9 mmol), acetic acid (4.16 mL), H_2SO_4 concentrated (4.16 mL) and H_2O (23.8 mL) was refluxed for 1 h. The reaction was cooled to 0 °C, and water was added and extracted with CH_2Cl_2 (2 × 40 mL). The organic fractions were combined, dried (anhydrous Na_2SO_4), filtered and the solvent was removed under reduced pressure. Compound **6** was purified by column chromatography (n-hexane and then, EtOAc/n-hexane 0.5:10) as a white solid, mp 29–31 °C, literature 28–31 °C [43]. Yield 60%. ^{1}H NMR (300.13 MHz, $CDCl_3$) δ 7.18 (dd, J_{HH} = 8.0, 1.8 Hz, 1H), 7.14–7.07 (m, 1H), 7.01 (m, J_{HH} = 8.1, 4.8, 1.5 Hz, 2H), 4.75 (s, OCH_2, 2H), 3.93 (s, SCH_2, 2H) ppm. HRMS (ESI-TOF) (m/z) calcd. for $C_9H_9O_2S$ $(M + H)^+$ 181.0323, found 181.0321.

3.5. *General Procedure for the Biotransamination of* **6** *Using ATAs Overexpressed in* Escherichia coli

The lyophilised cells of E. coli containing overexpressed transaminases (5 mg) were suspended in a 100 mM phosphate buffer pH 7.5 (475 μL) containing PLP (1 mM) and 2-propylamine (1 M). Then, a stock solution of ketone **6** in MeCN was added (25 μL of stock 0.4 M; final concentration 20 mM) and the mixture was shaken at 30 °C and 250 rpm for 20 h. After this time, the reaction was quenched by adding an aqueous 10 M NaOH solution (200 μL) and extracted with EtOAc (2 × 500 μL). The organic phases were combined and dried over anhydrous Na_2SO_4. The reaction crudes were analysed by GC to determine conversion values. Derivatisation were carried out in situ using acetic anhydride and K_2CO_3 for the measurement of the enantiomeric excesses through HPLC.

3.6. General Procedure for the Biotransamination of **6** Using Commercial ATAs

Transaminases from Codexis or Enzymicals AG (2 mg, 90% w/w) were suspended in a 100 mM phosphate buffer pH 7.5 (475 μL) containing PLP (1 mM) and 2-propylamine (1 M). Then, a stock solution of ketone **6** in MeCN was added (25 μL of stock 0.4 M; final concentration 20 mM) and the mixture was shaken at 30 °C and 250 rpm for 20 h. After this time, the reaction was quenched by adding an aqueous 10 M NaOH solution (200 μL) and extracted with EtOAc (2 × 500 μL). The organic phases were combined and dried over anhydrous Na_2SO_4. Reaction crude was analysed by GC to determine conversion values and in situ derivatisation was carried out using acetic anhydride and K_2CO_3 for the measurement of the enantiomeric excesses by HPLC.

3.7. Preparative Biotransamination of **6** under Optimised Conditions

Ketone **6** (225 mg, 1.25 mmol) was dissolved in MeCN (1.25 mL) and 100 mM phosphate buffer pH 7.5 (25 mL), containing PLP (0.5 mM) and 2-propylamine (1 M), and the TA-P1-G05 (75 mg, 33% w/w) were successively added. The mixture was shaken at 30 °C and 250 rpm for 22 h. The reaction was quenched by adding an aqueous NaOH 4 M solution (5 mL) and extracted with EtOAc (3 × 15 mL). The organic phases were combined, dried over anhydrous Na_2SO_4, combined and the solvent removed under reduced pressure, affording the (*R*)-**9** amine (220 mg).

(3*R*)-3,4-Dihydro-2*H*-1,5-benzoxathiepin-3-amine (*R*)-**9**. Yield: 220 mg (98%). ^{1}H NMR (300.13 MHz, $CDCl_3$): δ 7.37 (dd, J_{HH} = 7.7, 1.7 Hz, 1H), 7.15 (ddd, J_{HH} = 8.1, 7.3, 1.7 Hz, 1H), 7.02–6.92 (m, 2H), 4.12–4.08 (m, 2H), 3.50–3.42 (m, 1H), 3.18 (dd, J_{HH} = 14.2, 3.2 Hz, 1H), 2.80 (dd, J_{HH} = 14.2, 5.7 Hz, 1H), 1.89 (br s, 2H) ppm. For the free amine (*R*)-**9** in 93% *ee* $[\alpha]_D^{20}$ = +32.6 (c = 0.1, MeOH), and for the hydrochloride salt (*R*)-**9**·HCl in 93% *ee* $[\alpha]_D^{20}$ = +41.2 (c = 0.1, MeOH); literature $[\alpha]_D^{20}$ = +48.9 (c = 0.35, MeOH) for the (*R*)-**9**·HCl in >99% *ee* [4].

3.8. Computational Methods

Calculations were performed using the Gaussian 09 package [57] at the M06-2X/6-311++G(3df,2p) level [58]. Molecular geometries of the studied compounds were optimised with tight convergence criteria and the frequencies were computed in order to obtain the thermal correction to the energy (298.15 K).

The molecular electrostatic potential of 3,4-dihydro-2*H*-1,5-benzoxathiepin-3-one (**6**) was computed at M06-2X/6-311++G(3df,2p) level with tight SCF procedure and generating the density and potential cubes to plot the isodensity surface, colour-coded with the electrostatic potential.

4. Conclusions

The synthesis of the 3,4-dihydro-2*H*-1,5-benzoxathiepin-3-amine enantiomers has been possible by means of the stereoselective biotransamination of the 3,4-dihydro-2*H*-1,5-benzoxathiepin-3-one. A broad panel of commercially available amine transaminases were employed, finding after optimisation of parameters that affect the enzyme catalysis suitable reaction conditions for the access to both amine antipodes in high conversions and good selectivities. A scale-up experiment considering 50 mM substrate concentration was successfully achieved for the formation of the (*R*)-3,4-dihydro-2*H*-1,5-benzoxathiepin-3-amine (*R*-**9**), a valuable precursor of anti-proliferative agents.

Author Contributions: C.C, J.M.C. and V.G.-F. conceived the project; N.F.-S. performed the experiments for the chemical synthesis of the starting ketone; D.G.-M. performed the biotransamination experiments; D.G.-M. carried out analytical measurements and analysed the data; C.C., J.M.C. and V.G.-F. wrote the paper.

Funding: Financial support has been received from the Spanish Ministry of Economy and Competitiveness (MINECO, Projects CTQ2013-40855-R and CTQ2016-75752-R), Junta de Andalucía (Project CS2016.1) and Gobierno de Aragon-FEDER (Research group E19_17R).

Acknowledgments: D.G.-M. (Severo Ochoa predoctoral fellowship) thanks the Asturian regional government for personal funding.

Conflicts of Interest: The authors declare no conflict of interest.

References

1. Kimatrai, M.; Conejo-García, A.; Ramírez, A.; Andreolli, E.; García, M.Á.; Aránega, A.; Marchal, J.A.; Campos, J.M. Synthesis and Anticancer Activity of the (*R*,*S*)-Benzofused 1,5-Oxathiepin Moiety Tethered to Purines through Alkylidenoxy Linkers. *ChemMedChem* **2011**, *6*, 1854–1859. [CrossRef] [PubMed]
2. Le Grand, B.; Pignier, C.; Létienne, R.; Cuisiat, F.; Rolland, F.; Mas, A.; Vacher, B. Sodium late current blockers in ischemia reperfusion: Is the bullet magic? *J. Med. Chem.* **2008**, *51*, 3856–3866. [CrossRef] [PubMed]
3. Le Grand, B.; Pignier, C.; Létienne, R.; Cuisiat, F.; Rolland, F.; Mas, A.; Borras, M.; Vacher, B. Na^+ currents in cardioprotection: Better to be late. *J. Med. Chem.* **2009**, *52*, 4149–4160. [CrossRef] [PubMed]
4. Mahfoudh, N.; Marín-Ramos, N.I.; Gil, A.M.; Jiménez, A.I.; Choquesillo-Lazarte, D.; Kawano, D.F.; Campos, J.M.; Cativiela, C. Cysteine-based 3-substituted 1,5-benzoxathiepin derivatives: Two new classes of anti-proliferative agents. *Arab. J. Chem.* **2018**, *11*, 426–441. [CrossRef]
5. Vacher, V.; Brunel, Y.; Castan, C.F. An Improved Process for the Preparation of Benzoxathiepines and Their Intermediates. FR 2868779 A120051014, 14 October 2005.
6. Hudlicky, T.; Reed, J.W. Applications of biotransformations and biocatalysis to complexity generation in organic synthesis. *Chem. Soc. Rev.* **2009**, *38*, 3117–3132. [CrossRef] [PubMed]
7. Clouthier, C.M.; Pelletier, J.M. Expanding the organic toolbox: A guide to integrating biocatalysis in synthesis. *Chem. Soc. Rev.* **2012**, *41*, 1585–1605. [CrossRef] [PubMed]
8. Milner, S.E.; Maguire, A.R. Recent trends in whole cell and isolated enzymes in enantioselective synthesis. *Arkivoc* **2012**, 321–382.
9. Torrelo, G.; Hanefeld, U.; Hollmann, F. Biocatalysis. *Catal. Lett.* **2015**, *145*, 309–345. [CrossRef]
10. Albarrán-Velo, J.; González-Martínez, D.; Gotor-Fernández, V. Stereoselective Biocatalysis. A mature technology for the asymmetric synthesis of pharmaceutical building blocks. *Biocatal. Biotransf.* **2018**, *36*, 102–130. [CrossRef]
11. Höhne, M.; Bornscheuer, U.T. Biocatalytic Routes to Optically Active Amines. *ChemCatChem* **2009**, *1*, 42–51. [CrossRef]
12. Kroutil, W.; Fischereder, E.-M.; Fuchs, C.S.; Lechner, H.; Mutti, F.G.; Pressnitz, D.; Rajagopalan, A.; Sattler, J.H.; Simon, R.C.; Siirola, E. Asymmetric preparation of *prim*-, *sec*-, and *tert*-amines employing selected biocatalysts. *Org. Process Res. Dev.* **2013**, *17*, 751–759. [CrossRef] [PubMed]
13. Kohls, H.; Steffen-Munsberg, F.; Höhne, M. Recent achievements in developing the biocatalytic toolbox for chiral amine synthesis. *Curr. Opin. Chem. Biol.* **2014**, *19*, 180–192. [CrossRef] [PubMed]
14. Schrittwieser, J.H.; Velikogne, S.; Kroutil, W. Biocatalytic imine reduction and reductive amination of ketones. *Adv. Synth. Catal.* **2015**, *357*, 1655–1685. [CrossRef]
15. Gamenara, D.; Domínguez de María, P. Enantioselective imine reduction catalyzed by imine reductases and artificial metalloenzymes. *Org. Biomol. Chem.* **2014**, *12*, 2989–2992. [CrossRef] [PubMed]
16. Grogan, G.; Turner, N.J. InspIRED by nature: NADPH-dependent imine reductases (IREDs) as catalysts for the preparation of chiral amines. *Chem. Eur. J.* **2016**, *22*, 1900–1907. [CrossRef] [PubMed]
17. Mangas-Sánchez, J.; France, S.P.; Montgomery, S.L.; Aleku, G.A.; Man, H.; Sharma, M.; Ramsden, J.I.; Grogan, G.; Turner, N.J. Imine reductases (IREDs). *Curr. Opin. Chem. Biol.* **2017**, *37*, 19–25. [CrossRef] [PubMed]
18. Tufvesson, P.; Lima-Ramos, J.; Jensen, J.S.; Al-Haque, N.; Neto, W.; Woodley, J.M. Process Considerations for the Asymmetric Synthesis of Chiral Amines Using Transaminases. *Biotechnol. Bioeng.* **2011**, *108*, 1479–1493. [CrossRef] [PubMed]
19. Mathew, S.; Yun, H. ω-Transaminases for the production of optically pure amines and unnatural amino acids. *ACS Catal.* **2012**, *2*, 993–1001. [CrossRef]
20. Simon, R.C.; Richter, N.; Busto, E.; Kroutil, W. Recent developments of cascade reactions involving ω-transaminases. *ACS Catal.* **2014**, *4*, 129–143. [CrossRef]
21. Fuchs, M.; Farnberger, J.E.; Kroutil, W. The industrial age of biocatalytic transamination. *Eur. J. Org. Chem.* **2015**, 6965–6982. [CrossRef] [PubMed]
22. Guo, F.; Berglund, P. Transaminase biocatalysis: Optimization and application. *Green Chem.* **2017**, *19*, 333–360. [CrossRef]
23. Patil, M.D.; Grogan, G.; Bommarius, A.; Yun, H. Recent advances in ω-transaminase-mediated biocatalysis for the enantioselective synthesis of chiral amines. *Catalysts* **2018**, *8*, 254. [CrossRef]

24. Koszelewski, D.; Lavandera, I.; Clay, D.; Rozzell, D.; Kroutil, W. Asymmetric synthesis of optically pure pharmacologically relevant amines Employing ω-transaminases. *Adv. Synth. Catal.* **2008**, *350*, 2761–2766. [CrossRef]
25. Höhne, M.; Kühl, S.; Robins, K.; Bornscheuer, U.T. Efficient asymmetric synthesis of chiral amines by combining transaminase and pyruvate decarboxylase. *ChemBioChem* **2008**, *9*, 363–365. [CrossRef] [PubMed]
26. Mutti, F.G.; Fuchs, C.S.; Pressnitz, D.; Sattler, J.H.; Kroutil, W. Stereoselectivity of four (*R*)-selective transaminases for the asymmetric amination of ketones. *Adv. Synth. Catal.* **2011**, *353*, 3227–3233. [CrossRef]
27. Truppo, M.; Janey, J.M.; Grau, B.; Morley, K.; Pollack, S.; Hughes, G.; Davies, I. Asymmetric, biocatalytic labeled compound synthesis using transaminases. *Catal. Sci. Technol.* **2012**, *2*, 1556–1559. [CrossRef]
28. Pressnitz, D.; Fuchs, C.S.; Sattler, J.H.; Knaus, T.; Macheroux, P.; Mutti, F.G.; Kroutil, W. Asymmetric amination of tetralone and chromanone derivatives employing ω-transaminases. *ACS Catal.* **2013**, *3*, 555–559. [CrossRef]
29. Park, E.-S.; Malik, M.S.; Dong, J.-Y.; Shin, J.-S. One-pot production of enantiopure alkylamines and arylalkylamines of opposite chirality catalyzed by ω-transaminase. *ChemCatChem* **2013**, *5*, 1734–1738. [CrossRef]
30. Richter, N.; Simon, R.C.; Lechner, H.; Kroutil, W.; Ward, J.M.; Hailes, H.C. ω-Transaminases for the amination of functionalised cyclic ketones. *Org. Biomol. Chem.* **2015**, *13*, 8843–8851. [CrossRef] [PubMed]
31. Martínez-Montero, L.; Gotor, V.; Gotor-Fernández, V.; Lavandera, I. But-2-ene-1,4-diamine and but-2-ene-1,4-diol as donors for thermodynamically favored transaminase- and alcohol dehydrogenase-catalyzed processes. *Adv. Synth. Catal.* **2016**, *358*, 1618–1624. [CrossRef]
32. Gundersen, M.T.; Tufvesson, P.; Rackham, E.J.; Lloyd, R.C.; Woodley, J.M. A rapid selection procedure for simple commercial implementation of ω-transaminase reactions. *Org. Process Res. Dev.* **2016**, *20*, 602–608. [CrossRef]
33. Siirola, E.; Mutti, F.G.; Grischek, B.; Hoefler, S.F.; Fabian, W.M.F.; Grogan, G.; Kroutil, W. Asymmetric synthesis of 3-substituted cyclohexylamine derivatives from prochiral diketones via three biocatalytic steps. *Adv. Synth. Catal.* **2013**, *355*, 1703–1708. [CrossRef]
34. Tauber, K.; Fuchs, M.; Sattler, J.H.; Pitzer, J.; Pressnitz, D.; Koszelewski, D.; Faber, K.; Pfeffer, J.; Haas, T.; Kroutil, W. Artificial multi-enzyme networks for the asymmetric amination of *sec*- alcohols. *Chem. Eur. J.* **2013**, *19*, 4030–4035. [CrossRef] [PubMed]
35. Skalden, L.; Peters, C.; Dickerhoff, J.; Nobili, A.; Joosten, H.-J.; Weisz, K.; Höhne, M.; Bornscheuer, U.T. Two subtle amino acid changes in a transaminase substantially enhance or invert enantiopreference in cascade syntheses. *ChemBioChem* **2015**, *15*, 1041–1045. [CrossRef] [PubMed]
36. Monti, D.; Forchin, M.C.; Crotti, M.; Parmeggiani, F.; Gatti, F.G.; Brenna, E.; Riva, S. Cascade coupling of ene-reductases and ω-Transaminases for the stereoselective synthesis of diastereomerically enriched amines. *ChemCatChem* **2015**, *7*, 3106–3109. [CrossRef]
37. Skalden, L.; Peters, C.; Ratz, L.; Bornscheuer, U.T. Synthesis of (1*R*,3*R*)-1-amino-3-methylcyclon-n-hexane by an enzyme cascade reaction. *Tetrahedron* **2016**, *72*, 7207–7211. [CrossRef]
38. Liardo, E.; Ríos-Lombardía, N.; Morís, F.; Rebolledo, F.; González-Sabín, J. Hybrid organo- and biocatalytic process for the asymmetric transformation of alcohols into amines in aqueous medium. *ACS Catal.* **2017**, *7*, 4768–4774. [CrossRef]
39. Molinaro, C.; Bulger, P.G.; Lee, E.E.; Kosjek, B.; Lau, S.; Gauvreau, D.; Howard, M.E.; Wallace, D.J.; O'Shea, P.D. CRTH2 antagonist MK-7246: A synthetic evolution from discovery through development. *J. Org. Chem.* **2012**, *77*, 2299–2309. [CrossRef] [PubMed]
40. Richter, N.; Simon, R.C.; Kroutil, W.; Ward, J.M.; Hailes, H.C. Synthesis of pharmaceutically relevant 17-α-amino steroids using an ω-transaminase. *Chem. Commun.* **2014**, *50*, 6098–6100. [CrossRef] [PubMed]
41. Limanto, J.; Ashley, E.R.; Yin, J.; Beutner, G.L.; Grau, B.T.; Kassim, A.M.; Kim, M.M.; Klapars, A.; Liu, Z.; Strotman, H.R.; Truppo, M.D. A highly efficient asymmetric synthesis of Vernakalant. *Org. Lett.* **2014**, *16*, 2716–2719. [CrossRef] [PubMed]
42. Weiβ, M. S.; Pavlidis, I.V.; Spurr, P.; Hanlon, S.P.; Wirz, B.; Iding, H.; Bornscheuer, U.T. Protein-engineering of an amine transaminase for the stereoselective synthesis of a pharmaceutically relevant bicyclic amine. *Org. Biomol. Chem.* **2016**, *14*, 10249–10254.

43. Feng, Y.; Luo, Z.; Sun, G.; Chen, M.; Lai, J.; Lin, W.; Goldmann, S.; Zhang, L.; Wang, Z. Development of an Efficient and Scalable Biocatalytic Route to (3*R*)-3-Aminoazepane: A Pharmaceutically Important Intermediate. *Org. Process Res. Dev.* **2017**, *21*, 648–654. [CrossRef]
44. Sugihara, H.; Mabuchi, H.; Kawamatsu, Y. 1,5-Benzoxathiepin derivatives, I. Synthesis and reaction of 1,5-benzoxathiepin derivatives. *Chem. Pharm. Bull.* **1987**, *35*, 1919–1929. [CrossRef] [PubMed]
45. López-Iglesias, M.; González-Martínez, D.; Gotor, V.; Busto, E.; Kroutil, W.; Gotor-Fernández, V. Biocatalytic Transamination for the Asymmetric Synthesis of Pyridylalkylamines. Structural and Activity Features in the Reactivity of Transaminases. *ACS Catal.* **2016**, *6*, 4003–4009. [CrossRef]
46. López-Iglesias, M.; González-Martínez, D.; Rodríguez-Mata, M.; Gotor, V.; Busto, E.; Kroutil, W.; Gotor-Fernández, V. Asymmetric Biocatalytic Synthesis of Fluorinated Pyridines through Transesterification or Transamination: Computational Insights into the Reactivity of Transaminases. *Adv. Synth. Catal.* **2017**, *359*, 279–291. [CrossRef]
47. Cassimjee, K.E.; Branneby, C.; Abedi, V.; Wells, A.; Berglund, P. Transaminations with isopropyl amine: Equilibrium displacement with yeast alcohol dehydrogenase coupled to *in situ* cofactor regeneration. *Chem. Commun.* **2010**, *46*, 5569–5571. [CrossRef] [PubMed]
48. Green, A.P.; Turner, N.J.; O'Reilly, E. Chiral Amine Synthesis Using ω-Transaminases: An Amine Donor that Displaces Equilibria and Enables High-Throughput Screening. *Angew. Chem. Int. Ed.* **2014**, *53*, 10714–10717. [CrossRef] [PubMed]
49. Gomm, A.; Lewis, W.; Green, A.P.; O'Reilly, E. A New Generation of Smart Amine Donors for Transaminase-Mediated Biotransformations. *Chem. Eur. J.* **2016**, *22*, 12692–12695. [CrossRef] [PubMed]
50. Payer, S.E.; Schrittwieser, J.H.; Kroutil, W. Vicinal Diamines as Smart Cosubstrates in the Transaminase-Catalyzed Asymmetric Amination of Ketones. *Eur. J. Org. Chem.* **2017**, 2553–2559. [CrossRef]
51. Paul, C.E.; Rodríguez-Mata, M.; Busto, E.; Lavandera, I.; Gotor-Fernández, V.; Gotor, V.; García-Cerrada, S.; Mendiola, J.; de Frutos, Ó.; Collado, I. Transaminases applied to the synthesis of high added-value enantiopure amines. *Org. Process Res. Dev.* **2014**, *18*, 788–792. [CrossRef]
52. Kaulman, U.; Smithies, K.; Smith, M.E.B.; Hailes, H.C.; Ward, J.M. Substrate spectrum of ω-transaminase from *Chromobacterium violaceum* DSM30191 and its potential for biocatalysis. *Enzyme Microb. Technol.* **2007**, *41*, 628–637. [CrossRef]
53. Yamada, Y.; Iwasaki, A.; Kizaki, N.; Matsumoto, K.; Ikenaka, Y.; Ogura, M.; Hasegawa, J. Enzymic Preparation of Optically Active (*R*)-Amino Compounds with Transaminase of *Arthrobacter*. PCT Int. Appl. WO 9848030A1, 29 October 1998.
54. Pannuri, S.; Kamat, S.V.; Garcia, A.R.M. Methods for Engineering *Arthrobacter citreus* ω-Transaminase Variants with Improved Thermostability for Use in Enantiomeric Enrichment and Stereoselective Synthesis. PCT Int. Appl. WO 2006063336A2, 15 June 2006.
55. Savile, C.K.; Janey, J.M.; Mundorff, E.M.; Moore, J.C.; Tam, S.; Jarvis, W.R.; Colbeck, J.C.; Krebber, A.; Fleitz, F.J.; Brands, J.; et al. Biocatalytic Asymmetric Synthesis of Chiral Amines from Ketones Applied to Sitagliptin Manufacture. *Science* **2010**, *329*, 305–309. [CrossRef] [PubMed]
56. Koszelewski, D.; Göritzer, M.; Clay, D.; Seisser, B.; Kroutil, W. Synthesis of optically active amines employing recombinant ω-transaminases in *E. coli* cells. *ChemCatChem* **2010**, *2*, 73–77. [CrossRef]
57. Frisch, M.J.; Trucks, G.W.; Schlegel, H.B.; Scuseria, G.E.; Robb, M.A.; Cheeseman, J.R.; Scalmani, G.; Barone, V.; Mennucci, B.; Petersson, G.A.; et al. *Gaussian 09*; Revision D.01; Gaussian, Inc.: Wallingford, CT, USA, 2009.
58. Zhao, Y.; Truhlar, D.G. The M06 suite of density functionals for main group thermochemistry, thermochemical kinetics, noncovalent interactions, excited states, and transition elements: Two new functionals and systematic testing of four M06-class functionals and 12 other functionals. *Theor. Chem. Acc.* **2008**, *120*, 215–241.

Article

A Photo-Enzymatic Cascade to Transform Racemic Alcohols into Enantiomerically Pure Amines

Jenő Gacs [1], Wuyuan Zhang [1], Tanja Knaus [2], Francesco G. Mutti [2], Isabel W.C.E. Arends [1] and Frank Hollmann [1,*]

1 Department of Biotechnology, Delft University of Technology, Van der Maasweg 9, 2629HZ Delft, The Netherlands; gacsjeno@gmail.com or acsjeno@gmail.com (J.G.); w.zhang-1@tudelft.nl (W.Z.); I.W.C.E.Arends@uu.nl (I.W.C.E.A.)

2 Van't Hoff Institute for Molecular Sciences (HIMS-Biocat), University of Amsterdam, Science Park 904, 1098 XH Amsterdam, The Netherlands; T.Knaus@uva.nl (T.K.); F.Mutti@uva.nl (F.G.M.)

* Correspondence: f.hollmann@tudelft.nl; Tel.: +31-(0)152-781957

Received: 27 February 2019; Accepted: 20 March 2019; Published: 27 March 2019

Abstract: The consecutive photooxidation and reductive amination of various alcohols in a cascade reaction were realized by the combination of a photocatalyst and several enzymes. Whereas the photocatalyst (sodium anthraquinone-2-sulfonate) mediated the light-driven, aerobic oxidation of primary and secondary alcohols, the enzymes (various ω-transaminases) catalyzed the enantio-specific reductive amination of the intermediate aldehydes and ketones. The system worked in a one-pot one-step fashion, whereas the productivity was significantly improved by switching to a one-pot two-step procedure. A wide range of aliphatic and aromatic compounds was transformed into the enantiomerically pure corresponding amines via the photo-enzymatic cascade.

Keywords: photooxidation; cascade; alcohol; reductive amination; ω-transaminase

1. Introduction

In recent years, significant attention has been devoted to the synthesis of amines [1]. Especially, enantiomerically pure amines are of interest in natural product synthesis, as intermediates in pharmaceutical synthesis, and for a variety of other chemical products [2–4]. Direct amination of alcohols has been studied extensively [1,5]. The main advantage of prompt conversion of alcohols to amines is that both the substrate and the product are in the same oxidation state, and, thus, theoretically the additional use of any redox equivalents is not required [6,7]. Furthermore, many of the required alcohols are already available on an industrial scale, which facilitates the mass application of the technology [8]. On the other hand, most of the currently applied reactions either possess poor chemo-selectivity and require harsh reaction conditions (e.g., the reaction of alcohol with ammonia over various heterogeneous catalysts, such as tungsten, cobalt, nickel, chromium) [5] or have low efficiency and produce toxic intermediates [9]. It is also worth mentioning that the majority of the current methodologies either only accept symmetric substrates or produce racemic products. To synthesize chiral amine moieties, ketones are one of the most commonly used precursors [1].

Various approaches have been reported to selectively oxidize alcohols to the corresponding carbonyl products [10–13]. Compared to transition metal-based approaches [14], photooxidation provides an atom-economic and environmentally benign alternative [10]. Commercially available TiO_2 is used most commonly as a photocatalyst, but the applicability is restricted by its UV absorption [15]. However, efficient oxidation of 1-pentanol has been reported by irradiation up to 480 nm upon doping TiO_2 with Nb_2O_5 [16,17]. To further utilize visible light, numerous other heterogeneous catalysts have been tested [18]. By using graphitic carbon nitride (g-C_3N_4) [19,20] and vanadium-oxide grafted to numerous carriers [21,22], efficient alcohol oxidation mediated by visible light has been reported.

Besides heterogeneous systems, water-soluble sodium anthraquinone-2-sulfonate (SAS) is also known as an effective and inexpensive homogeneous photocatalyst [23–27]. Very recently we found that SAS is also very active for the oxidation of C–H bonds [27] and, therefore, can be implemented in photo-enzymatic cascades for the conversion of a very large substrate scope [28].

Even though photocatalysis provides an attractive approach to produce the carbonyl intermediate, typically, tedious protection/de-protection steps are required for the synthesis of chiral amines. These additional stages prolong the synthesis time and significantly decrease the atom efficiency of the reaction. To aid these shortcomings, biocatalytic approaches have been extensively studied [29–32]. Pyridoxal-5-phosphate (PLP)-dependent ω-transaminases (ω-TAs) provide an elegant way to convert a carbonyl group to an amine moiety at the expense of a sacrificial amine donor [31,33]. ω-TAs have been increasingly applied in industrial chemical synthesis, in particular for the manufacturing of active pharmaceutical ingredients (APIs) [29,34,35]. Other main applications deal with the production of surfactants [36], amino acids [37], and plastic fiber monomers [36,38,39]. To reach an acceptable conversion when L- or D-alanine is used as an amine donor, the reaction equilibrium of the transamination reaction must be shifted towards the product. Thus, one of the most common means to enhance the amine formation is to remove the pyruvate by-product through the addition of another enzyme, such as lactate dehydrogenase (LDH), alanine dehydrogenase (AlaDH), pyruvate decarboxylase (PDC), or an acetolactate synthase (ALS) [32]. On the other hand, the use of alternative amine donors provides a more favorable thermodynamic equilibrium. However, these molecules are either prohibitively expensive [39] or generate a co-product during the transamination that tends to polymerize, thus complicating product isolation and lowering product yield [40,41]. Therefore, it is not surprising that isopropylamine (IPA) has been used as the preferred amine donor, especially on an industrial scale. IPA is inexpensive and provides a much more favorable thermodynamic equilibrium compared to alanine [42], which can be further shifted via simple and selective evaporation of the low boiling acetone co-product [35]. Consequently, IPA was selected as the amine donor in this study.

Recently, numerous reports have been published about the effective conversion of a plethora of alcohols to (chiral) amines in multi-enzyme cascades [43,44]. Inspired by their results and the current advances in the selective photochemical oxidation of alcohols [18], herein we present a photo-enzymatic tandem reaction for the direct conversion of alcohols to (chiral) amines. Employing the consecutive oxidation and reductive amination steps in one pot, a further reduction of the environmental impact of the synthesis is possible (Scheme 1). As time-consuming intermediate isolation and purification steps are omitted, the required amount of organic solvents is minimized, which results in a lower E-factor (*Environmental* factor, defined as the mass of waste per mass of product) [45,46].

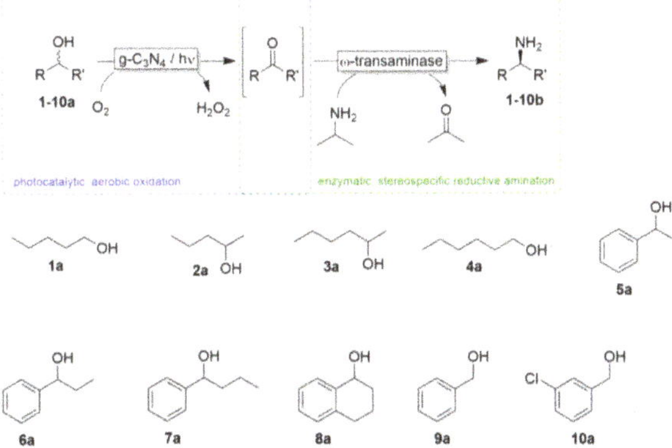

Scheme 1. Schematic representation of the photo-enzymatic cascade reaction and the examined substrates.

2. Results and Discussion

In the first set of experiments, we evaluated a range of homogeneous and heterogeneous photocatalysts available in our laboratory. More specifically, water-soluble sodium anthraquinone-2-sulfonate (SAS) and heterogeneous graphitic carbon nitride (g-C_3N_4), niobium(V)-oxide-doped titanium-dioxide (TiO_2-Nb_2O_5), and vanadium oxide (VO) deposited on various carriers, such as alumina (VO-Al_2O_3), zirconium-dioxide (VO-ZrO_2), and graphitic carbon nitride (VO-g-C_3N_4), were evaluated. As a model reaction, we chose the oxidation of *rac*-1-phenylethanol (**5a**) to acetophenone (**5b**) (Figure 1). Starting from 25 mM of the starting material, SAS mediated the full conversion of **5a** (TN = 30), while the best heterogeneous catalyst, g-C_3N_4, produced only 6.2 mM of product with 20% conversion. The discrepancy between conversion and product concentration in the case of SAS is probably due to the oxidative decomposition of the product. Since the other photocatalysts fell back significantly in their performance (Figure 1), we focused our attention on SAS and g-C_3N_4.

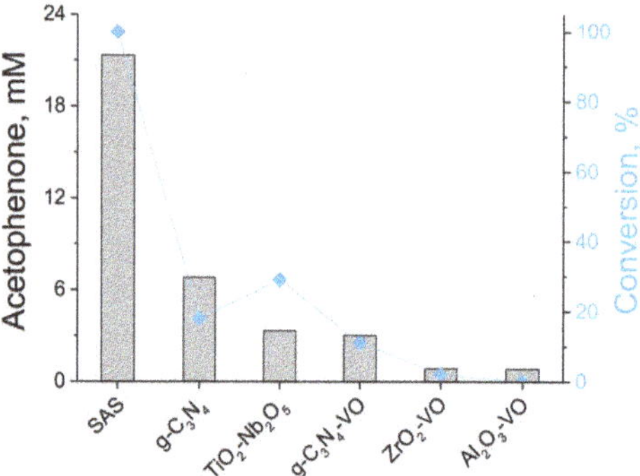

Figure 1. Alcohol oxidation using various photocatalysts. Reaction conditions: 1 mL MilliQ water with [**5a**] = 25 mM, [sodium anthraquinone-2-sulfonate (SAS)] = 0.75 mM, or [heterogeneous photocatalyst] = 10 mg/mL. Reactions were performed for 24 h at 30 °C under atmospheric conditions and irradiated under visible light ($\lambda > 400$ nm). These experiments have been performed as single experiments.

We performed the cascade reaction shown in Scheme 1 using SAS and g-C_3N_4 in a one-pot one-step fashion, i.e., adding the starting material and all catalysts at the beginning (Figure 2). For the reductive amination of the intermediate ketone, two ω-transaminases were applied: the (*R*)-selective ω-TA from *Aspergillus terreus* (*AT*ωTA) and the (*S*)-selective ω-TA from *Bacillus megaterium* (*BM*ωTA). In the case of g-C_3N_4, virtually no alcohol conversion was observed. The trace amounts of *rac*-**5c** originated from small acetophenone impurities within the commercial substrate. We attribute this lack of catalytic activity to the absorption of the biocatalysts to the g-C_3N_4 surface, thereby passivating it for the desired oxidation reaction.

The cascade reaction using SAS, however, produced 2.5 mM (*S*)- and 1.5 mM (*R*)-1-phenylethyl amine in 48 h. Even though enantiomerically pure amines were obtained, the performance of this system fell back behind our expectations. Therefore, we set out to investigate the limitations of the one-pot one-step procedure. Further characterization of the reaction conditions revealed some limitations of the one-pot one-step procedure to be as follows: (i) the catalytic activity decreased steadily in the presence of light and SAS. This may be attributed to an oxidative inactivation and degradation of the biocatalysts by photoexcited SAS and the used light (Figure S9); (ii) we found that the rate of the first step (SAS-mediated oxidation of phenylethanol) was significantly slower in the case of the one-pot one-step procedure. This may be attributed to SAS's activity on the stoichiometric amine donor and the product of the reductive amination step (Figures S11–S13).

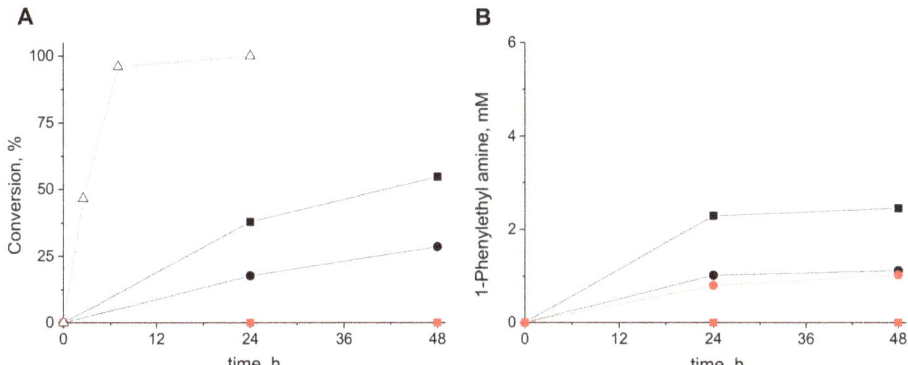

Figure 2. Results of the direct amination of 1-phenylethanol (**5a**) in a one-pot one-step fashion with SAS (−) and g-C_3N_4 (−) photocatalysts, employing *BM*ωTA (*Bacillus megaterium* ωTA ■/■) and *AT*ωTA (*Aspergillus terreus* ωTA •/•) enzymes. (**A**) Conversion of **5a**. As a comparison, the conversion of 20 mM **5a** in MilliQ water containing 0.75 mM SAS is also shown (△); (**B**) Production of 1-phenylethyl amine **5c**. Reaction conditions: 1 mL reaction mixture containing sodium phosphate (NaPi) buffer (100 mM, pH 9), [1-phenylethanol] = 15 mM, [SAS] = 0.75 mM, [isopropylamine, IPA] = 1 M, [Pyridoxal-5-phosphate, PLP] = 1 mM, [crude cell extract overexpressed with ω-TA enzyme] = 10 mg/mL. Reaction mixtures were incubated at 30 °C, irradiated with white light (λ > 400 nm).

Therefore, we turned our attention to a one-pot two-step procedure, wherein we first performed the photochemical alcohol oxidation followed by supplementation of the reaction mixture containing the reaction components needed for the reductive amination in a second step. Compared to the one-pot one-step case, the rate of the oxidation increased meaningfully, and the amount of product **5c** doubled with excellent enantiomeric excess (>99%) (Figure 3).

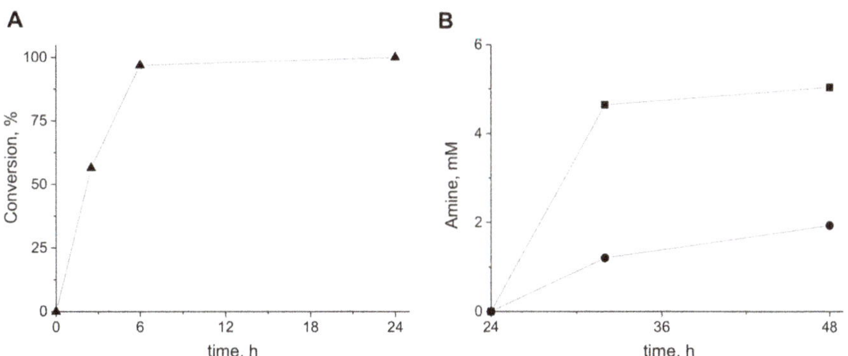

Figure 3. Results of the direct amination of 1-phenylethanol (**5a**) in a one-pot two-step fashion. (**A**) Conversion of **5a** (▲); (**B**) Production of **5c** with *BM*ωTA (■) and *AT*ωTA (•) enzymes; Reaction conditions: (**A**) 1 mL MilliQ water containing [**5a**] = 10 mM, [SAS] = 0.75 mM. Samples were incubated at 30 °C and irradiated under visible light (λ > 400 nm); (**B**) 0.5 mL pH 9 reaction mixture containing [NaPI] = 50 mM, [IPA] = 1 M, [PLP] = 1 mM, [crude cell extract overexpressed with ω-TA enzyme] = 10 mg/mL. Samples were incubated on a shaking plate at 30 °C at 600 rpm.

To gain more insight into the performance of the ω-transaminases, we varied the enzyme, amine donor, and cofactor concentrations. An increase in the enzyme concentration resulted in a higher initial reaction rate. However, approx. 2.9 mM of product was formed for all cases after 28 h using 1 M of IPA at pH 9 (Figure S23). Using the same concentration of IPA but increasing the pH to 10.2

led to a progressive increase in the yield above 6 mM (Figure S26). These data are in agreement with previous publications, which indicate that the deprotonated form of IPA in aqueous solution might act as the actual amine donor [35,47,48]. Testing different concentrations of IPA at pH 9 (0.5, 1, 2 M) resulted in both higher initial rate and final product concentration (Figure S24), as also the amount of free IPA amino donor in solution increased accordingly. We concluded that the final obtained conversion was determined by the actual concentration of free IPA in solution and the thermodynamic equilibrium constant of the reaction [42], and not by the stability of the biocatalyst. Therefore, the productivity of our system can be increased in the future by process engineering aimed at the selective removal of the acetone co-product. Finally, varying the concentration of the cofactor PLP did not show any noticeable influence on the product formation (Figure S25), which is also in agreement with previous findings. Conversely, an excessively high concentration of PLP is not only inconvenient economically but can also inhibit the ω-TA [49]. Based on the above findings, we chose the following parameters for our next investigation: [with ω-TA enzyme] = 10 mg/mL, [IPA] = 1 M, and [PLP] = 1 mM. The reactions were performed at 30 °C, an optimal temperature for both the photooxidation and reductive amination steps.

To further explore the scope of the one-pot two-step procedure, we investigated a broader scope of starting materials including aliphatic, aromatic, chiral, and non-chiral substrates (Scheme 1 and Table 1).

Table 1. Photocatalytic oxidation of various alcohols.

Substrate	Conversion [a] (%)	Yield [b] (%)
1a [c]	23	74
2a [c]	26	51
3a [c]	29	27
4a [c]	28	27
5a	99	94
6a	>99	74
7a	98	59
8a	88	28
9a	98	88
10a	>99	100

For analytical details see Supplementary Materials. [a] Conversion was determined by GC analysis and calculated as $X = (c_0 - c)/c_0$, where X is the conversion, c_0 is the in initial substrate concentration, and c is the final substrate concentration. [b] Yield was determined by GC analysis and calculated as $Y = (c_p - c_{p,0})/(c_0 - c)$, where Y is the yield, $c_{p,0}$ is the initial product concentration, c_p is the final product concentration c_0 is the initial substrate concentration, and c is the final substrate concentration. [c] 1 mL reaction mixture was used in a 2 mL glass vial and irradiated for 24 h. Reaction conditions: 2 mL MilliQ water in a 4 mL vial, containing [SAS] = 0.75 mM and [substrate] = 10 mM. Samples were incubated at 30 °C and irradiated under visible light (λ > 400 nm) for 9 h.

Aromatic substrates were oxidized with good (>80%) to excellent (>95%) conversion. However, by increasing the length of the aliphatic side chain from 1-phenylethanol (**5a**) to 1-phenylbutanol (**7a**), the yield decreased significantly. This is likely the result of oxidative decomposition, which is enhanced by the inductive effect of the aliphatic side chain [47]. This is also supported by the results for benzyl alcohol (**9a**) and 3-chlorobenzyl alcohol (**10a**). With the presence of an electron withdrawing group on the aromatic ring, which slightly reduces the activity of the hydroxyl group, the selectivity increased. It is also worth mentioning that besides oxidation, 1,2,3,4-tetrahydro-naphtol (**8a**) reacted with SAS, which was indicated by the absence of the characteristic yellow color of the photocatalyst after irradiation. In contrast to the aromatic substrates, aliphatic compounds demonstrated moderate conversion (>30%). Interestingly, C5 compounds (**1a, 2a**) showed better yield than C6 compounds (**3a, 4a**).

Following the photooxidation, the reductive amination step was carried out. The oxidized samples were diluted as such that the reaction mixture contained approx. 5 mM aromatic (except α-tetralone (**8b**) 1 mM) and 0.3–1.5 mM aliphatic substrate. In the reactions, the substrate scopes of five wild-type ω-TAs were examined, one (*R*)-selective from *Aspergillus terreus* (*AT*ωTA), and four (*S*)-selective ones from *Bacillus megaterium* (*BM*ωTA), *Chromobacterium violaceum* (*CV*ωTA), *Pseudomonas fluorescens* (*PF*ωTA), and *Vibrio fluvialis* (*VF*ωTA). For all ω-TAs tested, alkaline reaction conditions were found to be favorable for the reductive amination reaction (Figures S26–S30).

Amongst the prochiral compounds (Table 2), the entire range of substrates was transformed with excellent stereoselectivity (>99% *ee*). Acetophenone (**5b**) was accepted by all enzymes. In the case of aromatic compounds (**5b–7b**), the conversion decreased progressively with the increasing length of the alkyl chain attached to the aromatic ring. However, *VF*ωTA was an exception, as it similarly mediated the reductive amination of both propiophenone (**6b**) and butyrophenone (**7b**) with 35% conversion. Reductive amination of **8b** was also feasible using *BM*ωTA with excellent stereospecificity, which has been challenging for other systems [47,50]. Amongst the aliphatic substrates, 2-pentanone (**2b**) and 2-hexanal (**4b**) were converted to detectable levels of product only by *AT*ωTA and *VF*ωTA, respectively. However, the modest performances of other enzymes are likely to be the result of the mediocre initial concentration of the substrate, as both compounds have been reported to be efficiently converted to the corresponding chiral amine [47].

Table 2. Reductive amination of ketones.

		2b	3b	5b	6b	7b	8b
*AT*ωTA	c_{amine} (mM) [a]	0.66 (*R*)	n.d.	2.8 (*R*)	n.d.	n.d.	n.d.
	Conversion (%) [b]	36	n.d.	72	n.d.	n.d.	n.d.
	ee (%) [c]	>99	n.d.	>99	n.d.	n.d.	n.d.
*BM*ωTA	c_{amine} (mM)	n.d.	n.d.	4.93 (*S*)	10.75 (*S*)	n.c	0.42 (*S*)
	Conversion (%)	n.d.	n.d.	95	89	n.d.	20
	ee (%)	n.d.	n.d.	>99	>99%	n.d.	>99%
*CV*ωTA	c_{amine} (mM)	n.d.	n.d.	1.29 (*S*)	n.d.	n.d.	n.d.
	Conversion (%)	n.d.	n.d.	34	n.d.	n.d.	n.d.
	ee (%)	n.d.	n.d.	>99	n.d.	n.d.	n.d.
*PF*ωTA	c_{amine} (mM)	n.d.	n.d.	1.95 (*S*)	2.40	n.d.	n.d.
	Conversion (%)	n.d.	n.d.	43	35	n.d.	n.d.
	ee (%)	n.d.	n.d.	>99	>99	n.d.	n.d.
*VF*ωTA	c_{amine} (mM)	n.d.	2.25 (*S*)	3.73 (*S*)	2.44	1.20 (*S*)	n.d.
	Conversion (%)	n.d.	100	72	35	35	n.d.
	ee (%)	n.d.	>99	>99	>99 %	>99	n.d.

For analytical details, see Supplementary Materials. [a] Amine concentration was determined by GC analysis [b] Conversion was determined by GC analysis and calculated as indicated in Table 1. [c] Enantiomeric excess was determined by GC analysis on a chiral stationary phase (see Supplementary Materials). [d] n.d.: Not determined due to low concentration. Reaction conditions: 0.5 mL reaction mixture in a 2 mL glass vial containing [NaPi] 50 mM, [crude cell extract overexpressed with ω-TA enzyme] = 10 mg/mL, [IPA] = 1 M, [PLP] = 1 mM, and [aromatic ketone] ≈ 5 mM or [aliphatic ketone] ≈ 0.5 mM.

In the case of aldehydes, excellent conversions were achieved (Table 3). Besides VFωTA, all the enzymes completely converted hexanal (**3b**) to hexylamine (**3c**). However, it is also important to mention that similar to the other aliphatic compounds, the initial substrate concentration was low. Both benzaldehyde (**9b**) and 3-chlorobenzaldehyde (**10b**) were fully converted to the corresponding amine.

Table 3. Reductive amination of aldehydes.

		4b	9b	10b
*AT*ωTA	c_{amine} (mM) [a]	2.21	6.03	8.20
	Conversion (%) [b]	100	100	95
*BM*ωTA	c_{amine} (mM)	2.21	6.86	11.97
	Conversion (%)	100	100	98

Table 3. *Cont.*

		4b	9b	10b
*CV*ωTA	c_{amine} (mM)	2.21	7.00	12.82
	Conversion (%)	100	100	91
*PF*ωTA	c_{amine} (mM)	2.21	6.84	11.79
	Conversion (%)	100	100.00	91
*VF*ωTA	c_{amine} (mM)	n.d. [c]	7.19	11.83
	Conversion (%)	n.d.	100.00	98

For analytical details, see Supplementary Materials. [a] Amine concentration was determined by GC analysis [b] Conversion was determined by GC analysis and calculated as indicated in Table 1. [c] n.d.: Not determined due to low concentration. Reaction conditions: 0.5 mL reaction mixture in a 2 mL glass vial containing [NaPi] = 50 mM, [crude cell extract overexpressed with ω-TA enzyme] = 10 mg/mL, [IPA] = 1 M, [PLP] = 1 mM, and [aromatic ketone] ≈ 5 mM or [aliphatic ketone] ≈ 0.5 mM.

3. Materials and Methods

All chemicals were purchased Sigma-Aldrich (Zwijndrecht, The Netherlands), Fluka (Buchs, Switzerland), Acros (Geel, Belgium) or Alfa-Aesar (Karlsruhe, Germany) with the highest purity available and used without further treatment. The ω-transaminases were prepared via recombinant expression in *Escherichia coli* as described in detail in the Supplementary Materials.

3.1. Reaction Conditions for the One-Pot One-Step Cascade

In a 2 mL glass vial, 1 mL pH 9 reaction mixture was prepared, containing 100 mM sodium phosphate (NaPi), approx. 11 mM 1-phenylethanol (or one of the other alcohols investigated here), 1 mM pyridoxal-5-phosphate (PLP), 0.75 mM sodium anthraquinone-2-sulfonate (SAS), 1 M isopropylamine (IPA), and 10 mg lyophilized crude cell extract overexpressed with (*R*)- or (*S*)-selective ω-transaminase (ω-TA). Samples were irradiated under visible light (Osram Halolux CERAM 205W light bulb; λ > 400 nm) at 30 °C. The reaction mixture was stirred gently with a magnetic stirrer. At intervals, aliquots were taken, extracted with ethyl acetate, derivatized, and analyzed by gas chromatography (see Supplementary Materials).

3.2. Reaction Conditions for the One-Pot Two-Step Cascade

If not stated otherwise, in a 4 mL vial, 2 mL MilliQ water containing approx. 10 mM substrate and 0.75 mM SAS were irradiated with visible light (Osram Halolux CERAM 205W light bulb; λ > 400 nm) at 30 °C. The reaction mixture was gently stirred with a magnetic stirrer. At intervals, aliquots were taken, extracted with ethyl acetate, and analyzed by gas chromatography without derivatization (see Supplementary Materials). After the irradiation, 250 μL reaction mixture was diluted to 500 μL by using NaPi (100 mM) as such that the final samples contained NaPI (50 mM), IPA (1M), PLP (1 mM), and 10 mg/mL lyophilized whole cells overexpressed with (*R*)- or (*S*)-selective ω-TA. The pH of the system was adjusted to pH 9. Samples were incubated on a shaking plate at 30 °C with 600 rpm shaking speed. At intervals, aliquots were taken, extracted with ethyl acetate, derivatized, and analyzed by gas chromatography (see Supplementary Materials).

3.3. Derivatization of GC Samples

Samples (75 μL) were extracted with ethyl acetate (150 μL) dried ($MgSO_4$, 5 mg) and reacted with acetic anhydride (10 μL) in the presence of 4-(*N*,*N*-dimethylamino)pyridine (5 mg) at 45 °C for 45 min. The reaction was quenched with water (75 μL) and the samples were dried ($MgSO_4$, 5 mg) again.

4. Conclusions

In summary, a reaction sequence of consecutive photooxidation and reductive amination of various alcohols in one pot is reported. The tandem reaction with sodium anthraquinone-2-sulfonate

(SAS) as a photocatalyst was also feasible in a one-pot one-step approach; however, the yield was significantly inferior to the one-pot two-step case. The system was tested on a plethora of aromatic and aliphatic substrates. Even though all of the compounds were converted, the photooxidation of aromatic substrates proceeded much faster and with a higher yield. In the second step of the cascade reaction, recombinant transaminases originating from *Aspergillus terreus* (*AT*ωTA), *Bacillus megaterium* (*BM*ωTA), *Chromobacterium violaceum* (*CV*ωTA), *Pseudomonas fluorescens* (*PF*ωTA), and *Vibrio fluvialis* (*VF*ωTA) were used for the reductive amination. All of the non-chiral substrates were converted with high conversion and the prochiral substrates with excellent enantiomeric excess.

One issue en route to the preparative application is the inactivation of the biocatalysts by the photoexcited photocatalysts. This may be overcome by special separation of the photocatalytic oxidation step from the reductive amination step. For this, for example, immobilized catalysts in a flow chemistry setup may be a doable approach.

Supplementary Materials: The following are available online at http://www.mdpi.com/2073-4344/9/4/305/s1. Figure S1: SDS-Page for the expression of the ωTAs; Figure S2: FT-IR spectrum of SAS; Figure S3: FT-IR spectrum of g-C3N4; Figure S4: XRD spectrum of g-C3N4; Figure S5: FT-IR spectrum of g-C3N4; Figure S6: XRD spectrum of g-C3N4-VO; Figure S7: Effect of evaporation on the acetophenone concentration during the photo-oxidation of 1-phenylethanol; Figure S8: Control reaction of the photooxidation; Figure S9: Effect of the circumstances of the photooxidation on the efficiency of the reductive amination; Figure S10: Control reaction of the reductive amination; Figure S11: 1H NMR spectrum of 1-phenylethyl amine; Figure S12: 1H NMR spectrum of oxidation of 1-phenylethylamine; Figure S13: 1H NMR spectrum of oxidation of isopropylamine; Figure S14: Effect of light intensity on the photocatalytic activity of SAS; Figure S15: Effect of light composition on the photocatalytic activity of SAS; Figure S16: Effect of the reaction atmosphere on the photocatalytic activity of SAS; Figure S17: Effect of the amount of photocatalyst on the photocatalytic activity of SAS; Figure S18: Effect of pH on the photocatalytic activity of SAS; Figure S19: Effect of light composition on the photocatalytic activity of g-C3N4; Figure S20: Effect of the reaction atmosphere on the photocatalytic activity of g-C3N4; Figure S21: Effect of the photocatalyst amount on the photocatalytic activity of g-C3N4; Figure S22: Effect of the pH on the photocatalytic activity of g-C3N4; Figure S23: Effect of the enzyme concentration on the yield of the reductive amination; Figure S24: Effect of the amine donor concentration on the conversion of the reductive amination; Figure S25: Effect of the cofactor concentration on the yield of the reductive amination; Figure S26: pH dependency of ATω-TA; Figure S27: pH dependency of BMωTA; Figure S28: pH dependency of VFωTA; Figure S29: pH dependency of PFωTA; Figure S30: pH dependency of CVωTA; Figure S31: Representative GC chromatogram of SAS catalyzed photooxidation of 1-pentanol; Figure S32: Representative GC chromatogram of SAS catalyzed photooxidation of 2-pentanol; Figure S33: Representative GC chromatogram of SAS catalyzed photooxidation of 1-hexanol; Figure S34: Representative GC chromatogram of SAS catalyzed photooxidation of 2-hexanol; Figure S35: Representative GC chromatogram of SAS catalyzed photooxidation of 1-phenylethanol; Figure S36: Representative GC chromatogram of SAS catalyzed photooxidation of 1-phenylpropanol; Figure S37: Representative GC chromatogram of SAS catalyzed photooxidation of 1-phenylbutanol; Figure S38: Representative GC chromatogram of SAS catalyzed photooxidation of 1,2,3,4-tetrahydro-1-naphtol; Figure S39: Representative GC chromatogram of SAS catalyzed photooxidation of benzyl alcohol; Figure S40: Representative GC chromatogram of SAS catalyzed photooxidation of 3-chlorobenzyl alcohol; Figure S41: Representative GC chromatogram of derivatized n-pentylamine; Figure S42: Representative GC chromatogram of derivatized 2-aminopentane; Figure S43: Representative GC chromatogram of derivatized 1-hexanol, hexanal and 1-hexylamine mixture; Figure S44: Representative GC chromatogram of derivatized 2-aminohexane; Figure S45: Representative GC chromatogram of derivatized 1-phenylethanol, acetophenone and 1-phenylethyl amine; Figure S46: Representative GC chromatogram of derivatized 1-phenylpropanol, propiophenone (R) and (S) 1-phenylpropyl amine; Figure S47: Representative GC chromatogram of derivatized 1,2,3,4-tetrahydro-1-naphthol, α-tetralone, (R) and (S) 1,2,3,4-tetrahydro-1-naphthylamine; Figure S48: Representative GC chromatogram of derivatized benzyl alcohol, benzaldehyde, benzyl amine; Figure S49: Representative GC chromatogram of derivatized 3-chlorobenzyl alcohol, 3-chlorobenzaldehyde, 3-chlorobenzyl amine, Table S1: Used GC systems; Table S2: Details of GC analysis of the alcohol oxidation; Table S3: Details of GC analysis of the reductive amination.

Author Contributions: Conceptualization, F.H.; methodology, J.G., W.Z., T.K., F.G.M., and I.W.C.E.A; validation, J.G., W.Z., T.K., F.G.M., and I.W.C.E.A; formal analysis, J.G., W.Z., and F.H., writing—original draft preparation, J.G. and W.Z., writing—review and editing, W.Z. and F.H.

Funding: This research was funded by the European Research Council (ERC Consolidator Grant No. 648026).

Acknowledgments: We thank Remco van Oosten for help with GC method development.

Conflicts of Interest: The authors declare no conflict of interest.

References

1. Nugent, T.C.; El-Shazly, M. Chiral amine synthesis—Recent developments and trends for enamide reduction, reductive amination, and imine reduction. *Adv. Synth. Catal.* **2010**, *352*, 753–819. [CrossRef]
2. Welsch, M.E.; Snyder, S.A.; Stockwell, B.R. Privileged scaffolds for library design and drug discovery. *Curr. Opin. Chem. Biol.* **2010**, *14*, 347–361. [CrossRef]
3. Newman, D.J.; Cragg, G.M. Natural products as sources of new drugs over the 30 years from 1981 to 2010. *J. Nat. Prod.* **2012**, *75*, 311–335. [CrossRef] [PubMed]
4. Green, A.P.; Turner, N.J.; O'Reilly, E. Chiral amine synthesis using omega-transaminases: An amine donor that displaces equilibria and enables high-throughput screening. *Angew. Chem. Int. Ed.* **2014**, *53*, 10714–10717. [CrossRef]
5. Bähn, S.; Imm, S.; Neubert, L.; Zhang, M.; Neumann, H.; Beller, M. The catalytic amination of alcohols. *ChemCatChem* **2011**, *3*, 1853–1864. [CrossRef]
6. Imm, S.; Bähn, S.; Zhang, M.; Neubert, L.; Neumann, H.; Klasovsky, F.; Pfeffer, J.; Haas, T.; Beller, M. Improved ruthenium-catalyzed amination of alcohols with ammonia: Synthesis of diamines and amino esters. *Angew. Chem. Int. Ed.* **2011**, *50*, 7599–7603. [CrossRef] [PubMed]
7. Mutti, F.G.; Knaus, T.; Scrutton, N.S.; Breuer, M.; Turner, N.J. Conversion of alcohols to enantiopure amines through dual-enzyme hydrogen-borrowing cascades. *Science* **2015**, *349*, 1525–1529. [CrossRef] [PubMed]
8. Lawrence, S.A. *Amines: Synthesis, Properties and Applications*; Cambridge University Press: Cambridge, UK, 2004.
9. Fabiano, E.; Golding, B.T.; Sadeghi, M.M. A simple conversion of alcohols into amines. *Synthesis* **1987**, *1987*, 190–192. [CrossRef]
10. Xu, C.; Zhang, C.H.; Li, H.; Zhao, X.Y.; Song, L.; Li, X.B. An overview of selective oxidation of alcohols: Catalysts, oxidants and reaction mechanisms. *Catal. Surv. Asia* **2016**, *20*, 13–22. [CrossRef]
11. Mallat, T.; Baiker, A. Oxidation of alcohols with molecular oxygen on solid catalysts. *Chem. Rev.* **2004**, *104*, 3037–3058. [CrossRef]
12. Schultz, M.J.; Sigman, M.S. Recent advances in homogeneous transition metal-catalyzed aerobic alcohol oxidations. *Tetrahedron* **2006**, *62*, 8227–8241. [CrossRef]
13. Arends, I.W.C.E.; Sheldon, R.A. Modern oxidation of alcohols using environmentally benign oxidants. In *Modern Oxidation Method*; Bäckvall, J.-E., Ed.; Wiley-VCH: Weinheim, Germany, 2010; Volume 2, pp. 147–185.
14. Davis, S.E.; Ide, M.S.; Davis, R.J. Selective oxidation of alcohols and aldehydes over supported metal nanoparticles. *Green Chem.* **2013**, *15*, 17–45. [CrossRef]
15. Irie, H.; Miura, S.; Kamiya, K.; Hashimoto, K. Efficient visible light-sensitive photocatalysts: Grafting Cu(II) ions onto TiO(2) and WO(3) photocatalysts. *Chem. Phys. Lett.* **2008**, *457*, 202–205. [CrossRef]
16. Shishido, T.; Miyatake, T.; Teramura, K.; Hitomi, Y.; Yamashita, H.; Tanaka, T. Mechanism of photooxidation of alcohol over nb2o5. *J. Phys. Chem. C* **2009**, *113*, 18713–18718. [CrossRef]
17. Furukawa, S.; Shishido, T.; Teramura, K.; Tanaka, T. Photocatalytic oxidation of alcohols over TiO_2 covered with Nb_2O_5. *ACS Catal.* **2012**, *2*, 175–179. [CrossRef]
18. Chen, J.; Cen, J.; Xu, X.L.; Li, X.N. The application of heterogeneous visible light photocatalysts in organic synthesis. *Catal. Sci. Technol.* **2016**, *6*, 349–362. [CrossRef]
19. Su, F.Z.; Mathew, S.C.; Lipner, G.; Fu, X.Z.; Antonietti, M.; Blechert, S.; Wang, X.C. Mpg-C_3N_4-catalyzed selective oxidation of alcohols using O-2 and visible light. *J. Am. Chem. Soc.* **2010**, *132*, 16299–16301. [CrossRef]
20. Zhang, L.G.; Liu, D.; Guan, J.; Chen, X.F.; Guo, X.C.; Zhao, F.H.; Hou, T.G.; Mu, X.D. Metal-free g-C_3N_4 photocatalyst by sulfuric acid activation for selective aerobic oxidation of benzyl alcohol under visible light. *Mater. Res. Bull.* **2014**, *59*, 84–92. [CrossRef]
21. Verma, S.; Baig, R.B.N.; Nadagouda, M.N.; Varma, R.S. Selective oxidation of alcohols using photoactive vo@g-C_3N_4. *ACS Sustain. Chem. Eng.* **2016**, *4*, 1094–1098. [CrossRef]
22. Zavahir, S.; Xiao, Q.; Sarina, S.; Zhao, J.; Bottle, S.; Wellard, M.; Jia, J.F.; Jing, L.Q.; Huang, Y.M.; Blinco, J.P.; et al. Selective oxidation of aliphatic alcohols using molecular oxygen at ambient temperature: Mixed-valence vanadium oxide photocatalysts. *ACS Catal.* **2016**, *6*, 3580–3588. [CrossRef]

23. Clark, K.P.; Stonehil, H.I. Photochemistry and radiation-chemistry of anthraquinone-2-sodium-sulphonate in aqueous-solution Part 1. Photochemical kinetics in aerobic solution. *J. Chem. Soc. Faraday Trans. I* **1972**, *68*, 577–590. [CrossRef]
24. Clark, K.P.; Stonehil, H.I. Photochemistry and radiation-chemistry of 9,10-anthraquinone-2-sodium sulfonate in aqueous-solution Part 2. Photochemical products. *J. Chem. Soc. Faraday Trans. I* **1972**, *68*, 1676–1686. [CrossRef]
25. Wells, C.F. Hydrogen transfer to quinones Part 2. Reactivities of alcohols, ethers and ketones. *Trans. Faraday Soc.* **1961**, *57*, 1719–1731. [CrossRef]
26. Wells, C.F. Hydrogen transfer to quinones Part 1. Kinetics of deactivation of photo-excited quinone. *Trans. Faraday Soc.* **1961**, *57*, 1703–1718. [CrossRef]
27. Zhang, W.; Gacs, J.; Arends, I.W.C.E.; Hollmann, F. Selective photooxidation reactions using water soluble anthraquinone photocatalysts. *ChemCatChem* **2017**, *9*, 3821–3826. [CrossRef]
28. Zhang, W.; Fernandez Fueyo, E.; Hollmann, F.; Leemans Martin, L.; Pesic, M.; Wardenga, R.; Höhne, M.; Schmidt, S. Combining photo-organo redox- and enzyme catalysis facilitates asymmetric C-H bond functionalization. *Eur. J. Org. Chem.* **2019**, *10*, 80–84. [CrossRef]
29. Fuchs, C.S.; Farnberger, J.E.; Steinkellner, G.; Sattler, J.H.; Pickl, M.; Simon, R.C.; Zepeck, F.; Gruber, K.; Kroutil, W. Asymmetric amination of α-chiral aliphatic aldehydes via dynamic kinetic resolution to access stereocomplementary brivaracetam and pregabalin precursors. *Adv. Synth. Catal.* **2018**, *360*, 768–778. [CrossRef]
30. Hohne, M.; Bornscheuer, U.T. Biocatalytic routes to optically active amines. *ChemCatChem* **2009**, *1*, 42–51. [CrossRef]
31. Ward, J.; Wohlgemuth, R. High-yield biocatalytic amination reactions in organic synthesis. *Curr. Org. Chem.* **2010**, *14*, 1914–1927. [CrossRef]
32. Slabu, I.; Galman, J.L.; Lloyd, R.C.; Turner, N.J. Discovery, engineering, and synthetic application of transaminase biocatalysts. *ACS Catal.* **2017**, *7*, 8263–8284. [CrossRef]
33. Fuchs, M.; Farnberger, J.E.; Kroutil, W. The industrial age of biocatalytic transamination. *Eur. J. Org. Chem.* **2015**, *2015*, 6965–6982. [CrossRef]
34. Kelly, S.A.; Pohle, S.; Wharry, S.; Mix, S.; Allen, C.C.R.; Moody, T.S.; Gilmore, B.F. Application of omega-transaminases in the pharmaceutical industry. *Chem. Rev.* **2018**, *118*, 349–367. [CrossRef] [PubMed]
35. Savile, C.K.; Janey, J.M.; Mundorff, E.C.; Moore, J.C.; Tam, S.; Jarvis, W.R.; Colbeck, J.C.; Krebber, A.; Fleitz, F.J.; Brands, J.; et al. Biocatalytic asymmetric synthesis of chiral amines from ketones applied to sitagliptin manufacture. *Science* **2010**, *329*, 305–309. [CrossRef] [PubMed]
36. Palacio, C.M.; Crismaru, C.G.; Bartsch, S.; Navickas, V.; Ditrich, K.; Breuer, M.; Abu, R.; Woodley, J.M.; Baldenius, K.; Wu, B.; et al. Enzymatic network for production of ether amines from alcohols. *Biotechnol. Bioeng.* **2016**, *113*, 1853–1861. [CrossRef] [PubMed]
37. Mathew, S.; Yun, H. Ω-transaminases for the production of optically pure amines and unnatural amino acids. *ACS Catal.* **2012**, *2*, 993–1001. [CrossRef]
38. Sattler, J.H.; Fuchs, M.; Mutti, F.G.; Grischek, B.; Engel, P.; Pfeffer, J.; Woodley, J.M.; Kroutil, W. Introducing an in situ capping strategy in systems biocatalysis to access 6-aminohexanoic acid. *Angew. Chem. Int. Ed.* **2014**, *53*, 14153–14157. [CrossRef]
39. Wang, B.; Land, H.; Berglund, P. An efficient single-enzymatic cascade for asymmetric synthesis of chiral amines catalyzed by [small omega]-transaminase. *Chem. Comm.* **2013**, *49*, 161–163. [CrossRef] [PubMed]
40. Martínez-Montero, L.; Gotor, V.; Gotor-Fernández, V.; Lavandera, I. But-2-ene-1,4-diamine and but-2-ene-1,4-diol as donors for thermodynamically favored transaminase- and alcohol dehydrogenase-catalyzed processes. *Adv. Synth. Catal.* **2016**, *358*, 1618–1624. [CrossRef]
41. Gomm, A.; Lewis, W.; Green, A.P.; O'Reilly, E. A new generation of smart amine donors for transaminase-mediated biotransformations. *Chem. Eur. J.* **2016**, *22*, 12692–12695. [CrossRef]
42. Tufvesson, P.; Jensen, J.S.; Kroutil, W.; Woodley, J.M. Experimental determination of thermodynamic equilibrium in biocatalytic transamination. *Biotechnol. Bioeng.* **2012**, *109*, 2159–2162. [CrossRef] [PubMed]
43. Martinez-Montero, L.; Gotor, V.; Gotor-Fernandez, V.; Lavandera, I. Stereoselective amination of racemic sec-alcohols through sequential application of laccases and transaminases. *Green Chem.* **2017**, *19*, 474–480. [CrossRef]

44. Sattler, J.H.; Fuchs, M.; Tauber, K.; Mutti, F.G.; Faber, K.; Pfeffer, J.; Haas, T.; Kroutil, W. Redox self-sufficient biocatalyst network for the amination of primary alcohols. *Angew. Chem. Int. Ed.* **2012**, *51*, 9156–9159. [CrossRef]
45. Schrittwieser, J.H.; Sattler, J.; Resch, V.; Mutti, F.G.; Kroutil, W. Recent biocatalytic oxidation-reduction cascades. *Curr. Opin. Chem. Biol.* **2011**, *15*, 249–256. [CrossRef]
46. Sheldon, R.A. The E factor 25 years on: The rise of green chemistry and sustainability. *Green Chem.* **2017**, *19*, 18–43. [CrossRef]
47. Mutti, F.G.; Fuchs, C.S.; Pressnitz, D.; Sattler, J.H.; Kroutil, W. Stereoselectivity of four (R)-selective transaminases for the asymmetric amination of ketones. *Adv. Synth. Catal.* **2011**, *353*, 3227–3233. [CrossRef]
48. Mutti, F.G.; Kroutil, W. Asymmetric bio-amination of ketones in organic solvents. *Adv. Synth. Catal.* **2012**, *354*, 3409–3413. [CrossRef]
49. Cassimjee, K.E.; Humble, M.S.; Miceli, V.; Colomina, C.G.; Berglund, P. Active site quantification of an ω-transaminase by performing a half transamination reaction. *ACS Catal.* **2011**, *1*, 1051–1055. [CrossRef]
50. Fesko, K.; Steiner, K.; Breinbauer, R.; Schwab, H.; Schurmann, M.; Strohmeier, G.A. Investigation of one-enzyme systems in the omega-transaminase-catalyzed synthesis of chiral amines. *J. Mol. Catal. B Enzym.* **2013**, *96*, 103–110. [CrossRef]

MDPI
St. Alban-Anlage 66
4052 Basel
Switzerland
Tel. +41 61 683 77 34
Fax +41 61 302 89 18
www.mdpi.com

Catalysts Editorial Office
E-mail: catalysts@mdpi.com
www.mdpi.com/journal/catalysts

www.ingramcontent.com/pod-product-compliance
Lightning Source LLC
Chambersburg PA
CBHW051858210326
41597CB00033B/5946

* 9 7 8 3 0 3 9 2 1 7 0 8 3 *